Chemical Protective Clothing

Second Edition

Edited by Daniel H. Anna, PhD

AIHA PRESS

Disclaimers

Copyright

1st printing 2003

Available from:
American Industrial Hygiene Association
2700 Prosperity Avenue, Suite 250
Fairfax, VA 22031
(703) 849-8888
www.aiha.org

ISBN # 1-931504-46-6

Stock # 568-PC-03

Printed in the United States of America

Table of Contents

Chapter 1: Chemical Protective Clothing and the Skin: *Practical Considerations*

by Mark Boeniger

Chapter 2: Chemistry of Chemical Protective Clothing Formulations

By Jimmy L. Perkins

Chapter 3: Solvent-Polymer Interactions

By Jimmy L. Perkins

Chapter 4: Types of Chemical Protective Clothing

By Jeffrey O. Stull

Chapter 5: Test Methods & Standards

By Norman W. Henry III and Jeffrey O. Stull

Chapter 6: Full-Body Protection and Standards

By Jeffrey O. Stull

Chapter 7: Selection of Chemical Protective Clothing

By Jeffrey O. Stull

Chapter 8: *Full-Body Protection and Standards*

By Michael H. Ziskin, Jeff Behar, and S. Zack Mansdorf

CORRECTION

Chemical Protective Clothing, 2nd Edition

Chapter 8 was misidentified in the Table of Contents. It should read: **Decontamination of CPC**.

Chapter 9: Managing Heat Stress

by Thomas E. Bernard and Frank N. Dukes-Dobos

Chapter 10: Development of a CPC Program

By S. Zack Mansdorf

Authors

Jeff Behar, CIH, MS, MBA
Senior Industrial Hygienist
Jet Propulsion Laboratory
California Institute of Technology
Pasadena, CA

Thomas E. Bernard, PhD, CIH, PE
Professor
University of South Florida
College of Public Health
Tampa, FL

Mark F. Boeniger, CIH
Industrial Hygienist
NIOSH
Cincinnati, OH

Francis N. Dukes-Dobos, MD
University of South Florida
Tampa, FL

Norman W. Henry, III, CIH
Senior Research Chemist
DuPont Company Stine-Haskell Research Center
Newark, DE

S.Z Mansdorf, PhD, CIH, CSP, QEP
Safety, Industrial Health & Environment Director
L'Oreal
Aulnay sous Bois Cedex, France

Jimmy L. Perkins
Professor
University of Texas Health Science Center
School of Public Health
San Antonio, TX

Jeffrey O. Stull
Vice President of Operations
International Personnel Protection Inc.
Austin, TX

Michael H. Ziskin
President
Field Safety Corp.
North Branford, CT

Acknowledgments

The AIHA Protective Clothing and Equipment Committee is grateful to many people who were involved with the production of this second edition.

First, special thanks go to the authors for volunteering their valuable time and expertise. They have once again done an outstanding job in creating a concise and very informative resource for OEHS professionals. Thanks also go to the committee members who reviewed the technical content of each chapter.

Additionally, several people deserve individual recognition. Judy Foster, 2002 PC&E Committee Chair, shepherded the development process and maintained the project timeline. Even after her term as chair had ended, she stuck with the project to ensure continuity in committee leadership. Andry Nowosiwsky, 2003 PC&E Committee Chair, provided guidance and rallying support in the final stages of production. Fran Kuecker, AIHA staff, provided initial project support and championed making this book available by chapter prior to its completion. Katie Robert, AIHA staff, kept this project on track as project manager and provided cohesiveness and polish to this multi-authored publication as copy editor. Janet Scheren's design skills gave this book its sleek, finished look.

Finally, special thanks go to the authors of the first edition who were not able to participate this time, or who participated in a different capacity. Their early contributions were the basis for this dynamic publication.

Daniel H. Anna, PhD
Associate Professor
Occupational Safety & Environmental Health
Department of Industry and Technology
Millersville University of Pennsylvania

Introduction

The history of protective clothing and improvements in material technology and construction methods is well documented. From the early use of animal hides, to the use of natural materials, to the increasing number of synthetic materials available today, all protective clothing provides some level of protection. Matching the proper protective clothing to the hazard to take advantage of the protection offered is the goal for proper effective use. The concept of using protective clothing as the "last line of defense" implemented when engineering and administrative controls are not feasible continues to be the ideal standard of use. In jobs where personal protective clothing and equipment are the only available control option, the purpose of the clothing and equipment is to shield or isolate individuals from the chemical, physical, and biological hazards.

Exposure concerns have expanded throughout the years. In the 1970s and 1980s, emphasis was placed on reducing exposures to carcinogenic materials. In the 1980 and1990s, the prevention of exposure to bloodborne pathogens and the spread of diseases through contact with bodily fluids were added. Although world events in the last few years have quickly changed the perception, significance, and use of protective clothing, many of these changes were initiated by events of the past decade. The rapid introduction of terrorism, and chemical and biological threats into standard business risks, have, in many ways, elevated the significance of planning for non-standard conditions, where protective clothing and equipment plays a key role in the prevention of exposures. The ever-changing work environment in chemical, pharmaceutical, and other high tech industries often limits or prevents the effective implementation of engineering controls. As a result, the number of workers that rely on protective clothing to reduce exposures has increased.

Standards, Test Methods, and Performance Data

An increasing number of standards and practices emphasize the need for protective clothing to be part of an overall safety and health plan. OSHA requirements to complete and revise PPE assessments for all workplace hazards remain a significant challenge for many companies and workplaces. Even after completion of the hazard assessments, the selection of appropriate protective clothing becomes more challenging because of all of the available materials and models. Performance data for protective clothing has become widely available, but guidelines for the use and interpretation of the data continue to be developed. Test methods were developed to provide a basis for evaluating clothing and material effectiveness for protection against a specific hazard. Although methods are designed to allow comparisons between the relative performances of various products, the influence of other factors such as size and fit must be incorporated into the selection process. Without adequate interpretation guidelines, the selection process can be confusing for many safety and health professionals, particularly when the test method does not simulate the conditions of the actual exposure.

Product Development and the End User

The development of new clothing materials, and the sophisticated advances in polymer technology and laminate methods broadened the capabilities of available products. At the same time, there are additional variables to consider, which make selection more difficult.

In addition to the development of new materials, the focus has shifted to the factors that affect the end user. Protective clothing must not only be effective, but also be practical for use in the workplace. Improvements in manufacturing have increased the selection of garment sizes and models. As more attention is focused on understanding ergonomic issues, heat stress implications, and the relationship between the task and the clothing, improved fit and comfort will be necessary to help ensure proper and consistent use at all times. New materials, garments, and standards and test methods will need to be developed in a manner that will encourage the end user to wear the protective clothing as intended. Easy methods of identification, such as color-coding, will need to be incorporated to simplify the selection process. The development of field-based end of service indicators will help improve the usefulness and protection afforded by the clothing, and to minimize risks associated with the reuse of previously exposed clothing materials.

Risk Analysis

The combination of suits and other clothing and protective equipment, including respiratory protection, form ensembles for true full-body protective clothing. The increased reliance on protective clothing in emergency and disaster situations increases the need to determine the effective protection offered by the collective ensemble. As information about combinations of protective clothing components are developed, questions about "overprotection" of workers emerge. The realization of chemical and biological threats has shifted the emphasis onto preplanning, so that quick and timely selection decisions can be made. Risk based decisions about the appropriate level of protection may result in accepting some level of exposure.

With occupational skin disease being recognized as one of the leading work-related illnesses, the interest in preventing dermal exposures has risen considerably. Since dermal contact with hazardous materials can contribute significantly to a worker's overall exposure, knowledge of basic information about the skin, and the impact of protective clothing on the behavior of the skin as a protective mechanism, is essential to ensure proper selection of protective clothing materials. As future considerations are made regarding protective clothing, the issue of dermal exposures will need to be readdressed. It will be necessary to develop more sophisticated quantitative evaluation techniques that are more field-adaptable than those that currently exist. In addition, simple indicators of acceptable dermal exposures will need to be developed. Overall, this will mean an increased emphasis on examining multiple exposure routes, including the sources of dermal exposures from surface contamination as well as those from direct contact. Eventually, this may necessitate developing dermal occupational exposure limits and broadening the available biological exposure indices.

As information about the selection, use, and evaluation of protective clothing and equipment continue to evolve; development of a comprehensive program covering these aspects of protective clothing can enhance the level of protection afforded to the workers.

An understanding of the performance data in conjunction with the integration of risk assessment into the selection process can lead to better selections and therefore better protection for workers that rely on protective clothing to minimize or prevent exposures.

Chemical Protective Clothing, 2nd Edition – The Book

This updated book provides a comprehensive overview of the considerations necessary for implementation of a successful protective clothing and equipment program. The content is organized into three areas: Background, Fundamentals, and Theory; Manufacture, Design and Testing; and Considerations for Proper Selection and Comprehensive Program Development. Throughout the book, the discussion focuses primarily on the prevention or reduction of chemical exposures, but often the information can be applied in a broader manner to reduce exposures to biological and physical hazards.

Background, Fundamentals and Theory (Chapters 1-3)

As an introduction to protective clothing, the first chapters address the effects of chemical exposures on the skin, the theory behind the protective mechanisms of the clothing, and the interaction between the performance of protective clothing and the skin. The basics of polymer chemistry that are related to the manufacture of different protective clothing materials are presented. A review of permeation theory helps to explain the differences in protection levels and sets the stage for later discussion of the interpretation of testing data.

Manufacture, Design, and Testing (Chapters 4-6)

The second segment of the book covers the different types of protective clothing that are available and the basics of manufacturing processes. Differences in performance that results from the manufacturing methods and styles are discussed. A thorough review of the applicable test methods and standards, and a discussion of the test data and interpretation are provided.

Considerations for Proper Selection and Program Development (Chapters 7-10)

The last segment of the books is comprised of recommendations for developing an appropriate risk based selection strategy and additional considerations related to the proper use and handling of protective clothing. Issues related to proper decontamination of protective clothing, and prevention of heat related illnesses are also discussed. The book concludes with a chapter that covers the development of a comprehensive protective clothing and equipment program.

Without inclusion of all the elements of a good protective clothing program, the exposure risk to workers increases. The proper selection, use, and maintenance of protective clothing helps to ensure that the expected and actual protection levels are the same.

1

Chemical Protective Clothing and the Skin: *Practical Considerations*

By Mark Boeniger

Introduction

Many occupational safety and health professionals consider chemical protective clothing (CPC) as the "last line of defense" for protecting the skin from chemical and physical hazards. CPC is in several respects justly regarded this way. First, it physically is very close to the skin. Because of its intimate contact with the skin, its purpose is to form a physical barrier that separates the skin from the outside environment.

Aside from this rather obvious physical description of its proximity to the skin, CPC as a last line of defense can also be considered from a pragmatic viewpoint. There exists in industrial hygiene practice a hierarchy of controls that has evolved from the effectiveness, reliability, user comfort, and operating costs of each of the interventions. Generally, personal protective equipment (PPE) is considered a last resort for protecting workers, to be issued only when all other options for control are considered inadequate. This is clearly accepted to be true for inhalation hazards where use of respirators is acknowledged to have practical limitations. However with regard to protecting the skin, CPC is often considered first without giving much consideration to possible alternative approaches. Use of CPC should not be preferred over other options because there may be adverse consequences and costs associated with this choice. This chapter will provide some basic information about the skin, and focuses on the direct consequences of wearing CPC upon the skin and on the effectiveness of exposure reduction.

Skin Anatomy and Physiology

The skin is the largest organ of the body, constituting 10–15% of the normal body weight, with a surface area of approximately 1.8 m^2. After over a century of scientific study, much is now known about the seemingly amazingly effective function of the skin as a barrier to the outside environment and of helping to maintain homeostasis within. However, research is still intensely attempting to understand all the physicochemical aspects of the skin, especially among the cosmetic and pharmaceutical industries who venture to profit from improving the condition of the skin, and by delivery of drugs through the skin, respectively.

There are several excellent comprehensive books that are presently available that should be referred to if more information is needed on skin anatomy and function.[1-6] The purpose of this chapter is to provide only a basic broad background about the skin as it relates to occupational exposures. More specifically, this overview will provide the reader who is involved with the use of CPC with information that is relevant to the impact of wearing CPC on the skin. The close proximity of CPC to the skin, especially for gloves, and its potential health importance make this discussion essential.

For review, the skin constitutes three structurally and functionally distinct important regions: the stratum corneum, viable epidermis, and dermis. The stratum corneum is the uppermost and thinnest layer exposed to the outside world (see Figure 1). It is composed of flattened denucleated cells called corneocytes, which contain mainly highly cross-linked fibrous keratin proteins. Keratin has a high affinity for water, which is necessary for it to remain flexible. If dried, the stratum corneum will become brittle and crack.[7] The corneocytes are nonrespiring cells that are tightly connected forming a surprisingly rugged membrane. Each corneocyte is typically 25–40 mm (micrometers) in diameter and only 0.2 to 0.5 mm

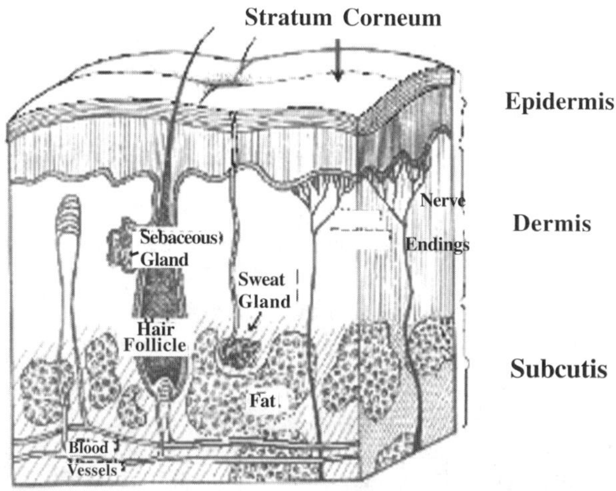

Figure 1. Simplified diagram of the skin including the upper stratum corneum consisting of flattened corneocytes, and the viable epidermin, dermis, and underlying subcutis tissue. The skin can be an important route for absorption of chemicals into the body, can be adversely affected resulting in serious medical conditions, and is an important immunologic organ of the body.

thick. These are randomly stacked some 15 to 25 layers thick over most human skin surfaces, which have an approximate total thickness of about 15 mm.[8] This varies somewhat by anatomical location, being thickest on the palmar skin and soles of the feet with a thickness of 400–600 mm. Figure 2 shows a scanning electron photomicrograph of the surface of the skin.

The intercellular space between the corneocytes has a diameter of between 0.01–0.1 µm and has been estimated to occupy approximately 20% of the total stratum corneum volume.[9-11] Within this space are well-organized lipophilic and hydrophilic domains that represent parallel channels (lamallar bilayers) of least resistance to diffusion of either nonpolar (lipophilic) or polar (hydrophilic) compounds, respectively.[12-14] These structured layers in turn occupy a space of only 0.5–1.0 µm.[15] These domains constitute an important, tortuous pathway for molecular transport through

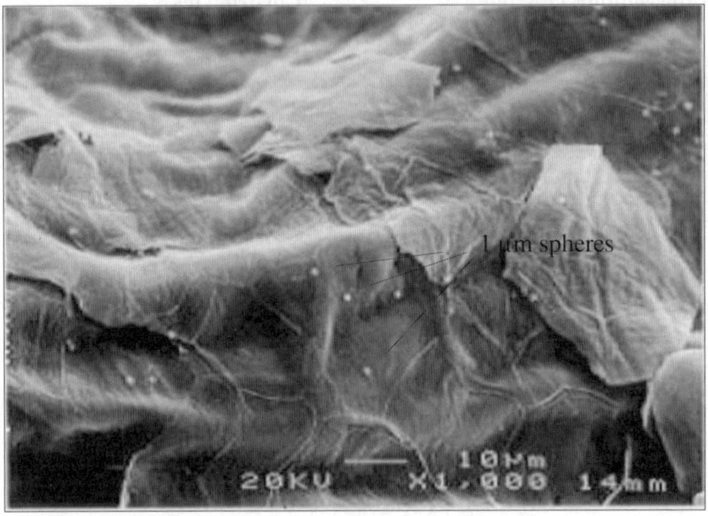

Figure 2. Scanning electron photomicrograph of the surface of the skin showing varied topography and plate-like corneocytes. Small particles of 1 μm are shown for comparison (courtesy of Dr. Sally Tinkle, NIOSH, Morgantown, WV).

the stratum corneum. Normally the intercellular fluid is predominantly lipid, consisting mainly of lipid classes called ceramides (~45%), cholesterol (~25%), and free fatty acids (~15%), which spontaneously form multilayer bilayers.[16] Lipids comprise approximately 8% of the total stratum corneum dry weight.[17] It has been found that long chain ceramides covalently bind to the surface proteins on the corneocytes, acting as anchors to which the other lipids adhere and also participate in intercellular adhesion.[18] Because the normal intercellular space consists predominantly of lipids, chemical solubility and diffusion of lipophilic (fat soluble) compounds through the skin occur more readily than for hydrophilic (water soluble) compounds. This will be discussed in more detail below. A simplified schematic of the stratum corneum construction with its dense corneocytes and intercellular fluid was first referred to as a brick and mortar model by Michaels et al. in

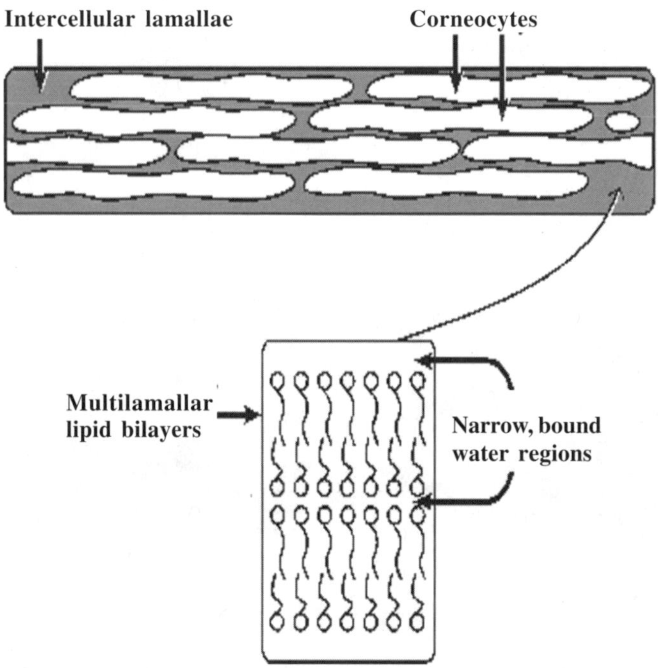

Figure 3. Organization of the stratum corneum and intercellular lipids.

1975 and is depicted in Figure 3.[19] We know at present that the skin barrier is probably far more complicated than was previously understood due to the organization of the intercellular lipids.[20]

The thickness of the stratum corneum varies by anatomical location on the human body. Table 1 shows the average skin thickness for several locations.[21] Overall, the stratum corneum is surprisingly thin. The stratum corneum on most of the body surface is only 10 to 16 µm thick, with limited areas of increased thickening related to the need for abrasion resistance, such as on the palms of the hands and soles of the feet where it is 400–600 µm thick. To appreciate how thin the majority of the stratum corneum is in respect to other familiar objects, the average human hair is about 50–70 µm in diameter.

Table 1:
Human Skin Thickness by Anatomical Site*

Skin Area	Stratum Corneum Thickness, mm
Abdomen	15
Volar Forearm	16
Back	10.5
Forehead	13
Scrotum	5
Back of Hand	49
Palm	400
Sole	600

*Adapted from Reference 21.

Polyester adhesive tape (3M Scotch Tape®) is about 25 µm thick. The thinnest synthetic glove is about 7 µm.

It is clear that the stratum corneum provides an important physical and chemical barrier. Because chemicals permeating the skin are not likely to pass directly through the corneocytes but must instead traverse the long distances of the intercellular spaces surrounding the flattened corneocytes, permeability is reduced by about 1000 times relative to a pure lipid phase.[19] As such, the apparent effective thickness of the stratum corneum has been calculated to be between 500 to 750 mm.[22,23] Another way to look at this is that it would require a homogeneous film at least 500 mm thick, impregnated with the same lipids as present in the skin, to provide equivalent protection from chemical permeation.

Beneath the stratum corneum is the viable epidermis. Within this layer is the location of the germinal stratum granulosum where new corneocytes are formed from keratinocytes. The viable epidermis is about 50 to 100 mm in thickness, depending on location. A complete replacement of the skin corneocytes that make up the stratum corneum occurs every 14 to 28 days in humans.[24] However, it has been shown that continuous occlusion increases this duration by about twofold.[25]

The viable epidermis provides an important and rather unique immune capability by providing both specific and nonspecific protection against pathogenic microorganisms and environmental antigens. The epidermis is unique in having enhanced capability over the respiratory or oral routes in recognizing foreign materials and mounting a immune response.[26,27] Respiratory sensitization may be effectively induced solely by cutaneous exposure to both low molecular weight compounds (i.e., haptens),[28,29] and high molecular weight organic substances, (i.e., allergens) such as natural rubber latex proteins.[30-33] The major immunologically active cellular constituents are keratinocytes, Langerhan's cells, skin-infiltrating T-lymphocytes, and postcapillary venule endothelial cells. Langerhans cells are mobile, can seek out foreign agents, and play an important role in immune processes, especially allergic contact dermatitis. It is important to recognize that this immune capability is just below the thin stratum corneum. Regional lymph nodes link the skin with the systemic immune system, which together are referred to as the skin-associated lymphoid tissue.[34,35]

The thickest layer of the skin is the dermis, which is about 500 mm to 3 μm thick, depending on anatomical region. This highly vascular layer contains the appendegeal structures like the hair follicles, eccrine, and sebaceous glands that originate in this layer. Within a cubic centimeter of human skin are 11–100 philosebacious glands with hair follicles and 100–400 eccrine glands. The total cross-sectional area of the appendages is probably 0.1–1% of the surface area of the skin, and the total volume available for percutaneous absorption (excluding hair diameter) is only about one-tenth of that.[36] Although these appendegeal structures represent less than 1/100th of the skin surface area, they may represent an important shunt through the stratum corneum for large hydrophobic compounds and small particles.[37]

Significance of Occupational Skin Exposure

Chemicals that contact the skin can cause adverse health effects in two ways. First and most obvious is when the skin itself is affected, and there are pathological changes. The most likely effects include allergic and irritant contact dermatitis, which comprise the bulk of occupational contact dermatitis (OCD). Another way skin contact can affect the worker, but is often much more obscure, is when potentially toxic chemicals are absorbed through the skin, adding to the systemic body burden and toxicity in internal organs.

The Bureau of Labor Statistics (BLS) estimates that skin disease currently accounts for 13% of all reported occupational disease.[38] According to the BLS the rate of occupational skin diseases was 81 cases per 100,000 workers in 1997. The estimated annual cost may be upward of $1 billion.[39] According to the latest available data, dermatitis is the third most common cause of compensable temporary total and partial disability and the sixth most common cause of permanent partial disability in the United States.[40] Although both major causes of OCD are highly preventable through avoidance of exposure, the prognosis of untreated OCD is generally poor.[41,42]

The extent skin absorption contributes to other possible causes of disability such as systemic poisoning, neurotoxic effects, and ill-defined conditions, is presently poorly documented. A full understanding of the health significance of skin exposure leading to systemic toxicity is far less clear because of the difficulty in objectively determining the role of skin absorption to an adverse health outcome, especially if the illness were the result of chronic exposures. If skin absorption contributed to a fraction of the estimated total annual 60,000 deaths and 860,000 occupational illnesses attributed to workplace exposures, it could be a substantial number.[43]

Occupational Dermatitis

Occupational skin disease includes any abnormality of the skin induced or aggravated by the work environment. The term dermatitis relates only to skin conditions having an inflammatory component involved with its pathogenesis, while dermatosis relates to skin disease from any cause and with any pathologic outcome.[44]

Causes of occupational dermatoses include (1) mechanical, caused by friction, pressure, and mechanical disruption, (2) chemical, (3) physical, caused by extremes in temperature and radiation (principally ultraviolet), and (4) biological, caused by microbiological and parasitic organisms.[44,45] One estimate is that about 75% of occupationally related skin disease seen in the infirmaries of industrial plants was attributed to mechanical trauma. It was noted that while this type of injury is usually relatively minor, it can predispose the skin to more serious dermatoses due to the skin's compromised mechanical barrier.[44] About 90–95% of all work-related dermatoses, not including those that are caused by mechanical trauma, are considered OCD or eczema.[46] Occupational contact dermatitis is typically characterized by inflammation and erythema (reddening), itching, or the formation of scales as a result of contact with external chemicals or substances. The occurrence of pustules (small pus-containing superficial lesions) is rare and usually is a sign of secondary infection. Contact dermatitis can be further divided into two etiological classes—allergic and irritant.

Allergic contact dermatitis is a delayed-type immunological reaction in response to contact with an allergen in sensitized individuals. This reaction is also referred to as Type 4, or T-cell-mediated immune reaction, since there is a procession of cellular events within the body leading up to the inflammatory response. In predisposed individuals, initially in the induction period allergenic chemicals penetrate the intact skin as small molecules (usually <400 MW), and they are incompletely allergenic (haptens) until they bind to protein and form a complete allergen.[47] Langerhans cells are specialized cutaneous immune effector cells that bind with the allergen and direct the allergen to a regional lymph node. Within the lymph node, interaction with T lymphocytes is followed by replication of sensitized T lymphocytes and completion of the induction phase. Sensitization can occur after just one single exposure, but

Table 2:
Some Chemicals Causing Allergic Dermatitis Among Workers

Chemical	Occurrence
acrylates	paint plasticizer, plastics
bacampicillin	pharmaceutical
benzocaine	pharmaceutical
chloroacetamide	water-base preservative in paints, glues, cosmetics
colophony	electronic solder flux, adhesives
cobalt metal, fume, and dust	metal smelting
diglycidyl ether of bisphenol A	epoxy, product fabrication with resin
ethylenediamine	solvent and chemical intermediate
formalin	textiles, embalming
hydrazine	soft solder flux, chemical intermediate, metal cleaning
d-limonene	cleansers, degreasers
mercaptobenzothiazole	rubber, PPC, antimicrobial agent
methacrylate compounds	dental laboratory denture technicians
methylene diisocyanate	rigid polyurethane
neomycin sulfate	pharmaceutical antibacterial
nickel	stainless steel, metal products
p-phenylenediamine	oxidative hair dyes, cosmetology
parabens mixture	preservative in skin medication, cosmetics, cleansers
phenyl glycidyl ether	epoxy resin
picric acid	battery manufacture, colored glass, explosives
poison ivy	outdoor work
potassium dichromate	histology, leather, matches, spackle compound, photography
substilisins	detergent manufacture
thiram	rubber manufacture, PPC, food disinfectant, lub oils
toluene 2,4-diisocyanate	polyurethane foam manufacture

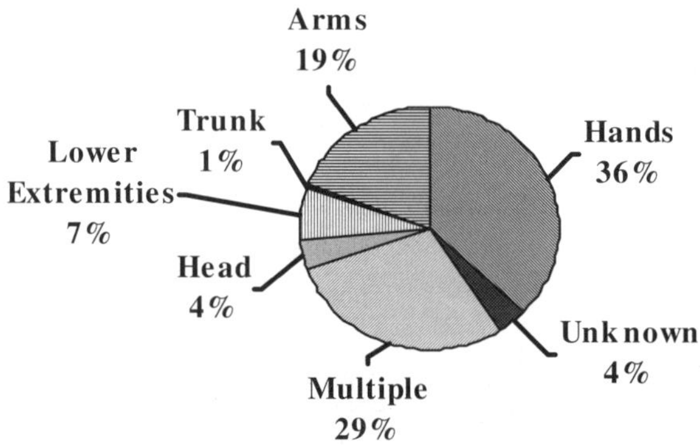

Figure 4. Locations on the body affected by occupational contact dermatitis resulting in worker's compensation in Oregon, 1988–1992. Data from Reference 56.

requires a lag period of a few days to a couple of weeks for induction to be complete. Once sensitized, and upon re-exposure to the allergen (the elicitation phase), normally it takes from 12 to 96 hours for a reaction to occur, but more usually 48 to 72 hours after exposure.[48] Table 2 lists some common chemicals known to cause allergic contact dermatitis in industry. Allergic contact dermatitis accounts for about 30 to 50% of the cases of contact dermatitis in the workplace.[49]

Induction of allergic contact dermatitis is known to depend on the concentration of the allergen on the skin surface. If a sensitizing dose of the allergen is spread over a larger surface area, the likelihood of sensitization appreciably declines. It is believed that sensitization is dependent on the number of allergen molecules per Langerhans cell—a small number of cells bearing many molecules being more effective than having many cells bearing a few molecules.[50] There appears to be a threshold surface concentration for induction of all sensitizers, and the range of induction concentrations is quite large. Caution is warranted in strictly interpreting experimental laboratory data since influences in the workplace might alter percutaneous absorption of chemicals. Also, repeating the exposure over a period of time seems more effective in inducing sensitization than is a single dose.

Genetic disposition plays a prominent role in determining individual susceptibility.[51] Although there appears to be a fairly linear dose-response to sensitizing compounds among humans, once sensitized there is wide individual variability in the provocation (elicitation) threshold concentration. The concentration necessary for eliciting an individual response in sensitized individuals can span at least a 100-fold range.[52]

It has been reported that 90% of all occupational allergic contact dermatitis was found on the back of the hands and the forearms.[53] However, contact dermatitis among homemakers occurred in almost 50% of the cases on the palms, whereas 15% of the time it affected the back of the hands and fingers.[54] In a different study of dental laboratory technicians, the fingertips were primarily involved in allergic contact dermatitis (93%), whereas in irritant contact dermatitis, the dorsum of the fingers were affected in 80% of the cases.[55] Figure 4 depicts the locations on the body for occupational contact dermatitis based on 879 recent Oregon workers' compensation cases for the period 1988–1992.[56]

Contact urticaria (Type I immune reaction) is an immediate immunological response in the skin resulting from circulating chemical-specific antibodies coming into contact, most commonly, with exogenous proteinaceous molecules (e.g., animal dander, natural rubber latex proteins, foodstuffs, industrial enzymes). The reaction becomes apparent as a wheal, or hive, which is a firm, rounded or flat-topped elevated lesion that results from edema (swelling) of the dermis. Wheals are often pink in color, priuritic, and the individual lesion persists fewer than 24 hours.

In type I reactions, in addition to the skin, the respiratory and gastrointestinal tracts, as well as the cardiovascular system, may respond after cutaneous exposure to allergens. Much less frequently, contact urticaria can result from exposure to low molecular weight chemicals forming chemical–protein conjugates (e.g., 2-ethylhexyl acrylate) in the epidermis. Response is mediated by allergen-provoked release of histamine from cutaneous mast cells. In Finland, a recent survey of OCD cases from 1990–1994 found that almost 30% of all occupational immunologically mediated dermatoses were

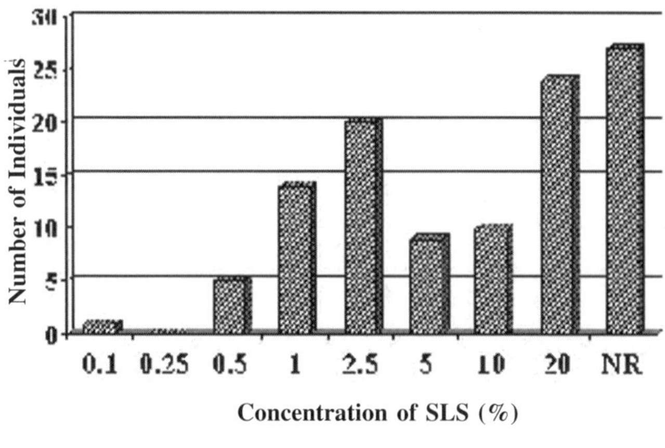

Figure 5. Acute irritant response thresholds among individuals with six skin types. Minimum SLS concentrations inducing irritation from 110 individuals with all skin types showing wide variation in susceptibility. Data from Reference 66.

due to contact urticaria, while the remainder were allergic contact dermatitis.[57]

Contact dermatitis from irritants constitutes about 50 to 80% of all OCD cases.[58] There are several forms of response common to irritant exposures that are dependent on the chemical substance, concentration, duration, and the individual exposed. The first type is caused by a single application of a strong compound, which results in a toxic, acute reaction primarily associated with reddening (erythema) and possibly a burning or stinging sensation. The second type results from repeated exposure, which results in erythema, chapping, and fissures in the skin. The third type also results from repeated exposures, but develops into a chronic dermatitis that is characterized by erythema and scaling, with frequent fissuring of the stratum corneum.[59] A compensatory process of tolerance for irritants, sometimes called "hardening," can occur that results in lichenified (thickened) skin. A subcategory of irritant dermatitis manifests only after a lag time of 8 to 24 hours or longer, and is thus referred to as being a delayed type.[59] Some industrial chemicals known to cause delayed effects include epichlorohydrin, ethylene oxide, hydrofluoric acid, some acrylates like hexanediol and butanediol diacrylate, and propane sulfone. In agriculture, the pesticide triphenyl tin hydroxide can cause delayed skin effects. Such compounds that cause a delayed irritant reaction characteristically penetrate the stratum corneum slowly and are cytotoxic to the viable epidermis.

Usually acute irritant response is rapid and begins to subside in 24 to 72 hours but as mentioned above some compounds can take longer, responding similarly to delayed allergic reactions.[60] However, in experimental studies with sodium lauryl sulfate (SLS), complete functional skin recovery after a single 24-hour exposure had not completely occurred 12 days following exposure. [61] The prognosis for complete resolution of both the first and second type of acute irritation is good if exposure is quickly discontinued. The prognosis for chronic irritation of the skin is variable but generally poor.[41,62-64]

Substantial individual range of susceptibility has been found to exist among individuals when exposed in the same way to a model irritant, such as the detergent SLS.[65,66] The range of threshold concentrations necessary to induce an irritant response to SLS is indicated in data depicted in Figure 5 for a 4-hour occluded patch test. One hundred and ten individuals representing all skin types ranging from very fair skin (burns easily, never tans) to deep normal pigmentation were included. There seem to be little relationship between skin pigmentation type, ability to sunburn, or gender as predictors of skin sensitivity to irritants.[67] Using sensitive measures that do not include erythema, black skin may be more sensitive than white skin.[68] Young skin seems more susceptible than older skin in spite of the fact that older skin tends to thin with advancing age.[69] People with respiratory atopy without outward signs of cutaneous atopy are generally no more susceptible to irritants than nonatopics but persons with a history or current atopic dermatitis seem several fold more prone.[70-72] The probability of a given concentration to induce irritation is also dependent on the season. This is because of seasonal differences in ambient humidity and hence skin hydration. In one test group, 45% reacted to 20%

SLS in summer, whereas 91% reacted to this concentration in the winter.[73] For other compounds, like alkalies and powders, irritant response seems more likely in the summer.[60]

Chemical Protective Clothing and Occupational Dermatitis

The recent increase in occupational sensitization to natural rubber latex proteins due to use of latex gloves has received much attention. This is a type I allergy involving high molecular weight allergens resulting in immediate type urticarial responses. Fortunately steps have been taken to reduce the incidence of this problem and latex gloves should not be commonly used in occupations outside of health care industries where bloodbourne pathogen protection is the primary intent.[74] Another type of allergic response due to small molecular weight chemicals, commonly used as accelerators in the polymerization of natural and synthetic rubber, is responsible for type IV, delayed response sensitization (allergic contact dermatitits). Most of the cases of glove-related contact dermatitis involve rubber gloves whereas cases reported from use of synthetic rubber gloves are relative few.[75] These allergen compounds belong to classes of chemicals including thiazoles, thiurams, dithiocarbamates, guanadines, thioureas, and amine aldehydes.[76] Other potential allergen additives are antioxidants, typically derivatives of *p*-phenylenediamine, that are primarily used in rubber products. These compounds may leach out of gloves and into the wearer's skin.[77] Because damaged skin is more likely to absorb chemicals, it should be kept in mind that if a person with dermatitic skin begins using gloves that contain allergens, this could be the cause of more serious problems later.[78,79]

In Finland, 1991, the eight most common types of chemicals responsible for allergic contact dermatitis were rubber chemicals, nickel, epoxy resins, formaldehyde, thiuram sulfides, chromate (hexavalent), isothiazolinones, and colophony. It is likely that the principal contact with rubber chemicals, thiuram sulfides, and isothiazolinones occurred from wearing chemical protective gloves.[80] In a later report, the most common causes of allergic contact dermatitis were rubber chemicals (26%), synthetic resins, plastics, glues, and paints (19%), and metals (16%). While the sources of the rubber chemical allergies were most often attributed to protective gloves, no cases of occupational allergic dermatoses were detected in the rubber industry during 1993.[81] Paradoxically hand cleansers and gloves were among the top causes of occupational dermatitis in most other studies as well.[82]

In the United Kingdom, the most frequent agents for contact dermatitis were rubber chemicals and materials (14.1% of cases reported by dermatologists), soaps and cleaners (12.7%), nickel (11. 9%), wet work (11.1%), personal protective equipment (6.2%), petroleum products (6.3%), cutting oils and coolants (5.6%), and epoxy and other resins (6.1%).[83]

In addition to allergens, the wearing of gloves may cause or exacerbate skin irritation (also see section on occlusion below). According to Taylor, irritant contact dermatitis is probably the most frequent adverse glove reaction, especially among those that wear latex gloves.[84] Wilkinson recently reported the type of dermatitis among 44 patients with glove intolerance and found that 75% had irritant contact dermatitis, 20% had contact urticaria, and 8% had allergic contact dermatitis.[85] Contributors to skin irritation may include powders that are added to enhance donning and doffing, trapping of irritating hand cleaners or other chemicals underneath an occlusive covering, or just the simple friction of glove to skin contact that occurs when flexing.[86] The degree to which each individual variable or combined causes exacerbates skin irritation has not been well characterized, but there are limited studies relating to this in the literature. For instance, a questionnaire survey of dental surgeons revealed that 29% of the respondents indicated that they experienced skin irritation. This was more evident among those who wore gloves most often.[87] In another study of hospital and dental care workers, 37% reported skin problems associated with frequent glove use.[88]

Cornstarch powder is presently commonly used as a donning agent on most powdered gloves and replaced talc for the most part, although talcum powder may still used on a limited basis as a mold release agent in glove manufacturing.[89] Thus, even powder-free gloves are likely to contain some powder, albeit at considerably lower levels.[90] More common presently is the use of calcium carbonate as a mold release agent. Also in use are silicone oils and chlorination, which enhance the slipperiness or smoothness of the inside surface. Brehler et al. found the glove irritation effect variable depending upon the glove and powder used.[91] They also found some gloves with alkaline surface pH conditions (up to 10.2) that reportedly could alter the skin surface pH for hours afterward. Although in their experimental studies unpowdered gloves seemed less irritating than powdered gloves, other user acceptance surveys indicated little consistency with this.[87] It is likely that several factors acting in combination may contribute to glove irritancy.

Permeation and Penetration of Substances Through the Skin

Certain chemicals will be absorbed through the skin much more readily than others. In intact skin, the main means of absorption is through permeation. Permeation is the molecular diffusion of chemicals within a matrix like the skin. The solubility of chemicals in the stratum corneum plays a big part in determining the concentration gradient achieved between the outside and inside. If one knows the skin absorption rate, and it is combined with the toxicity of the chemical once absorbed, one can estimate the potential risk of skin exposure.

It has been found from extensive study of the experimental permeation data for humans that the two principal factors that determine a chemical's likelihood to permeate are its solubility and molecular size. Solubility is typically expressed as water solubility or preferentially as the octanol–water partition coefficient, the latter being a representation of the chemical's solubility in a nonpolar phase versus a polar phase. Because of the wide numerical range of octanol–water coefficient values, or K_{ow}, the log of the K_{ow} is usually reported. The majority of chemicals have a log K_{ow} of less than -1 to 6, representing indeed a very wide range of solubilities. Because the intercellular channels contain bipolar moieties, chemicals with a log K_{ow}

between 1 and 3 most readily permeate the skin. Chemicals that are very lipophilic will not readily diffuse through the hydrophilic epidermis, and therefore may remain in the stratum corneum for a considerable time. Very hydrophilic compounds may also remain in the skin for prolonged periods because of their poor penetration of the stratum corneum, possibly exceeding one week in duration.[92] In addition, some compounds appear to have an affinity for proteins, reversibly interacting with specific chemical sites.[93] Prolonged residence in the stratum corneum is referred to as chemical substantivity. Another term referring to this phenomenon is the skin depot.

In one evaluation of the chemical features of 17 compounds with a wide range of octanol–water solubilities, the strongest positive correlations were found between partition coefficient (K_{ow}) and protein binding, a marginal correlation was found between protein binding and concentration in the epidermis, while partition coefficient was unable to predict concentration in the epidermis.[94] The second chemical-specific property of importance is the molecular size of the compound. Above a molecular weight of 500 daltons, permeation flux through the intact skin is usually negligible for most chemicals. Within this range, however exist a large number of chemicals of commercial importance.

In addition to permeation that occurs through molecular diffusion, compounds and small particles may penetrate the skin through discontinuities in the stratum corneum or perhaps through physical processes that occur during skin flexing. For instance, contact urticaria can result from cutaneous contact with natural protein allergens. Contact urticaria is a quite common problem among workers exposed to animal products, enzymes, and food proteins.[95] Typically these high molecular weight (HMW) proteins are usually >10,000 MW (>10 kilodalton), but are apparently able to traverse the stratum corneum to first induce sensitization and then result in a subsequent elicitation of response. Mice, for instance, have been sensitized through topical application on intact skin by ovalbumin and other allergens.[96,97] The molecular weight of ovalbumin (OVA) is 45 kilodalton. These experiments resulted in statistically significant elevations in OVA-IgE specific antibody in the dosed group. Also found was that the

Table 3:
Comparative Relative Permeability of Human Skin to Topical ^{14}C Hydrocortisone, Parathion, Malathion, and Water

Regional Variation	Parathion[1]	Malathion[1]	Hydrocortisone[1]	Water[2]
Forearm (ventral)	1	1	1	1
Palm	1.3	0.9	0.8	3.7
Ball of foot	1.6	1	—	12.6
Abdomen	2.1	1.4	—	1.1
Back of hand	2.4	1.8	—	1.8
Scalp	3.7	—	3.5	—
Angle of jaw	3.9	—	13	—
Forehead	4.2	3.4	6	2.7
Axilla	7.4	4.2	3.6	—
Scrotum	11.8	—	42	5.5

[1]Data from Reference 100.
[2]Data from Reference 21.

concentration required to induce sensitization decreased with repeated epicutaneous exposures. Natural rubber latex proteins have also been shown to penetrate intact human skin with much greater penetration if the skin is damaged, and topical dosing has produced respiratory sensitivity in mice.[98,99]

Factors Affecting the Skin Barrier

The following section describes some of the main variants that affect individual skin barrier function and health. Understanding these influences is essential to better protecting workers' skin and determining the prudent use of CPC.

Table 4:
Individual Range in Alveolar Air Concentration Following a 30-Minute Dermal Exposure*

Solvent	Subjects	End of Exposure
carbon tetrachloride	3	0.11–0.83
trichloroethylene	3	0.033–0.76
tetrachloroethylene	5	0.17–0.17
methylene chloride	3	2.3–3.6
1,1,1-trichloroethane	6	0.19–1.02

*Adapted from Reference 110.

Anatomical Differences

All skin over the human body is not exactly the same. In addition to differences in thickness, as shown in Table 1, there are compositional differences in the main pathways of diffusion. Permeation of chemicals through the stratum corneum was stated above to be primarily through the intercellular bipolar lipid channels. It has been found that there are regional variations in skin permeability that correspond to the differing amounts of intercellular lipids therein. Table 3 shows the relative difference in skin permeation of three organic compounds as well as water through various regions of the human body.[21,100] Parathion is the least water soluble while hydrocortisone is the most. The increase in hydrocortisone permeation in some regions of the body appears to correspond to eccrine sweat production. Note how the palm, even though about 27 times thicker than the forearm (refer to Table 1), is almost equivalent in its barrier function. The planar stratum corneum (palms and soles of the feet), although thicker and able to resist physical abrasion better, has less intercellular lipid and is a poor diffusion barrier to chemicals, especially by those compounds that are more hydrophilic. The percentage weight of lipids in the planar stratum corneum is 1.3%, whereas lipids account for 7.2% by weight in the forehead skin.[101] Comparison of the water loss through each site (planar: 20–40 g/m^2/hr versus forehead: 4-7

Table 5:
Effect of Type of Physical Damage on Skin Absorption of Nicotinic Acid in Human Skin*

Condition of Skin	% Absorbed	
	In Vivo	In Vitro
Normal	7	5
Abraded	47	51
Tape stripped		58
UV irradiated[1]		
1.5 minutes	22	7
6 minutes	51	13

[1]Application was made 3 days after irridation.

*Adapted from Reference 118.

g/m²/hr) also correlates with the difference in intercellular lipid content.[102]

Just as there are regional anatomical differences in skin permeability, there are also differences in regional susceptibility to skin irritants. Permeability and irritant response are probably at least partly related, although biological mediators of response also play a role.[103] Cua et al. found the thigh to be the most sensitive to the irritant sodium lauryl sulfate while the palm and ankle were least responsive.[104] This is fortunate given the increased likelihood of contamination of these sites.

Inter-Individual Differences

Inter-individual differences in persons with apparently healthy skin appear common and can be appreciable. According to Feldman and Maibach, the standard deviation of expected skin permeabilities should be expected to be one-third to one-half the mean value.[105] Assuming a normal distribution, 1 person in 10 will absorb twice the mean value, while 1 in 20 will absorb three times this amount. Up to tenfold differences in interpersonal skin absorption rate have been seen within small group studies that have been reported in the literature for such compounds as hydrocortisone and parathion.[106-108] The apparent transdermal absorption rate of nitroglycerin in healthy volunteers, using the same site, resulted in variations from 21% to 78% within six subjects.[109] Individual variation in alveolar air concentrations following a 30-minute skin exposure to five different chlorinated solvents is shown in Table 4. A large range in the apparent uptake was seen, assuming minor confounding from differences of *in vivo* metabolism of the absorbed compound.[110] Similar large inter-individual differences were seen by Lauwerys et al. when 11 male volunteers were asked to immerse both hands into pure m-xylene for 20 minutes.[111] The skin of all volunteers was free of lesions. The total amount of m-xylene absorbed ranged from 16 to 110 mg (6.9-fold range). Same person differences tested twice during a one-week period were twofold or less. The range of skin permeability seen in healthy individuals may also account for the large range of susceptibilities of response to irritants.[65]

Gender and race, as these relate to the skin, have only been marginally studied to date. However, significant gender-related differences have not been found after repeated, daily application of an irritant.[112,113] Black skin was found more responsive to irritation than white skin using objective techniques such as transepidermal water loss and increased cutaneous blood flow, but no differences in erythema or in diagnosed cases of dermatitis have been noted.[114-116] Age decreases the thickness of the stratum corneum, the lipid content, and transepidermal water loss, but visual and objective measures of responsiveness to both irritants and allergens appear to decline with age.[69,117]

Physical Damage

The barrier properties of the stratum corneum, given its thinness, make it quite unique. The important practical aspect of this knowledge is the realization that when the stratum corneum is healthy, it can perform an outstanding job, relative to its thickness, of resisting chemical insults. But if this thin barrier is physically damaged, or the intercellular lipids are altered, the stratum corneum becomes much less of a barrier. Table 5 presents some examples of damage by abrasion and

Table 6:
Effect of Chemical Irritation on Percutaneous Penetration*

Compound	Log Partition Coefficient	Percent Enhancement
HC	1.6	260
M	3.1	160
IB	3.5	190
AC	6	140

*Adapted from Reference 129.

ultraviolet light irradiation to the stratum corneum and the effect on absorption that different types of damage might have.[118]

Abrasions and cuts are probably the most common insults to workers' skin. A nonintact stratum corneum offers little protection against permeation. There have been numerous documented cases where exposure to chemicals, not normally absorbed through the skin in sufficient amounts to cause even mild effects, have actually resulted in death when a few scratches were present. In one case, a women pruning orchard trees that had been sprayed earlier with paraquat developed scratches on her unprotected arms and hands during her work. Normally, paraquat is not a dermal exposure hazard due to its poor permeation through healthy skin. In her case, death ensued from respiratory failure a few days later.[119] Experimentally removing the stratum corneum by tape stripping resulted in 2.6- to 8.5-fold increase in absorption of seven different pesticides.[120] Some dangerous radionuclides, like cobalt-chloride, which are poorly absorbed through intact skin (<0.1%) will rapidly penetrate abraded skin (52%) after 60 minutes.[121]

Dermatoses

Dermatoses constitute a broad range of skin conditions that can result from a variety of chemical and physical traumas

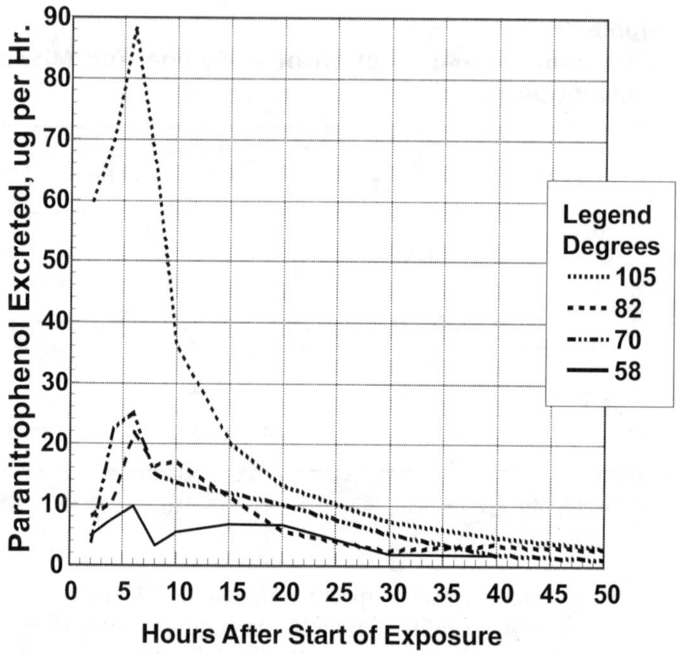

Figure 6. Affect of ambient temperature on percutaneous absorption of parathion dust from the forearm skin. Data from Reference 132.

to the skin, and as pointed out before, damaged skin presents a less effective barrier. Dermatitis is certainly a form of damage. This is especially true when the dermatitic skin is in the acute stage of response.

For instance, percutaneous absorption of the pesticide lindane in patients with severe scabies was 10- to 40-fold greater than in normal persons.[122] In other studies, measurable differences were seen in metabolite excretion of carbon disulfide exposed workers and in dimethylformamide exposed workers who had skin irritation or skin disease.[123,124] Percutaneous absorption of xylene vapor appeared to be about three times greater in a volunteer with atopic dermatitis.[125] Finally, a higher blood concentration of a pesticide was seen in a formulator as compared to his coworkers, which was attributable to the presence of scleroderma in the worker with the

Table 7:
Percutaneous Absorption of M-Xylene and Mixed Solvent Effects in In Vivo Human Skin*

Solvent	Symptoms	Methylhippuric Acid Excretion
Pure Xylene	Erythema No "tightness"	1.0
1:1 Xylene + Isobutanol	Erythema No "tightness"	0.4
10:10:1.5	Mild wrinkling; skin dry and oily	1.1

*Adapted from Reference 125.

high blood levels.[126] Hyperproliferation of the stratum corneum, as in psoriasis, exfoliative dermatitis, or ichtyosis, also results in increased permeability of the skin.[127,128] Rapidly generated stratum corneum, while thicker, is usually a poorer barrier.

Irritation Effects

Irritation of the skin can also result in increased percutaneous absorption. The data in Table 6 shows an enhancement in percutaneous absorption that is related to the water solubility of four model compounds.[129] These compounds were applied to the upper back of hairless guinea pigs after 0.5% sodium lauryl sulfate in water was applied in an occlusive chamber for 24 hours. Because the resistance of the stratum corneum to diffusion is greater to polar compounds than to nonpolar compounds, any disruption of the barrier should enhance penetration of hydrophilic compounds to a greater extent than would occur for lipophilic compounds. In this cited experiment, enhancement was 2.6-fold for the most hydrophilic compound but only 1.3-fold for the most lipophilic.

In another study where sodium lauryl sulfate was used to induce skin irritation, response to nickel salts in sensitized persons and animals was appreciably increased with the addition of the irritant.[130] Irritation, as well as several other factors that can influence sensitization, are reviewed elsewhere.[131]

Temperature and Relative Humidity

Molecular diffusion and dissolution will increase within matrices as the temperature increases. It should be no surprise that as temperature increases, the permeation of chemicals through membranes such as the skin will also increase. Normally, the skin surface temperature is between 32–35°C, but this can increase or decrease with heightened or lowered ambient temperature, or if the temperature of a liquid contacting the skin were different.

Figure 6 shows how the ambient temperature affected the excretion of p-nitrophenol in the urine of human volunteers after 5 grams of 2% parathion dust had been applied topically for 2 hours. Only one hand and forearm were exposed and placed into a temperature-controlled chamber. Following the 2-hour exposure, the skin surface area was decontaminated by scrubbing with soap and water for 5 minutes, followed by two washes with ethyl alcohol. The average excretion rate increased from 4.9 mm/hr at 58°F (14°C) to 19.6 mm/hr at 105°F (40 °C) over a total period of 41 hours.[132] Chemical vapor penetration also has been experimentally determined to increase with ambient temperature. Percutaneous absorption of aniline vapor, for example, increases about 20% for each 5°C increase in air temperature.[133] Percutaneous uptake was also significantly increased for 2-butoxyethanol vapor when the air temperature (but also humidity) was increased from 23°C to 33°C.[134]

Increasing relative humidity has been shown to increase percutaneous absorption. Chang and Reviere found under constant temperature the percutaneous absorption of parathion increased by two- to fourfold when the humidity was increased from 20% to 90%.[135]

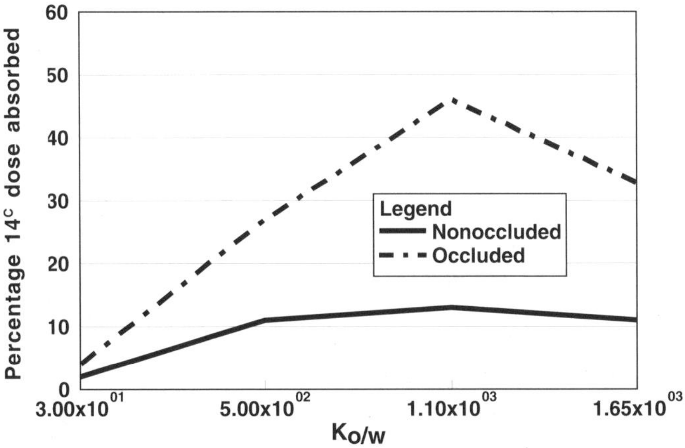

Figure 7. Percutaneous absorption of 4 steroidal compounds with and without occlusion. A single application of 4 µm/cm2 was applied to the ventral forearm. Occlusion was continuous for 24 hours, after which the site was washed. Mean and SD shown. Data from Reference 135.

Vehicle

A vehicle in this context is commonly a liquid in which another contaminant, which may be more toxic, is contained. Vehicles are important to percutaneous absorption because they may enhance the absorption rate by either disrupting the stratum corneum barrier function, or encouraging partitioning toward the skin, or a combination of these processes. As will be shown in the next examples, vehicles, or the co-components of a mixture, can have significant effects on how much of a toxic chemical enters the skin.

Another example of vehicle effects on percutaneous absorption, particularly where the vehicle is a mixture of solvents, is demonstrated in the data in Table 7. In this experiment, xylene absorption as measured by methylhippuric acid excretion, was compared to exposures of pure xylene, 1:1 mixture of xylene plus isobutanol, or 10:10:1.5 mixture of xylene, isobutanol, and water to saturation.[125] In each case volunteers immersed both hands up to the wrists in the solvents for 15 minutes at room temperature. Interestingly, the addition of a small amount of water to the mix dramatically increased xylene permeation rate so that it was almost three times greater than the xylene–isobutanol mix, and slightly more than pure xylene alone. With only xylene and isobutanol, there appeared to be a conspicuous dehydration of the skin by isobutanol and a delay in the absorption of xylene evident in the methylhippuric acid excretion (data not shown). The result seen by adding a small amount of water might be due to the increased hydration of the stratum corneum provided by the addition of water, or increasing the partitioning of xylene out of the saturated aqueous mixture towards the skin. This example may have relevance to wearing CPC as perspiration buildup is commonly associated with this.

More typically, the effect of the vehicle is compared among different pure solvents. The rule of thumb that "likes dissolve likes" applies to permeation in that hydrophilic solutes will tend not to partition to the skin if in a hydrophilic solvent, and lipophilic solutes will tend not to partition as well into the skin if present in a lipophilic solvent. This principle is exemplified by the 10-fold difference in the skin flux rate of benzocaine when in a water (lipophobic) vehicle or in polyethylene glycol (PEG) 400 (lipophilic) vehicle. The solubility of benzocaine in water is 1.26 g/L whereas the solubility in PEG 400 is 435 g/L. The skin flux rates were 0.1 $mg/cm^2/hr$ and 0.01 $mg/cm^2/hr$, respectively.[136]

Occlusion

Occlusion is one of the most effective means of increasing chemical absorption through the skin and the adverse effects on the skin of occlusion are probably the most important outcomes of wearing CPC. It should be realized that in referring to the skin permeation effects of occlusion, the workplace scenario most similar to this is when the contaminant gets underneath CPC. Polymer gloves provide excellent occlusive coverings. Because of the enhanced permeation attributed to occlusion, situations have been reported where increased absorption occurred because the personal protective clothing became contaminated on the inside (see Performance of CPC below).

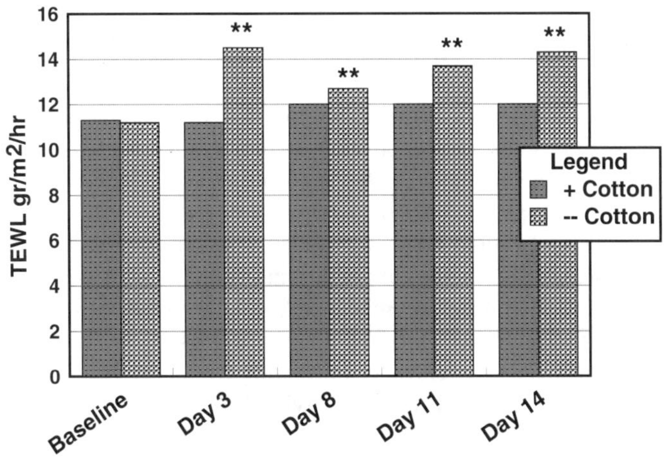

Figure 8. Effect of long-term experimental occlusion on skin integrity. A significant detrimental effect of wearing a glove for 6 hours per day for 14 days was detected by measuring transepidermal water loss, which is indicative of barrier integrity. It was concluded that gloves may be a substantial factor in the pathogenesis of cumulative irritant contact dermatitis, but that with proper use the risk might be minimized. (Adapted from Reference 144.)

Covering the skin with a moisture impenetrable barrier not only prohibits evaporative loss of any chemical that may have contaminated the surface of the skin, but increased skin hydration and increased temperature enhances the absorption rate of the contaminant. This effect has long been known, as demonstrated by Burckhardt in 1939, who successfully used occlusion to experimentally promote sensitization to turpentine.[137] With occlusion, nearly all animals were sensitized; without it none were sensitized. In this experiment, the site of exposure was occluded for 8 to 12 hours.

The stratum corneum normally contains between 5–15% water, but this can be increased to as much as 50% by external factors.[138] Occlusion is an effective way to hyperhydrate the skin. The mechanisms by which hydration can influence percutaneous absorption include altering the partitioning between the surface chemical and the skin due to the increasing presence of water, swelling the corneocytes and possibly altering the intercellular lipid phase organization, increasing the skin surface temperature, and increasing the subcutaneous blood flow. It has been postulated that with hyperhydration of the normally lipophilic stratum corneum, the effective partition coefficient of the penetrant between the stratum corneum and viable hydrophilic epidermis is reduced since the two tissue phases now appear more similar.[139] Thus occlusion has been shown to enhance absorption from several percent to several fold in experimental studies.

Occlusion does not appear to influence equally the percutaneous absorption of all compounds. Rather, the impact of occlusion is influenced by the polarity of the chemical. It most increases the absorption of moderately lipophilic molecules but is less effective on the absorption rate of highly lipophilic or highly hydrophilic compounds. Figure 7 illustrates this process for four steroids with and without occlusion.[140] There may be some chemical-specific factor that also affects occlusive penetration, since the same degree of enhancement per the K_{ow} is not always seen with all compounds.[141-142] Nevertheless, occlusion appears to enhance percutaneous absorption to some extent for all compounds tested.

Just as the permeability of human skin was shown to vary from location to location, it seems that there are anatomical differences in the extent of enhancement of percutaneous absorption due to occlusion. In an experiment by Qiao et al. using weanling pigs, occlusion greatly increases parathion absorption in most sites tested but to a lesser degree in the skin of the shoulder region.[143] In terms of human relevance, the pig abdominal skin absorption value matches the absorption value for human forearm skin in nonoccluded tests with parathion. These data also underscore the importance of skin site selection when conducting animal experimentation and comparing results among laboratories.

The above studies demonstrate the increase in permeation due to a one-time occlusion of contaminants on normal skin. In addition, prolonged occlusion has been found to produce chronic impairment of the barrier function of the skin. Figure 8 shows the results of an experiment to determine the long-term effect on barrier function by prolonged wearing of an occlusive glove.[144] The protocol involved having 18 volunteers wear a nonlatex hypoallergenic glove for a minimum of 6 hours per day for 14 days while sleeping at night. It was noted that during sleep skin temperature and friction are reduced (a best-case situation). Half of the volunteers wore a cotton glove underneath the occlusive glove. The results indicated increased transepidermal water loss at each of the measurement days up to day 14 for the skin with the occlusive glove only. However, wearing of the cotton glove prevented this damage. The authors concluded that prolonged occlusion can damage the barrier function of the skin and might enhance susceptibility to irritants and sensitizing agents. Minimizing the wearing of gloves and wearing of a cotton glove underneath are two ways to reduce the occlusive effects on the skin.

Lasting susceptibility to irritants has been demonstrated by Graves et al. after the skin was occluded.[145] In their experiment, the skin was occluded for 4 hours or 8 hours. After the covering was removed, skin permeability was evaluated by measuring the time to onset of hyperthermia due to topical application of the irritant hexyl nicotinate. After 4 hours of occlusion, the time to onset was reduced to 59% of its preocclusion value. After 8 hours of occlusion, this was further reduced to 38% of the preocclusion value. After occlusion, transepidermal water loss was also significantly reduced.

Direct physiological damage to the skin due to skin occlusion is evident. Within two days of occlusion marked cytotoxic damage to Langerhans cells, melanocytes, and keratinocytes occurs.[146] Intercellular edema and marked swelling of corneocytes are also prominent. Occlusion inhibits production of all key intercellular lipids after barrier disruption, prolonging the normal barrier restoration process.[147-149] The damaging effects of prolonged occlusion can quickly lead to dermatitis.

Interesting questions asked by Graves et al. include the following: (1) at what duration of wearing a glove continuously would there be a significant impairment of the stratum

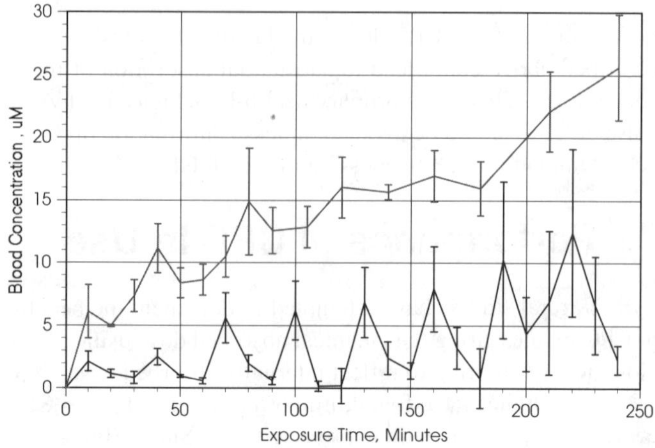

Figure 9. Concentration of n-butanol in blood of guinea pigs during continuous (-•-) and intermittent (-■-) skin exposure. Evaporative loss of volatile chemicals from uncovered skin will prevent large amounts of absorption from occurring, but continuous exposure may occur if solvents penetrate holes in CPC. (Adapted from Reference 163.)

corneum barrier function?; (2) how frequent and how long should nonoccluded periods be to prevent a cumulative effect?; (3) is the impairment of barrier function seen in these experiments sufficient to increase the incidence of irritant contact dermatitis?; and (4) does this impairment correspond to an increased susceptibility to mechanical trauma?[145] These are all relevant and practical questions, and further research will be needed to provide the answers.

Decontamination of CPC and Reuse

Decontamination of used CPC remains a perplexing problem due to great variability of success depending on the CPC material and the chemical of interest. Furthermore, reliable test methods to validate decontamination effectiveness are needed in order to make an informed judgment. If internal contamination of CPC occurs due to incomplete decontamination, wearing of this CPC due to the factors of occlusion described above could lead to substantial absorption of the contaminant. This is the primary health-based reason why reuse of CPC should be avoided, unless confirmation of decontamination effectiveness can be assured.

Performance of CPC In-Use

Several studies have attempted to document the benefit of CPC by measuring the chemical absorbed dose using biological monitoring to reflect internally absorbed dose.[150-154] These studies have found little or no benefit of using CPC compared to working with unprotected skin. Since CPC is expensive to purchase and uncomfortable for the user to wear, this is an unepected and undesired outcome. Assuming that the CPC should provide some barrier protection, there are two possible explanations for this outcome.

One possibility for failing to demonstrate benefit of wearing CPC is that because of poor work practices chemical contamination is allowed inside the user's CPC. Seldom is the contamination that might occur inside CPC checked, which can occur by repeated doffing and donning of the same CPC. Workers' unprotected skin may contact contaminated surfaces when the CPC is not worn, even if for a brief time, or when removing contaminated CPC incorrectly.[155-158] The effect of occlusion, as described above, can significantly enhance percutaneous absorption. The second possibility is that previous chemical contamination residing within the skin depot from previous days may be mobilized by the occlusion caused by CPC. This skin depot mobilization could significantly confound the significance of exposure reduction intervention, especially if the pre-intervention consisted of unprotected exposure and the intervention being evaluated consisted of wearing CPC. All of the cited CPC intervention effectiveness studies performed biological monitoring soon after introduction of the CPC. Both scenarios, consisting of enhancing absorption of albeit lesser amounts of exposure, or mobilizing residing skin depot, are both effects of occlusion that can come from wearing CPC. To reiterate, the outcome was among these CPC intervention effectiveness studies was that no difference or an increase in absorbed dose occurred after the introduction of CPC.

On the other hand, workers who have been regularly wearing gloves for a period of time, compared to workers who typically do not wear gloves, appear to benefit.[159,160] For instance, in one study of greenhouse workers who wore new thin nitrile gloves each day there was a statistically significant difference of serum cholinesterase activity compared to workers who worked bare-handed.[161] Only in future studies, where good work practices are ensured and where the worker is allowed an adjustment period that will factor out this second scenario, will the real benefit of using CPC be known.

In the workplace contact with chemicals may be intermittent and sporadic. A key characteristic of the chemical that will determine the residence time of a single contact is the volatility of the chemical. For uncovered skin, highly volatile chemicals will evaporate before appreciable percutaneous absorption occurs.[162] Volatility inversely correlates with the persistence of a chemical on the skin surface and the amount of chemical ultimately absorbed systemically. Figure 9 presents the results of continuous dermal exposure to n-butanol (vapor pressure = 6 mm Hg) and 1-minute exposure every 30 minutes. The periodic exposure results in a blood concentration that is lower than for continuous exposure. For more volatile compounds, such as toluene (v.p. = 21 mm Hg) and 1,1,1-trichloroethane (v.p. 100 mm Hg), no such cumulative effect occurs.[163] The absorbed dose of highly volatile compounds from repeated short exposures is appreciably less when compared to continuous immersion.[110] Presumably, the systemic absorption of less volatile compounds than n-butanol would eventually approach continuous contact conditions as evaporative loss decreases.

The relevance of this is that liquid chemicals entering pinholes or cuts in gloves will likely remain in contact with the skin for much longer periods of time if not allowed to evaporate. Liquid chemicals are typically low molecular weight compounds and are readily absorbed through the skin. Pinholes in certain new gloves have been reported in a substantial proportion of the time and in used gloves the chance of cuts increases with duration of use.[164-165] Double gloving has been shown to be effective in reducing the probability of skin contact due to the occurrence of tears or chemical permeation. Furthermore, it is important to monitor the condition of gloves periodically and not to use them for extended durations.[166-170]

Summary

Human skin is a marvelous creation that should be treated with care. The skin evolved long before the existence of chemicals that are encountered in the 21st century. It also evolved to be exposed to the open air. Therefore, chemical exposures and the use of CPC are to the skin unnatural conditions. There must be balance and judicious use of CPC in order for it to be most effective. Having a good understanding of the skin and its limits by interaction with the environment and to chemical exposures is important to optimizing the benefits of CPC and for reducing absorption of toxic chemicals or deleterious effects on the skin itself. The goal of this chapter was to highlight those factors that the occupational health and safety person responsible for a CPC program should be aware of in order to meet those aims.

References

1. Kanerva, L., P. Elsner, J.E. Wahlberg and H.I. Maibach, *Handbook of Occupational Dermatology* (Springer, Berlin, 2000).
2. Marzulli, F.N., and H.I. Maibach, *Dermatotoxicology*, 5th Edition (Taylor & Francis, Bristol, PA, 1996).
3. Harris, M.K., *Essential Resources for Industrial Hygiene: A Compendium of Current Practice Standards and Guidelines*, (American Industrial Hygiene Association, Fairfax, VA, 2000).
4. Mellström, G.A., J.E. Wahlberg and H.I. Maibach, *Protective Gloves for Occupational Use* (CRC Press, Boca Raton, FL, 1994).
5. Lodén, M., and H.I. Maibach, *Dry Skin and Moisturizers: Chemistry and Function* (CRC Press, Boca Raton, FL, 2000).
6. Freinkel, R.K., and D.T. Woodley, *The Biology of the Skin* (Parthenon Publishing, New York, 2001).
7. Blank, I.H., Further Observations on Factors which Influence the Water Content of Stratum Corneum. *J. Invest. Dermatol.* 21, 259-271 (1953).
8. Flynn, G.L., and B. Stewart, Percutaneous Drug Penetration: Choosing Candidates for Transdermal Development. *Drug Develop. Res.* 13, 169-185 (1988).
9. Elias, P.M., and D.S. Friend, The Permeability Barrier in Mammalian Epidermis. *J. Cell Biology* 65, 180-191 (1975)
10. Elias, P.M., S. Grayson, M.A. Lampe, M.A. Williams and B.E. Brown, The Intercorneocyte Spaces. In *The Stratum Corneum*, edited by R. Marks and G. Plewig (Springer Verlag, Berlin, 1983, pp. 53-67).
11. Jackson, S.M., and P.M. Elias, Skin as an Organ of Protection. In *Dermatology in General Medicine*, Vol. 1, 4th ed., edited by T.B. Fitzpatrick, A.Z. Eisen, K. Wolff et al. (McGraw Hill, New York, 1993, pp. 241-245).

12. Elias, P.M., J. Goerke and D.S. Friend, Mammalian Epidermal Barrier Layer Lipids: Composition and Influence on Structure. *J. Invest. Dermatol.* 69, 535-546 (1977).

13. Swartzendruber, D.C., P.W. Wertz, and D.J. Kitko, Molecular Models of the Intercellular Lipid Lamellae in Mammalian Stratum Corneum. *J. Invest. Dermatol.* 92, 251-257 (1989).

14. Froslind, B., A Domain Mosaic Model of the Skin Barrier. *Acta Derm. Venereol (Stockh)* 74, 1-6 (1994).

15. Hou, S., A.K. Mitra, S.H. White et al., Membrane Structures in Normal and Essential Fatty Acid-deficient Stratum Corneum: Characterisation by Ruthenium Tetroxide Staining and X-ray Defraction. *J. Invest. Dermatol.* 96, 215-223 (1991).

16. Yardley, H.J., Epidermal Lipids. In *Biochemistry and Physiology of the Skin*, edited by L.A. Goldsmith (Oxford University Press, New York, 1991, p. 363).

17. Gray, G.M., and H.J. Yardley, Lipid Compositions of Cells Isolated From Pig, Human, and Rat Epidermis. *Journal of Lipid Research* 16, 434-440 (1975).

18. Wertz, P.W., D.C. Swartzendruber, D.J. Kitko, K.C. Madison and D.T. Downing, The Role of the Corneocyte Lipid Envelope in Cohesion of the Stratum Corneum. *J. Invest. Dermatol.* 93, 169-172 (1989).

19. Michaels, A.S., S.K. Chandrasekaran and J.E. Shaw, Drug Permeation Through Human Skin. Theory and in Vitro Experimental Measurement. *Am. Inst. Chem. Eng.* 21, 985-996 (1975).

20. Forslind, B., A Domain Mosiac Model of the Skin Barrier. *Acta Derm. Venereol (Stochh.)* 74, 1-6 (1994).

21. Scheuplein, R.J., and I.H Blank, Permeability of the Skin. *Physiol. Rev.* 51, 702-747 (1971).

22. Potts, R.O., and M.L. Francoeur, The Influence of Stratum Corneum Morphology on Water Permeability. *J. Invest. Dermatol.* 96, 495-499 (1991).

23. Potts, R.O., and R.H. Guy, Predicting Skin Permeability. *Pharmaceut. Res.* 9, 663-669 (1992).

24. Treherne, J.E., The Permeability of Skin to Some Electrolytes. *J. Physiology (London)* 133, 171-180 (1965).

25. Takahashi, M, Y. Machida and R. Marks, Measurement of Turnover Time of Sstratum Corneum Using Dansyl Chloride Fluorescence. *J. Soc. Cosmet. Chem.* 38, 321-331 (1987).

26. Dearman, R.J., and P.A. Botham, Inhalation Exposure to Respiratory Sensitizing Chemicals Down-Regulates Guinea Pig IgE and Pulmonary Responses. *Int. Arch. Allergy Appl. Immunol.* 92, 425-432 (1990).

27. Kimber, I., The Role of the Skin in the Development of Chemical Respiratory Hypersensitivity. *Toxicology Letters* 86, 89-92 (1996).

28. Karol, M.H., B.A. Hauth, E.J. Riley and C.M. Magreni, Dermal Contact with Toluene Diisocyanate (Tdi) Produces Respiratory Tract Hypersensitivity in Guinea Pigs. *Tox Appl Pharm* 58, 221-230.

29. Rattray, N.J., P.A. Botham, P.M. Hext, D.R. Woodcock, I. Fielding, R.J. Dearman and I. Kimber, Induction of Respiratory Hypersensitivity to Diphenylmethane-4,4'-diisocyanate (MDI) in Guinea Pigs: Influence of Route of Exposure. *Toxicology* 88, 15-30 (1994).

30. Wang, L-F., J-Y. Lin, K-H. Hsieh and R-H. Lin, Epicutaneous Exposure of Protein Antigen Induces a Predominant Th2-like response with High IgE Production in Mice, *J. Immunology* 156, 4079-4082 (1996).

31. Arakawa, H., J. Lotvall, I. Kawikova, A. Morikawa, C. G. Lofdahl and B. E. Skoogh, Airway Responses Following Intradermal Sensitization to Different Types of Allergens: Ovalbumin, Trimellitic Anhydride and Dermatophagoides farinae. *Internat. Arch. Allergy Immunology.* 108, 274-280 (1995).

32. Woolhiser, M.R., A.E. Munson and B.J. Meade, Role of Sensitization Routes in the Development of Type 1 Hypersensitivity to Natural Rubber Latex in Mice. *Am. J. Industr. Med. (Suppl.)* 1, 139-141 (1999).

33. Saloga, J., H. Renz, G.L. Larsen and E.W. Gelfand, Increased Airways Responsiveness in Mice Depends on Local Challenge with Antigen. *Am. J. Respir. Crit. Care Med.* 149, 65-70 (1994).

34. Leung, D. Y. M. Atopic Dermatitis: The Skin as a Window into the Pathogenesis of Chronic Allergic Diseases. J. Allergy Clin. Immunol. 96:302-318 (1995).

35. Kimber, I., The Skin Immune System. In *Dermatotoxicology, 5th Ed.*, edited by F.N. Marzulli and H.I. Maibach (Taylor and Francis, Bristol, PA, 1996).

36. Blank, I.H., and R.J. Scheuplein, Transport Into and Within the Skin. *Br. J. Dermatol.* 81 (suppl. 4), 4-10 (1969).

37. Lauer, A.C., L.M. Lieb and C. Ramachandran, Transfollicular Drug Delivery. *Pharm. Res.* 12, 179-186 (1995).

38. Bureau of Labor Statistics (BLS), *Occupational Injuries and Illnesses in the United States.* U.S. Department of Labor, BLS, published annually since 1972; 1997 data published in Bulletin 2518, October 1999.

39. Mathias, C.G.T., The Cost of Occupational Skin Diseases. *Arch. Dermatol.* 121, 332-334 (1985).

40. Leigh, J.P., and T.R. Miller, Job-Related Diseases and Occupations Within a Large Workers' Compensation Data Set. *Am. J. Indust. Med.* 33, 197-211 (1998).

41. Hogan, D.J., The Prognosis of Occupational Contact Dermatitis. *Oc. Med.: State of the Art Rev.* 9, 53-58 (1994).

42. Birmingham, D.J., Prolonged and Recurrent Occupational Dermatitis. *Oc. Med.: State of the Art Rev.* 1, 349-355 (1986).

43. Leigh, J.P., S.B. Markowitz, M. Fahs, C. Shin and P.J. Landrigan, Occupational Injury and Illness in the United States: Estimates of Costs, Morbidity, and Mortality. *Arch. Intern. Med.* 157, 1557-1568 (1997).

44. Tucker, S.B., and M.M. Key, Occupational Skin Disease. In *Environmental and Occupational Medicine, 2nd Edition*, edited by W.N. Rom (Little, Brown & Co., Boston, 1992).

45. Harvey,D.T., and D.J. Hogan, Common Environmental Dermatoses. In *Environmental Medicine*, edited by S.M. Brooks (Mosby, St. Louis, 1995, pp. 263-281).

46. Lushniak, B.D., Epidemiology of Occupational Contact Dermatitis. *Dermatol. Clinics* 13, 671-680 (1995).

47. Marzulli,F.N., and H.I. Maibach, Allergic Contact Dermatitis. In *Dermatotoxicology, 5th edition*, edited by F.N. Marzulli and H.I. Maibach (Taylor & Francis, Bristol, PA, 1996).

48. Magnusson, B., and A.M. Kligman, *Allergic Contact Dermatitis in the Guinea Pig. Identification of Contact Allergens* (Thomas, Springfield, IL, 1970).

49. Holness, L., Characteristic Features of Occupational Dermatitis: Epidemiologic Studies of Occupational Skin Disease Reported by Contact Dermatitis Clinics. *Oc. Med: State* of *the Art Rev.* 9, 45-52 (1994).

50. Upadyke, M.R., and H.I. Maibach, Influence of Area of Application of Allergens on Sensitization in Contact Dermatitis. *Contact Dermatitis* 27, 281-286 (1992).

51. Magnusson, B., and A.M. Kligman, Factors Influencing Allergic Contact Sensitization. In *Dermatotoxocology, 2nd Edition*, edited by F.N. Marzulli (Hemisphere Publ., Washington D.C., 1983).

52. Basketter, D.A., G. Cookman and G.F. Gerberick, Skin Sensitization Thresholds: Determination in Predictive Models. *Food Chem. Toxicol.* 35, 417-425 (1997).

53. Meneghini, C.L., and G. Angelini, Primary and Secondary Sites of Occupational Contact Dermatitis. *Dermatosen.*

32, 205-207 (1984).

54. Cronin, E., Clinical Patterns of Hand Eczema in Women. *Contact Dermatitis* 13, 153-161 (1985).

55. Rustemeyer, T., and P.J. Frosch, Occupational Skin Diseases in Dental Laboratory Technicians. *Contact Dermatitis 34*, 125-133 (1996).

56. NIOSH, Sentinel Event Notification System for Occupational Risk (SENSOR) Program, National Institute for Occupational Safety and Health (NIOSH) Cincinnati, Ohio (1997).

57. Kanerva, L., J. Toikkanen and R. Jolanki, Statistical Data on Occupational Contact Urticaria. *Contact Derm.* 35, 229-233 (1996).

58. Holness, L., Characteristic Features of Occupational Dermatitis: Epidemiologic Studies of Occupational Skin Disease Reported by Contact Dermatitis Clinics. *Oc. Med: State of the Art Rev.* 9, 45-52 (1994).

59. Weltfriend, S., M. Bason, K. Lammintausta and H.I. Maibach, Irritant Dermatitis (Irritation). In *Dermatotoxicology, 5th Edition*, edited by F.N. Marzulli and H.I. Maibach (Taylor & Francis, Bristol, PA, 1996, pp. 87-118).

60. Bjornberg, A., Irritant Dermatitis. In *Occupational and Industrial Dermatology, 2nd Ed.*, edited by H.I. Maibach (Year Book Medical Publishers, Chicago, 1987).

61. Patil, S.M., P. Singh and H.I. Maibach, Cumulative Irritancy in Man to Sodium Lauryl Sulfate: the Overlap Phenomenon. *Internat. J. Pharmaceutics* 110, 147-154 (1994).

62. Fitzgerald, D.A., and J.S.C. English, The Long-Term Prognosis in Irritant Contact Hand Dermatitis. In *Irritant Dermatitis, New Clinical and Experimental Aspects*, edited by P. Elsner and H.I. Maibach (Basel, Karger, 1995, pp. 73-76).

63. Wall, L., and K. Gerbauer, A Follow-Up Study of Occupational Skin Disease in Western Australia.

Contact Dermatitis 524, 241-243 (1991).

64. Birmingham, D.J., Prolonged and Recurrent Occupational Dermatitis. *Oc. Med.: State of the Art Reviews* 1, 349-355 (1986).

65. Judge, M.R., H.A. Griffiths and D.A. Basketter, Variation in Response of Human Skin to Irritant Challenge. *Contact Dermatitis* 34, 115-117 (1996).

66. McFadden, J.P., S.H. Wakelin and D.A. Basketter, Acute Irritation Thresholds in Subjects with Type I–Type VI Skin. *Contact Dermatitis* 38, 147-149 (1998).

67. Wilhelm, K.P., and H.I. Maibach, Factors Predisposing to Cutaneous Irritation. *Dermatological Clinics* 8, 17-22 (1990).

68. Berardesca, E., and H. Maibach. Racial Differences in Skin Pathophysiology. *J. Am. Acad. Derm.* 34, 667-672 (1996).

69. Cua, A.B., K.P. Wilhelm and H. I Maibach, Cutaneous Sodium Lauryl Sulphate Irritation Potential: Age and Regional Variability. *British J. Derm.* 123, 607-613 (1990).

70. Hannuksela, A., and M. Hannuksela, Irritant Effects of a Detergent in Wash, Chamber and Repeated Open Application Tests. *Contact Dermatitis* 34, 134-137 (1996).

71. Seidenari, S., B. Belletti, B and M.E. Schiavi, Skin Reactivity to Sodium Lauryl Sulfate in Patients with Respiratory Atopy. *J. Am. Acad. Derm.* 35, 47-52 (1996).

72. Coenraads, P.-J., and T.L. Diepgen, Risk of Hand Excema in Employees with Past or Present Atopic Dermatitis. *Int.* Arch. *Occup. Environ. Health* 71, 7-13 (1998).

73. Basketter, D.A., H.A. Griffiths, X.M. Wang et al.: Individual, Ethnic and Seasonal Variability in Irritant Susceptibility of Skin: The Implications for a Predictive Human Patch Test. *Contact Dermatitis* 35, 208-213 (1996).

74. National Institute for Occupational Safety and Health:

Alert - Preventing Allergic Reactions to Natural Rubber Latex in the Workplace. DHHS (NIOSH) Publication No. 97-135, Cincinnati, Ohio (1997).

75. Eslander, T., R. Jolanki and L. Kanerva, Allergic Contact Dermatitis from Rubber and Plastic Gloves. In *Protective Gloves for Occupational Use*, edited by G.A. Mellström, J.E. Wahlberg and H.I. Maibach (CRC Press, Boca Raton, FL, 1994).

76. Belsito, D.V., Rubber. In *Handbook of Occupational Dermatology*, edited by L. Kanerva, P. Elsner, J.E. Wahlberg and H.I. Maibach (Springer, Berlin, 2000).

77. Knudsen, B.B., E. Larsen, H. Egsgaard and T. Menné, Release of Thiurams and Carbamates from Rubber Gloves. *Contact Derm.* 28, 63-69 (1993).

78. Fregert, S., Occupational Dermatitis in a 10-year Material, *Contact Dermatitis* 1, 96-107 (1975).

79. Themido, R., and F. Menezes Brandao, Contact Allergy to Thiurams. *Contact Derm.* 10, 251 (1984).

80. Kanerva, L., R. Jolanki and J. Toikkanen, Frequencies of Occupational Allergic Diseases and Gender Differences in Finland. *Int. Arch. Occup. Environ. Health* 66, 111-116 (1994).

81. Kanerva, L., R. Jolanki and J.K. Toikkanen, Statistics on Occupational Dermatoses in Finland. In *Irritant Dermatitis, New Clinical and Experimental Aspects*, edited by P. Elsner and H.I. Maibach (Basel, Karger, 1995, pp. 28-40).

82. Halkier-Sorensen, L., B.H. Petersen and K. Thestrup-Pedersen, Epidemiology of Occupational Skin Diseases in Denmark: Notification, Recognition and Compensation. In *The Irritant Contact Dermatitis Syndrome*, edited by P.G.M. van der Valk and H.I. Maibach (CRC Press, Boca Raton, FL, 1996, pp. 23-52).

83. Cherry, N., J.D. Meyer, A. Adisesh, R. Brooke, V. Owen-Smith, C. Swales and M.H. Beck, Surveillance of Occupational Skin Disease: EPIDERM and OPRA. *Br. J.*

Dermatol. 142, 1128-1134 (2000).

84. Taylor, J.S., *Protective Gloves for Occupational Use* (CRC Press, Boca Raton, FL, 1994, pp. 255-265).

85. Wilkinson, S.M., Presented at the Jadassohn Centenary Congress, European Society of Contact Dermatis/ American Contact Dermatitis Society, London, abstract 111 (October 1996).

86. Wrangsjo, K., K. Osterman and M. van Hage-Hamsten, Glove-Related Skin Symptoms Among Operating Theatre and Dental Care Unit Personnel (II). Clinical Examination, Tests and Laboratory Findings Indicating Latex Allergy. *Contact Dermatitis*, 30, 139-143 (1994).

87. Burke, F.J., N.H. Wilson and S.W. Cheung, Factors Associated with Skin Irritation of the Hands Experienced by General Dental Practitioners. *Contact Dermatitis* 32, 35-38 (1995).

88. Wrangsjèo, K., K. Osterman and M. Van Hage-Hamsten, Glove-Related Skin Symptoms Among Operating Theatre and Dental Care Unit Personnel (I). Interview Investigation. *Contact Dermatitis* 30, 102-107 (1994).

89. Woods, J.A., R.F. Morgan, F.H. Watkins and R.F. Edlich, Surgical Glove Lubricants: From Toxicity to Opportunity. *J Emerg Med* 15, 209-220 (REF: 59) (1997).

90. Phillips, M.L., C.C. Meagher and D.L. Johnson, What is "Powder Free?" Characterization of Powder Aerosol Produced During Simulated Use of Powdered and Powder Free Latex Gloves. *Oc. Environ. Med.* 58, 479-481 (2001).

91. Brehler, R., W. Voss and S. Mèuller, Glove Powder Affects Skin Roughness, One Parameter of Skin Irritation. *Contact Dermatitis* 39, 227-230 (1998).

92. Bucks, D.A.W., Predictive Approaches II: Mass-Balance Procedure. In *Topical Drug Bioavailability, Bioequivalence, and Penetration*, edited by V.P. Shah and H.I. Maibach (Phenum Press, 1993, pp. 183-195).

93. Artuc, M., C. Reinhold, G. Stuttgen and J. Gazith, A

Rapid Measurement for Measuring Drug Enrichment in Epidermis. *Arch. Dermatol. Res.* 268, 129-140 (1980).

94. Miselnicky, S.R., J.L. Lichtin and A. Sakr, The Influence of Solubility, Protein Binding, and Percutaneous Absorption on Reservoir Formation in Skin. *J. Soc. Cosmet. Chem.* 39, 169-177 (1988).

95. Kanerva, L., J. Toikkanen and R. Jolanki, Statistical Data on Occupational Contact Urticaria. *Contact Derm.* 35, 229-233 (1996).

96. Wang, L.-F., J.Y. Lin, K.H. Hsieh and R.H. Lin, Epicutaneous Exposure of Protein Antigen Induces a Predominant Th-2-like Response with High Ige Production in Mice. *J. Immunology* 156, 4079-4082 (1996).

97. Saloga J., and J. Knop, Does Sensitization Through the Skin Occur? *Allergy* 55, 905-909 (2000).

98. Hayes, B.B., A. Afshari, L. Millecchia, P.A. Willard, S.P. Povoski and B.J. Meade, Evaluation of Percutaneous Penetration of Natural Rubber Latex Proteins. *Toxicol. Sci.* 56, 262-270 (2000).

99. Woolhiser, M.R., A.E. Munson and B.J. Meade, Immunological Responses of Mice Following Administration of Natural Rubber Latex Proteins by Different Routes of Exposure. *Toxicol. Sci.* 55, 343-351 (2000).

100. Maibach, H.I., R.J. Feldmann, T.H. Milby and W.F. Serat, Regional Variation in Percutaneous Penetration in Man: Pesticides. *Arch. Environ. Health* 23, 208-211 (1971).

101. Elias, P.M., E.R. Cooper, A. Korc and B.E. Brown, Percutaneous Transport in Relation to Stratum Corneum Structure and Lipid Composition. *J. Invest. Dermatol.* 76, 297-301 (1981).

102. Lotte, C., A. Rougier, D.R. Wilson and H.I. Maibach, In Vivo Relationship Between Transepidermal Water Loss and Percutaneous Penetration of Some Organic Compounds in Man: Effect of Anatomic Site. *Arch.*

Dermatol. Res. 279, 351-356 (1987).

103. Harvell, J.D., and H.I. Maibach, Percutaneous Absorption and Inflammation in Aged Skin: A Review. *J. Am. Acad. Dermatol.* 31, 1015-1021 (1994).

104. Cua, A.B., K.-P. Wilhelm and H.I. Maibach, Cutaneous Sodium Lauryl Sulphate Irritation Potential: Age and Regional Variability. *Br. J. Dermatol.* 123, 607-613 (1990).

105. Feldman, R.J., and H.I. Maibach, Percutaneous Penetration of Some Pesticides and Herbicides in Man. *Toxicol. Appl. Pharmacol.* 28, 126-132 (1974).

106. Feldman, R.J., and H.I. Maibach, Penetration of ^{14}Chydrocortisone Through Normal Skin. *Arch. Dermatol.* 91, 661-666 (1965).

107. Maibach, H.I., In Vivo Percutaneous Penetration of Corticoids in Man and Unresolved Problems in Their Efficacy. *Dermatologica* 152 (suppl. 1), 11-25 (1976).

108. Dary, C.C., J.N. Blancato and M. Castles et al., Dermal Absorption and Disposition of Formulations of Malathion in Sprague-Dawley Rats and Humans. In *Biomarkers of Human Exposure to Pesticides,* edited by M.A. Saleh, J. Blancato and C.H. Nauman (ACS Symposium Series No. 542, American Chemical Society, 1994, pp. 231-263).

109. Wester, R.C., and H.I. Maibach, Clinical Considerations for Transdermal Delivery. In *Transdermal Delivery of Drugs, Vol. 1*, edited by R.C. Wester (CRC Press, Baton Rouge, 1987, pp. 72-78).

110. Stewart, R.D., and H.C. Dodd, Absorption of Carbon Tetrachloride, Trichloroethylene, Tetrachloroethylene, Methylene Chloride, and 1,1,1-trichloroethane Through the Human Skin. *Ind. Hyg. J.* 25, 439-446 (1964).

111. Lauwerys, R.R., T. Dath, J.M. Lachapelle, J.P. Buchet and H. Roels, The Influence of Two Barrier Creams on the Percutaneous Absorption of m-Xylene in Man. *J. Occup. Med.* 20, 17-20 (1978).

112. Lammintausta, K., H.I. Maibach and D. Wilson, Irritant Reactivity in Males and Females. *Contact Dermatitis* 17, 276-280 (1987).

113. Goh, C.L., and S.E. Chia, Skin Irritability to Sodium Lauryl Sulfate - as Measured by Skin Water Vapour Loss - by Sex and Race. *Clin. Exp. Dermatol.* 13, 16-19 (1988).

114. Basketter, D.A., H.A. Griffiths, X.M. Wang,, K.P. Wilhelm and J. McFadden, Individual, Ethnic and Seasonal Variability in Irritant Susceptibility of Skin: the Implications for a Predictive Human Patch Test. *Contact Dermatitis* 35, 208-213 (1996).

115. Anonymous, North American Contact Dermatitis Group: Epidemiology of Contact Dermatitis in North America. *Arch. Dermatol.* 108, 537-540 (1973).

116. Behrens, V., P. Seligman, L. Cameron, C.G. Mathias and L. Fine, The Prevalence of Back Pain, Hand Discomfort, and Dermatitis in the U.S. Working Population. *Am. J. Publ. Health* 84, 1780-1785 (1994).

117. Harvell, J.D., and H.I. Maibach, Percutaneous Absorption and Inflammation in Aged Skin: A Review. *J. Am. Acad. Dermatol.* 31, 1015-1021 (1994).

118. Bronaugh, R.L., and R.F. Stewart, Methods for in Vitro Percutaneous Absorption Studies, V. Permeation Through Damaged Skin. *J. Pharmaceut. Sci.* 74, 1062-1066 (1985).

119. Newhouse, M., D. McEvoy and D. Rosenthal, Percutaneous Paraquat Absorption: An Association with Cutaneous Lesion and Respiratory Failure. *Arch. Dermatol.* 114, 1516-1519 (1978).

120. Maibach, H.I., and R. Feldman, Systemic Absorption of Pesticides Through the Skin of Man. In *Occupational Exposure to Pesticides.* Report to the Federal Working Group on Pest Management from the Task Group on Occupational Exposure to Pesticides (1974).

121. Kusama, T., S. Itoh and Y. Yoshizawa, Absorption of Radionuclides Through Wounded Skin. *Health Physics*

51, 138-141 (1986).

122. Lange, M., K. Nitzsche and A. Zesch, Percutaneous Absorption of Lindane in Health Volunteers and Scabies Patients: Dependency of Penetration Kinetics in Serum upon Frequency of Application, Time and Mode of Washing. *Arch. Dermatol. Res.* 271, 387-399 (1981).

123. Drexler, H., T.H. Goen and J. Angerer, Carbon Disulphide. II. Investigations on the Uptake of CS2 and the Excretion of Its Metabolite 2-Thiothiazolidine-4-Carboxylic Acid after Occupational Exposure. *Int. Arch. Occup. Environ. Health* 67, 5-10 (1995).

124. Wrbitzky, R., and J. Angerer, N,N-dimethylformamide - Influence of Working Conditions and Skin Penetration on the Internal Exposure of Workers in Synthetic Textile Production. *Int. Arch. Occup. Environ. Health* 71, 309-316 (1998).

125. Riihimaki, V., Percutaneous Absorption of M-xylene from a Mixture of M-xylene and Isobutyl Alcohol in Man. *Scand. J. Work Environ. Health* 5, 143-150 (1979).

126. Starr, J.G., and N.J. Clifford, Absorption of Pesticides in a Chronic Skin Disease. *Arch. Environ. Health* 22, 396-400 (1971).

127. Solomon, A.E., and N.J. Lowes, Percutaneous Absorption in Experimental Epidermal Disease. *Br. J. Dermatol.* 100, 717-722 (1979).

128. Scheuplein, R.J., and R.L. Bronaugh, Percutaneous Absorption. In *Biochemistry and Physiology of the Skin*, edited by L.A. Goldsmith, Ed., (Oxford University Press, 1983, pp. 1255-1295).

129. Wilhelm, K.-P., C. Surber and H.I. Maibach, Effect of Sodium Lauryl Sulfate-induced Skin Irritation on in Vivo Percutaneous Penetration of Four Drugs. *J. Invest. Dermatol.* 97, 927-932 (1991).

130. Samitz, M.H., and H. Pomerantz, Studies of the Effects on the Skin of Nickel and Chromium Salts. *A.M.A. Arch.*

Indust. Heatlh, 18, 473-479 (1958).

131. Klingman, A.M., The Identification of Contact Allergens by Human Assay. Ii. Factors Influencing the Induction and Measurement of Allergic Contact Dermatitis. *J. Invest. Dermatol.* 47, 375-392 (1966).

132. Funckes, A.J., G.R. Hayes and W.V. Hartwell, Urinary Excretion of Paranitrophenol by Volunteers Following Dermal Exposure to Parathion at Different Ambient Temperatures. *J. Agricul. Food Chem.* 11, 455-457 (1963).

133. Dutkiewicz, T., The Absorption of Aniline Vapors Through the Skin, *Medycyna Pracy* 8, 25-26 (1957).

134. Johanson, G., and A. Boman, Percutaneous Absorption of 2-butoxyethanol Vapour in Human Subjects. *Br. J. Industr. Med.* 48, 788-792 (1991).

135. Chang, S.K., and J.E. Reviere, Percutaneous Absorption of Parathion in Vitro in Porcine Skin: Effects of Dose, Temperature, Humidity, and Perfusate Composition on Absorption Flux. *Fund. Appl. Toxicol.* 17, 494-504 (1991).

136. Proceedings of the Pharmacokinetics and Topical Applied Cosmetics Symposium, Cosmetics Toiletries and Fragrance Association, CTFA Monograph 2, Research Triangle Park (1983).

137. Burckhardt, W., Experimentelle Sensibilisierung Des Meerschweinchens Gegen Terpentinol. *Acta Derm. Venereol.* 19, 359-364 (1939).

138. Bird, M.G., Industrial Solvents: Some Factors Affecting Their Passage into and Through the Skin. *Ann. Occup. Hyg.* 24, 235-244 (1981).

139. Bucks, D.A.W., H.I. Maibach and R.H. Guy, Occlusion Does Not Uniformly Enhance Penetration In Vivo. In *In Vitro Percutaneous Absorption: Principles, Fundamentals, and Applications*, edited by R. Bronaugh, and H.I. Maibach (CRC Press, Boca Raton, FL, 1996, pp. 77-93).

140. Guy, R.H., Kinetics of Drug Absorption Across Human Skin In Vivo. In *Pharmacology and the Skin, Vol. 1.*, edited by B. Shroot and Schaefer (Karger, Basel, 1987, pp. 70-76).

141. Makki, S., Percutaneous Absorption of Three Psoralens Commonly Used in Therapy: Effect of Skin Occlusion (In Vitro Study). *Internat. J. Pharmaceutics* 133, 245-252 (1996).

142. Behl, C.R., G.L. Flynn, T. Kurihara, N. Harper, W. Smith, W. I. Higuchi, N.F. Ho and C.L. Pierson, Hydration and Percutaneous Absorption I. Influence of Hydration on Alkanol Permeation Through Hairless Mouse Skin. *J. Invest. Dermatol.* 75, 346-351 (1980).

143. Qiao, G.L., S.K. Chang and J.E. Riviere, Effects of Anatomical Site and Occlusion on the Percutaneous Absorption and Residue Pattern of 2,6 [Ring-^{14}c] Parathion in Vivo in Pigs. *Toxicol. Applied Pharmacol.* 122, 131-138 (1993).

144. Ramsing, D.W., and T. Agner, Effect of Glove Occlusion on Human Skin (II). Long-Term Experimental Exposure, *Contact Dermatitis* 34, 258-262 (1996).

145. Graves, C.J., C. Edwards and R. Marks, The Occlusive Effects of Protective Gloves on the Barrier Properties of the Stratum Corneum. In *Irritant Dermatitis, New Clinical and Experimental Aspects,* edited by P. Elsner and H.I. Maibach (Basel, Karger, 1995, pp. 87-94).

146. Klingman, A.M., Hydration Injury to Human Skin. In *The Irritant Contact Dermatitis Syndrome*, edited by P.G.M. van der Valk and H.I. Maibach (CRC Press, Boca Raton, FL 1996).

147. Menon, G.K., K.R. Feingold and P.M. Elias, Lamellar Body Secretory Response to Barrier Disruption. *J. Invest. Dermatol.* 98, 279-289 (1992).

148. Grubauer, G., K.R. Feingold and P.M. Elias, Relationship of Epidermal Lipogenesis to Cutaneous Barrier Function," *J. Lipid Res.* 28, 746-752 (1987).

149. Holleran, W.M., K.R. Feingold, M.Q. Man, W.N. Gao, J.M. Lee and P.M. Elias, Regulation of Epidermal Sphingolipid Synthesis by Permeability Barrier Function, *J. Lipid Res.* 32, 1151-1158 (1991).

150. Kusters, E., Biological Monitoring of MDA. *Br. J. Indust. Med.* 49, 72 (1992).

151. Niven, K.J.M., A. Robertson, E.R. Waclawski, S. Hagen, A.J. Scott, B. Cherrie, R. Topliss, H. Wilkins, M. Lovett, P.L. Bodsworth and M. McWilliam, Occupational Hygiene Assessment of Exposure to Insecticides and the Effectiveness of Protective Clothing During Sheep Dipping Operations. Institute of Occupational Medicine report TM/94/04 (August 1996).

152. Brouwer, D.H., S.A.F. De Vreede, W.J.A. Meuling and J.J. van Hemmen, Determination of the Efficiency for Pesticide Exposure Reduction with Protective Clothing: A Field Study Using Biological Monitoring. In *Worker Exposure to Agrochemicals*, edited by H.C. Honeycutt. ACS Symposium Series (CRC, Lewis Publishers, Baton Rouge, LA, 1999, pp. 65-86).

153. Aprea, C., G. Sciarra, P. Sartorelli, E. Desideri, R. Amati and E. Sartorelli, Biological Monitoring of Exposure to Organophosphorus Insecticides by Assay of Urinary Alkylphosphates: Influence of Protective Measures During Manual Operations with Treated Plants. *Int. Arch. Occup. Environ. Health* 66, 333-338 (1994).

154. Hogstedt, C., and R. Stahl, Skin Absorption and Protective Gloves in Dynamite Work. *Am. Industr. Hyg. Assoc. J.* 41, 367-372 (1980).

155. Garrod, A.N., M. Martinez, J. Pearson, A. Proud and D.A. Rimmer, Exposure to Preservatives Used in the Industrial Pre-treatment of Timber. *Ann. Occup. Hyg.* 43, 543-555 (1999).

156. Kusters, E., Biological Monitoring of MDA. *Br. J. Ind. Med.* 49, 72 (1992).

157. Sanderson, W.T., V. Ringenburg and R. Biagini, Exposure

to Commercial Pesticide Applicators to the Herbicide Alachlor. *Am. Indust. Hyg. Assoc. J.* 56, 890-897 (1995).

158. Garrod, A.N., A.M. Phillips and J.A. Pemberton, Potential Exposure of Hands Inside Protective Gloves - A Summary of Data from Aon-agricultural Pesticide Surveys. *Ann Occup. Hyg.* 45, 55-60 (2001).

159. Ghosh, S.K., V.N. Gokani, P.B. Doctor and J.R. Parikh, Intervention Studies Against "Green Symptoms" among Indian Tobacco Harvesters. *Arch. Environ. Health* 46, 316-317 (1991).

160. Lauwerys, R.R., A. Kivits, M. Lhoir, P. Rigolet, D. Houbeau, J.P. Buchet and H.A. Roels, Biological Surveillance of Workers Exposed to Dimethylformamide and the Influence of Skin Protection on Its Percutaneous Absorption. *Int. Arch. Occup. Environ. Health* 45, 189-203 (1980).

161. Nielsen, J.B., and H.R. Andersen, Cholinesterase Activity in Female Greenhouse Workers - Influence of Work Practices and Use of Oral Contraception. 10th Annual Conference of the International Society of Exposure Analysis, Asilomar Conference Center, Abstract 1A-04o, submitted to *Am. J. Indust. Med.* (2000).

162. Reifenrath, W.G., and T.S. Spencer, Evaporation and Pentration from Skin. In *Percutaneous Absorption: Mechanisms-Methodology-Drug Delivery, Second Edition*, edited by R.L. Bronaugh and H.I. Maibach (Marcel Dekker, 1989, pp. 313-334).

163. Boman, A., G. Hagelthorn and K. Magnusson, Percutaneous Absorption of Organic Solvents During Intermittent Exposure in Guinea Pigs. *Acta Derm. Venereol. (Stockh)* 75, 114-119 (1995).

164. Otis, L.L., and J.A. Cottone, Prevalence of Perforations in Disposable Latex Gloves During Routine Dental Treatment. *J. Am. Dental Assoc.* 118, 321-324 (1989).

165. Paulssen, J., T. Eidem and R. Kristiansen, Perforations in

Surgeons' Gloves. *J. Hospital Infection* 11, 82-85 (1988).

166. Matta, H. , A.M. Thompson and J.B. Rainey, Does Wearing Two Pairs of Gloves Protect Operating Theatre Staff from Skin Contamination? *Brit. Med. J.* 297, 597-598 (1988).

167. Caillot, J.L., C. Cote, H. Abidi and J. Fabry, Electronic Evaluation of the Value of Double Gloving. *Brit. J. Surgery* 86, 1387-1390 (1999).

168. Avery, C.M.E., J. Taylor and P.A. Johnson, Double Gloving and a System for Identitying Glove Perforation in Maxilofacial Trauma Surgery. *Brit. J. Oral Maxillofacial Surgery* 37, 316-319 (1999).

169. Canning, K.M., W. Jablonski and P.B. McQuinlan, Quantification of Surface Defects on Chemically Protective Gloves Following Their Use in Agriculture. *Ann. Agric. Environ. Med.* 5, 45-56 (1998).

170. National Institute for Occupational Safety and Health, Health Hazard Evaluation Report of Dirty Bird, Inc., HETA 95-0248-2562 (1995).

2

Chemistry of Chemical Protective Clothing Formulations

By Jimmy L. Perkins

Introduction

The science of chemical protective clothing (CPC) materials is really the science of polymers. An understanding of the properties of the polymeric materials that CPC items are made of facilitates understanding of the strengths and weaknesses of the clothing. This chapter presents a brief review of salient aspects of polymer chemistry.

Many types of polymers are used for protective clothing. No two products are identical, though they may be the same generic type. For example, nitrile gloves made by two different manufacturers will have somewhat different constituents. Many different chemicals go into a CPC product and its formulation is referred to as a "recipe." The way a polymer product is made has very important consequences for its protective properties. Minor changes in the manufacturing process can produce significant changes in the protective properties.

Chemical protective products are naturally expected to resist chemical permeation, penetration, and degradation. However, many other qualities considered by the manufacturer are relevant, including fit, tactility, tensile strength, heat resistance, and comfort. Often, compromises must be made in some of these characteristics to optimize others.

Some major variations in polymer systems result from the use of copolymers, laminates, and blends. The degree of crosslinking, crystallinity, and the amount of plasticizer and fillers in a polymer greatly affect the properties of the final product. This chapter will discuss these variables, including the chemistry, and how they affect the structure of the polymer product, which in turn will affect protective properties. Further information is available in references that deal with the science of polymer chemistry. The content of this chapter is taken largely from the texts listed in the suggested readings that can be found at the chapter's end.

Polymer Types

Polymers are made from monomer molecules bonded together in chains through a chemical reaction. Details of all polymerization reactions are not important, but a few examples are provided with a focus on condensation reactions and addition reactions.

Condensation is one of the most important reactions in the manufacture of many polymers, including polyesters. Polyester is the condensation product of dialcohol and diacid, as illustrated in Figure 1a.

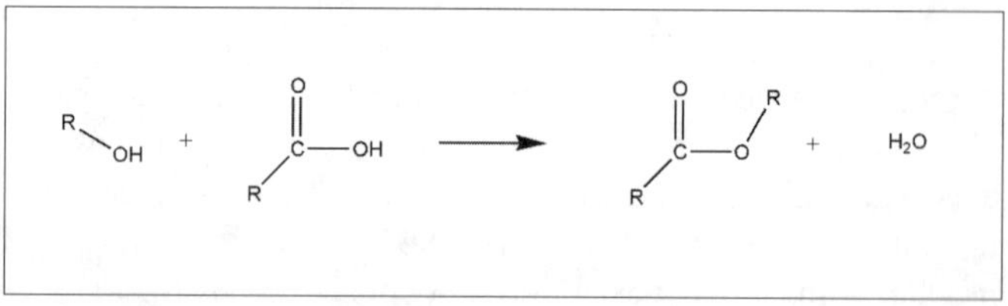

Figure 1a. A condensation reaction from two monomers that yields a polymer and water.

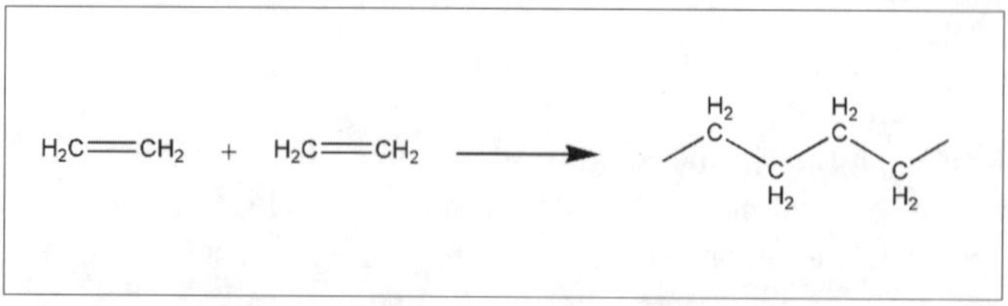

Figure 1b. An addition reaction from one monomer, ethylene, yielding polyethylene.

The hydrogen from the alcohol and the OH group from the acid are cleaved, forming water and an ester. The alcohol and the acid must be bifunctional (reactive in two places) so that once the initial reaction occurs, the ends of the new product may react, continuing the process.

Another important polymer reaction is the addition reaction. In this reaction, no product is produced other than the polymer. In contrast, in a condensation reaction, the repeating unit is somewhat different from that of the monomer(s) and a byproduct (water) is formed. Hence, the repeating unit in an addition reaction has the same composition as the original monomer. In addition reactions, the important step is the opening of a double bond, as in Figure 1(b) where polyethylene is made from ethylene. In the simplest reaction to form polyethylene, no hydrogen is lost from the ethylene groups. Only the double bond is opened, leaving a pair of free radicals to react with other monomers. This same process is used in the manufacture of polyvinyl chloride (PVC) and other polymers made from double-bonded monomers including styrene, propylene, and acrylonitrile. The Appendix lists polymer types and structures used in protective clothing.

When two differing monomers react by addition, a copolymer is formed. A random copolymer does not have a specific order for the constituent monomers. The Appendix includes examples of copolymers used in protective clothing, such as Viton® (a copolymer of hexafluoropropylene and tetrafluoroethylene), and nitrile rubber (a copolymer of acrylonitrile and butadiene). In both cases, the two different monomers that form the polymer are found randomly throughout the polymer chain. The condensation reaction of dialcohol and diacid in Figure 1(a) may appear to yield a copolymer. However, even though two distinct monomers are used in making the polymer, these molecules must be arranged in a specific order for the condensation reaction to occur, i.e., there is only one repeating unit. Figure 2 provides examples of polymer types.

In some cases, block copolymers, as opposed to random copolymers, are used. In block copolymers, the reaction is not random with respect to the monomer; each monomer is reacted with itself to give chains of a certain length. These two chain types then react.

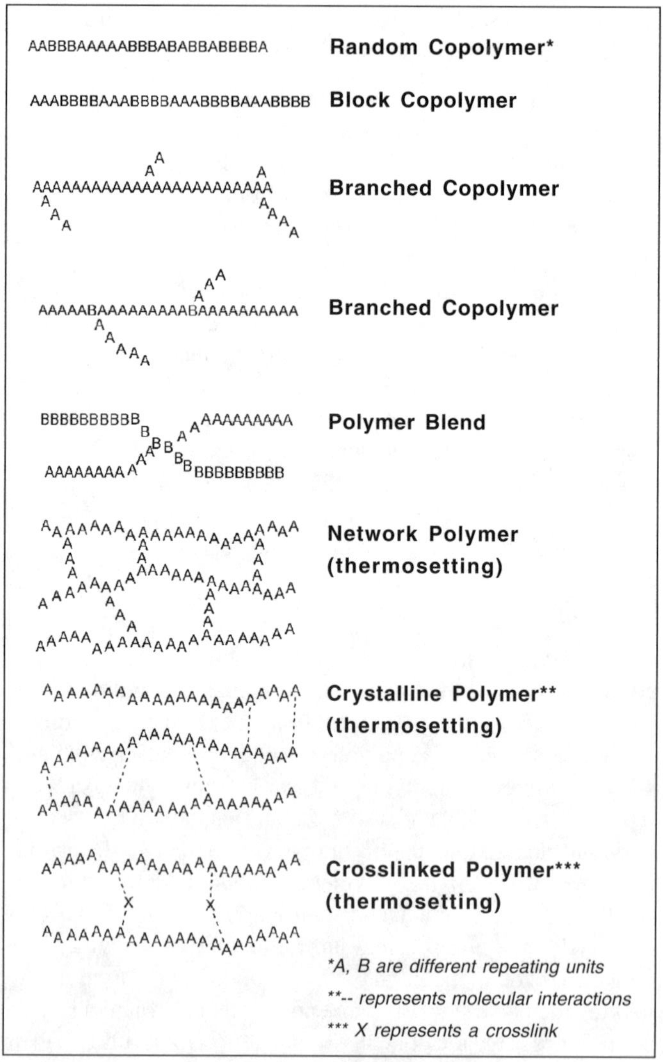

Figure 2. Basic Types of Polymer Structures

A polymer blend is a mixture of two polymer resins, not a copolymer. In a blend, one chain formed by the "A" monomers may lie next to a chain formed by the "B" monomers. Chemical bonds do not attach the two, unless a specific chemical known as a crosslinking agent is added to accomplish interpolymer bonding.

A laminate is simply a multilayer of two or more polymer sheets that are adhered together. Each layer can be comprised of any of the types of polymers (condensation, addition, block, blend). Extra layers may be added to a laminated construction to improve adhesion between layers.

Polymers may be classified further as being linear, branched, or crosslinked. Linear polymers have no side chains. Polyethylene, which is made up of a zigzag chain of carbon atoms, is nevertheless considered a linear polymer because, in its simplest form, there are no side chains attached to the polymer strand.

Branched polymers have side chains appearing along the backbone of the main polymer strand. These side chains develop because of an occasional reaction of the polymer backbone with monomer molecules at locations other than the ends. If a trifunctional monomer—reactive in three places—is introduced in an exact place or randomly along a linear chain, two of its functional groups are used as the polymerization process proceeds along the backbone chain, leaving only one functional group to react to form a branch. These branching reactions can occur by chance (if a hydrogen is lost in the formation of polyethylene then a branch will occur there), but in some cases a certain amount of trifunctional monomer is added to a polymer batch to induce branching. Branching has a significant influence on the properties of polymers.

The crosslinked or networked polymer is an extreme case of a branched polymer (Figure 2). As the length and frequency of branches on a polymer chain increase, the branches will eventually reach from chain to chain and become crosslinked. As the chains become more connected in three dimensions, the entire object becomes one large molecule. An extreme example, a bowling ball, is one large networked polymer molecule.

Crosslinks may form if sufficient amounts of trifunctional monomers are added. The process known as curing (heating after forming an object to final shape) generates crosslinks. Charles Goodyear first used sulfur to cure natural rubber, causing it to harden because the added sulfur linked the polymer chains together. These bridges hold the chains in place and decrease slipping and sliding against one another. Without this curing process, natural rubber loses its shape and becomes a sticky liquid when it is warmed.

Different sorts of polymers respond differently to heat. A cured polymer becomes hard and cannot be melted without forcing a decomposition of the crosslinked structure. This type of polymer behavior is known as thermosetting. To decompose, a thermosetting polymer requires extended exposures to heat or chemical reagents, and decomposition is not reversible. By analogy, an egg is a thermosetting substance.

Other polymers are thermoplastic. These polymers will soften or melt when heated, change shape if stressed, and when cooled, reverse and regain solid state again. Chocolate is a thermoplastic substance. In thermoplastic polymers, the energy holding the polymer in a solid or rubbery form results from molecular interactions rather than crosslinks or electrostatic attractions, not chemical bonds. These molecular interaction forces influence the degree of crystallinity of the polymer. These forces can be influenced by the introduction of additives such as plasticizers that change attractions in the polymer. Molecular interactions and crystallinity are discussed later in this chapter.

Some plastic substances can be shaped or molded and will hold their forms over a fairly wide temperature range. They do not need the heat that is required to change the shape of a thermoplastic, of which waxes and moist clay are examples.

A rubber substance, or elastomer, has the ability to withstand and recover from deformation or elongation under stress. The difference between a plastic and an elastomer is partially a function of the molecular attractions between the polymer chains. Substances used to coat fibers or form gloves are usually elastomeric and thermosetting. One notable exception is the thermoplastic PVC. PVC is a plastic, but through the addition of plasticizers, can be made flexible. However, PVC still does not have the elastic properties of an elastomer.

An additional variable affecting the characteristics of CPC materials is the selected manufacturing method. Solvent dipping, milling, and latex-suspension methods are all very important in determining the characteristics of the final product. These manufacturing processes and effects on permeation characteristics are discussed in Chapters 3 and 4.

Polymer Chemistry

Bonds and Interactions

Understanding the characteristics that make a polymer appropriate for CPC requires understanding of polymer structures and the various types of chemical bonds holding polymers together. Table 1 lists the different types of chemical bonds and molecular interactions and contrasts these according to the amount of energy necessary to break them.

Carbon atoms form covalent bonds with one another and with several other elements. As shown in Table 1, these are strong bonds with relatively large disassociation energies required to break the bonds. The monomers in a polymer are generally held together by covalent bonds, as are the links in a crosslinked polymer (although the latter in some cases is held together by ionic bonds). For CPC characteristics, the covalent bond is the most important bond in the polymer backbone and in the crosslinking of polymers.

Table 1:
Types of Bonds and their Disassociation Energies in kcal/mole

Type	Disassociation Energy
Covalent	50–200
Ionic	10–20
Hydrogen	3–7
Polar	1.5–3
Dispersion	0.5–2

Adapted from Rosen.

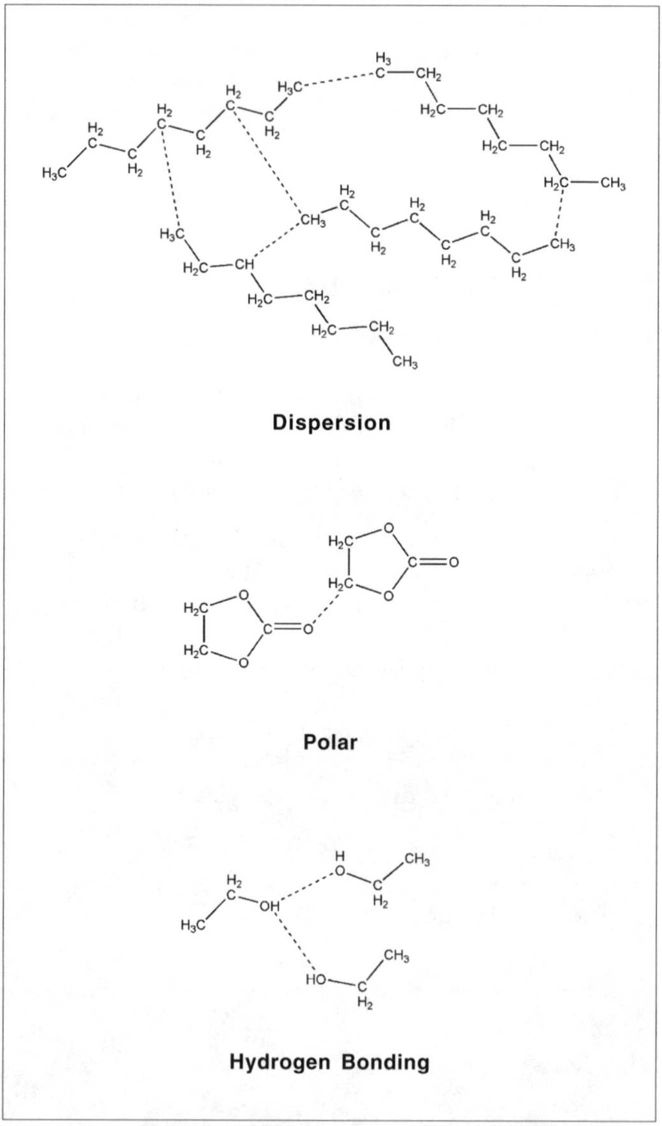

Figure 3. Examples of Molecular Interactions: dispersion forces (heptane), polar (ethylene carbonate), and hydrogen bonding (ethanol).

The three significant molecular interactions that occur between non-crosslinked polymer chains are dispersion forces, polar bonds or dipole interactions, and hydrogen bonds. All three are formed by electrostatic attractions. These are sometimes referred to as secondary bonds, but they are not bonds in the true chemical sense because they can be formed and broken without changing the characteristics of the substance. This breaking and reforming behavior is not true for covalent bonds. Ionic bond characteristics are somewhere between secondary and covalent characteristics. As shown in Table 1, the disassociation energies of these three types of molecular interactions are much smaller than those for covalent bonds.

These molecular interactions also represent the cohesive forces responsible for holding molecules together in a liquid. Examples of the three molecular interactions are shown in Figure 3.

Dispersion forces are solely responsible for holding together non-polar-bonding or non-hydrogen-bonding liquids, such as pentane, hexane, and other straight chain hydrocarbons. Dispersion forces arise from temporary dipole moments that fluctuate about a molecule, depending on positional shifts in its electron cloud. Thus, the molecules in pentane are constantly being attracted to one another and then repelled as the electron cloud shifts.

Polar bonding is present between many molecules that contain oxygen, fluorine, or nitrogen atoms, such as ketones, fluorinated hydrocarbons, and amines. Polar bonding is caused by the permanent dipoles present in these molecules. These attractions are stronger than dispersion forces as they are more permanent and involve greater energies of attraction.

Hydrogen bonding is a special case of polar bonding, present in molecules that contain two polar ends of opposite charge, one of which is an "exposed" hydrogen atom. The best example of hydrogen bonding occurs in water where hydrogen from one molecule readily interacts with oxygen from another. Hydrogen bonding is important in alcohols and organic acids. Hydrogen bonded to a strongly electronegative atom like oxygen is necessary for good hydrogen bonding. Electronegativity of the oxygen shifts electrons in the oxygen-hydrogen bond away from the hydrogen, leaving a very positive proton.

Crosslinking and Crystallinity

The various bond types determine the final structure of the polymer. If a polymer is not cured or otherwise crosslinked, only secondary bonds hold the polymer chains together. By raising the temperature, the chains become more mobile, just as molecules in a liquid become more mobile (volatile) when the temperature is raised. Consequently, the structure of a solid polymer where the chains are held together only by secondary bonds, at room temperature, will melt as the temperature rises. In contrast, a cured thermosetting polymer has covalent bonding between the polymer chains and will maintain characteristics until degraded.

During the curing process, covalent bonding between polymer chains is usually provided by a crosslinking agent–also known as an accelerator–or by natural crosslinking between the polymer chains due to the presence of trifunctional monomers. A thermosetting polymer will not melt until it reaches degradation temperature, the point when covalent bonds break and degrade into smaller molecules. Typically, a considerably higher amount of energy is required to degrade a crosslinked polymer than to melt a thermoplastic polymer.

Molecular interactions are extremely important as variables that influence the ability of foreign molecules to permeate through polymers. A high degree of molecular interaction leads to greater crystallinity of the polymer. Although no polymers are purely crystalline, some including high-density polyethylene, approach this state. Crystalline polymers constitute a metal-like lattice, with few spaces left for permeant molecules to pass between the polymer chains. Crystallinity can be induced in a polymer by stretching it during cooling. Stretching effectively helps to align the chains, increasing the opportunity for molecular interaction across the chain.

Crosslinking can inhibit the passage of permeant molecules since the polymer chains are held together by much stronger bonds than those involved in crystallinity. In both crystallinity and crosslinking, the random movement of the polymer chains is restricted. Consequently, the probability that holes of a certain size will open between the polymer chains is lower and permeation is reduced.

The inhibition of permeation is likely to be strongest for crystalline polymers. The forces in a highly oriented material may interact between every atom in one polymer chain and adjacent atoms in the next. The covalent crosslinks in a cured polymer act only between the branching points, of which there are far fewer. The total energy of crystallization may be higher than the crosslink energy of a cured polymer, even though the individual crosslinks of the cured polymer are much stronger than any one secondary bond in the crystalline material. At most, secondary bonds of crystallization act over relatively short distances of a few angstroms. A covalent link may extend over several atoms and 12 or more angstroms. This means it is much easier for a permeating molecule of a hazardous chemical to find enough room between two crosslinked polymer molecules to get past, as compared to two polymer molecules linked by multiple secondary bonds of crystallization.

Dissolution of the permeant molecules into the polymer through molecular interactions is important to the permeation process. In a crosslinked polymer, the polymer-solvent secondary bonds cannot break the primary polymer-polymer covalent crosslink bonds. When an organic solvent is applied that will interact with these polymers, swelling occurs. The polymers usually will not decompose unless swelling is severe. However, non-crosslinked polymers will easily dissolve when the appropriate solvent is used. An appropriate solvent for such a polymer has molecular interaction forces very similar to those between the polymer-repeating units.

Plasticizers and Physical Properties

Plasticizers are agents added to polymers to reduce crystallinity. Commonly used plasticizers are phthalic and adipic acid esters, e.g., di(2-ethyl hexyl) phthalate, commonly called DOP or dioctyl phthalate. Plasticizers form weak secondary bonds between the polymer chains and serve to spread the chains apart, increasing the polymer's flexibility. Fewer molecular interactions can take place between the chains.

Plasticizers have much lower molecular weights and therefore much higher solubilities than polymers. Since the bonds between plasticizer and polymer are not covalent, the appropriate solvents can dissolve the plasticizer from the material. When a plasticizer is dissolved from the polymer material, the polymer shrinks and becomes hard and inflexible. Some of the space in the polymer film that was occupied by the plasticizer is likely to remain open and act as permeation channels for any chemicals. The chemical resistance of a plastic that has been exposed to an incompatible chemical is likely to be ruined.

Another way of plasticizing a polymer is to incorporate the plasticizer directly into the polymer chain through covalent bonding. A second monomer is used as a plasticizer. Small amounts of these "soft" monomers are added to the recipe and incorporated at random positions in the polymer chain. Solvent cannot remove these internal plasticizers.

Geometric differences in polymers can affect permeation characteristics. Molecules having the same atoms in different geometric arrangements are called isomers. In polymers containing symmetric monomers, isomerism is not important. However, if the monomer is not symmetrical, such as vinyl chloride, then the specific positions of the chlorine atoms along the polymer chain can be important and can change the properties of the polymer.

In general, if chains formed by the asymmetrical monomer are somewhat regular in structure, the degree of crystallinity or interaction between the chains is increased, leading to a harder, tougher material. If the chains are twisted or irregular, then a softer, more flexible polymer is usually formed because of the decrease of molecular interaction between the chains. The molecular weight of a polymer is an average molecular weight based on the probable distribution of chain lengths present and is important to the polymer's final form and resistance to permeation.

At the ends of each polymer chain the functional groups may have chemical characteristics different from the repeating units of the polymer chain. For example, a condensation reaction typically produces linking groups that have much lower polarity than either of the end groups on the monomers. If the molecular weight is small, then the chain length is small, and the concentration of the end groups is greater. If the end groups are polar and the polymer repeating groups are non-polar, then a low molecular weight may increase the permeability of the polymer to polar solvents. Analogously, if repeating groups interact more with each other than with the end groups, then fewer chain interactions and less crystallinity will occur as molecular weight decreases. In general, the higher the molecular weight, the greater the interactions between the chains and the lower the solubility of a polymer to solvents and other challenges.

Polymer Transitions

Three transitions occur in polymers upon heating from a relatively cool state. When cooled to a low enough temperature, any polymer will assume the characteristics of glass– hardness, stiffness, and brittleness. At this temperature, also called the glass-transition temperature, large-scale molecular motion probably does not take place. Instead, only atoms and small groups of atoms are able to move against the local restraints of secondary bond forces. As the polymer is heated, plasticity returns again at the glass transition temperature, and large-scale molecular motion starts again.

While all polymers have a glass-transition temperature, in highly crystalline polymers, the glass transition temperature is difficult to detect. In these polymers, a second transition, known as melting, is approached at about the same time as the glass transition. In highly crystalline polymers, this temperature is comparable to the melting point in low molecular-weight compounds, i.e., the point at which sufficient heat is added to the system to overcome most molecular interactions caused by secondary bonding. The melting temperature of a polymer increases with degree of crystallinity.

The third transition is degradation. After a polymer is heated past the melting point, eventually the polymer undergoes a final change and degrades. This temperature is referred to as the degradation temperature.

In summary, a polymer is glassy at low temperatures. With increasing heat, the polymer becomes plastic or perhaps rubbery at the first transition. At the second transition, it melts or becomes liquid-like; and finally degrades at the third transition. In polymer chemistry, the glass-transition temperature and the factors that influence it are extremely important in manufacturing for desirable characteristics.

Free volume within the polymer is the amount of space the polymer occupies that isn't actually filled by the polymer chains. The greater the free volume, the more room the molecules will have for displacement, and the lower the glass-transition temperature will be. With increased free volume, there is more room for foreign chemicals to enter the polymer.

The mobility or freedom of polymer chains to rotate internally, affects the glass-transition temperature. If the individual polymer chains have difficulty in rotating because pendant (side) groups interfere with one another, then the glass-transition temperature will be increased. For chains to rotate freely, additional energy often is needed to allow the pendant groups to "pass by" one another during the rotation. If the energy requirement is too large and the rotation does not occur, then the glass-transition temperature is usually higher than it is for a polymer with a similar structure (isomers or near isomers) that does not possess the rotation effect.

Crosslinking in small amounts does not appreciably affect the glass-transition temperature. However, a high degree of crosslinking restricts chain mobility and thereby raises the glass-transition temperature.

Summary

The subject of polymer chemistry is very complex. This chapter provides an introduction to polymer chemistry—and to the various relationships necessary for a complete understanding of chemical permeation through protective clothing. The most important relationships follow:

- CPC materials may be pure polymers, copolymers, laminates of polymers, or blends of polymers.
  - **Pure polymers** are "pure" because they contain only one repeating unit.
  - **Copolymers** contain at least two distinct types of repeating units.
  - **Blended polymers** have two or more previously polymerized polymer chains. These chains are mixed in the final manufacturing process to form a polymer product that is a blend of two or more polymer chain types.
  - **Laminates** are cemented, or otherwise joined, polymer sheets composed of different individual polymer types.
- **Crosslinking** and **crystallinity** are determined by the types of bonds in the polymer.
  - **Crosslinking** is dependent upon the same type of bonding (covalent) that holds repeating units together in the backbone of a polymer.
  - **Crystallinity** is dependent upon secondary bonding or molecular interactions between the monomers and the polymer chain.
- Through **secondary bonding**, plasticizers interact with the polymer chains, holding them apart to increase the polymer's flexibility.
- **Polymer transitions** are important in the manufacture of a final polymer product.
  - **Glass-transition temperature** is the temperature

below which a polymer is glass-like. Above this temperature, the polymer is plastic if it contains a relatively high degree of crystallinity; it is rubbery if it is amorphous.

- **Melting temperature** is the point above which the polymer melts or doesn't hold a shape due to the breakage of the molecular interaction forces between the polymer strands.
- **Degradation temperature** is the temperature at which covalent bonds begin to break.

Suggested Readings:

F.W. Billmeyer, *Textbook of Polymer Science*, 3rd edition. (Wiley Interscience, New York, 1984). [Complex, complete reference.]

J. Brandrup, and E.H. Immergut, Eds., *Polymer Handbook*, 2nd edition. (Wiley Interscience, New York, 1975). [Physical-chemical constants for polymers.]

H.G. Elias and R.A. Pethrick, *Polymer Yearbook*, 1st edition. (Harwood Academic, New York, 1984). [Yearly reviews of important subjects.]

S.L. Rosen, *Fundamental Principles of Polymeric Materials* (Wiley Interscience, New York, 1982). [Short, concise, easy to understand.]

R.B. Seymour, *Modern Plastics Technology* (Reston Publishing, Reston, VA, 1975). [Good reference for polymer structures.]

Appendix

Common Names and Repeating Units for Polymers in Chemical Protective Clothing

Name (constituents)	*Repeating Units*
Flexible Films	
Butyl (97% isobutylene/3% isoprene copolymer)	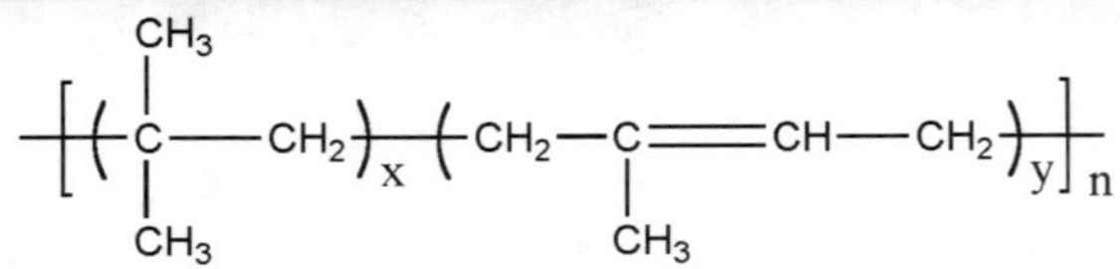
Chlorobutyl (butyl rubber with chlorine atoms substituted randomly for hydrogen atoms)	*See Butyl*
Chlorinated polyethylene (polyethylene with 36–45% by weight chlorine atoms substituted randomly for hydrogen atoms)	*See Polyethylene*

EVA/PE
(blend of 14% ethylene/vinyl acetate copolymer; 86% polyethylene)

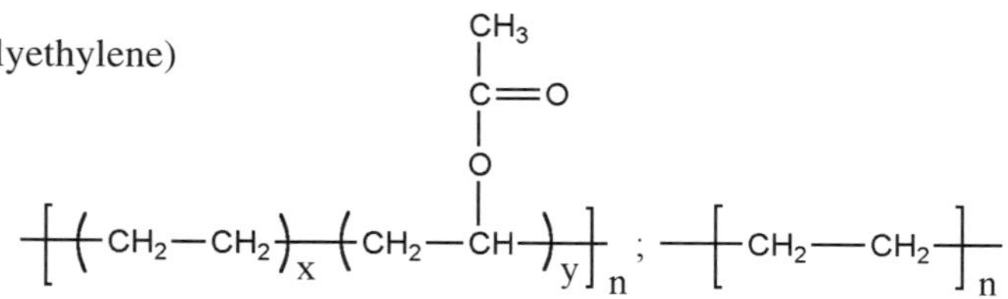

EVOH
(ethylene/vinyl alcohol copolymer)

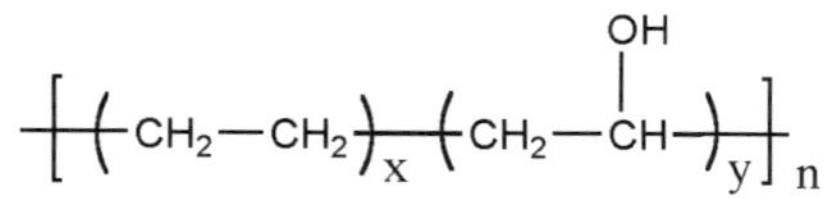

Fluorine/Chloroprene (Viton, chloroprene laminate)

See Viton, Neoprene

FEP, fluorinated ethylene propylene resin
(Hexafluoropropylene/ tetrafluoroethylene copolymer)

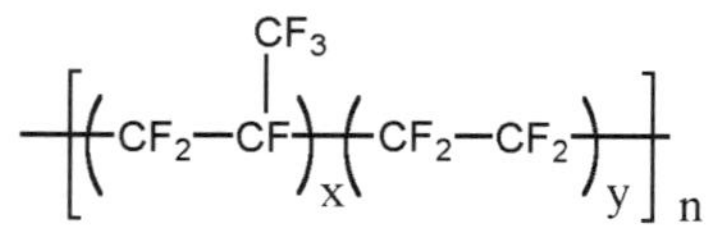

FEP/TFE (FEP; tetrafluoroethylene blend)	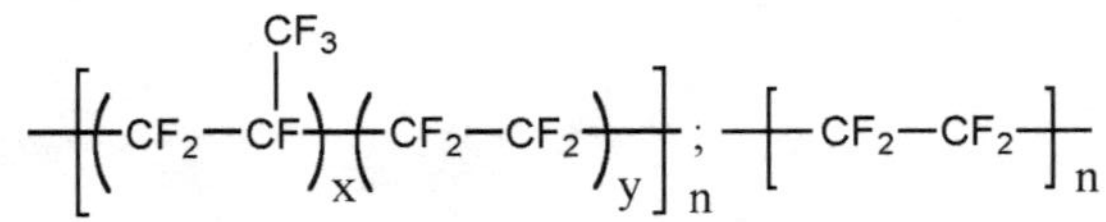
Natural rubber (isoprene)	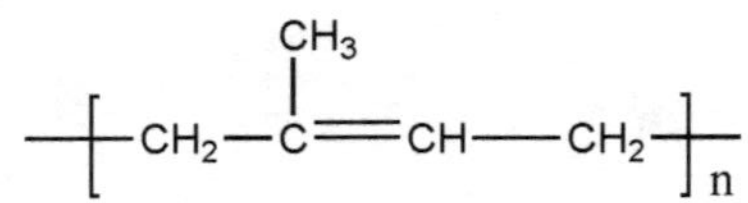
NBR, Nitrile (random or alternating acrylonitrile/butadiene copolymer)	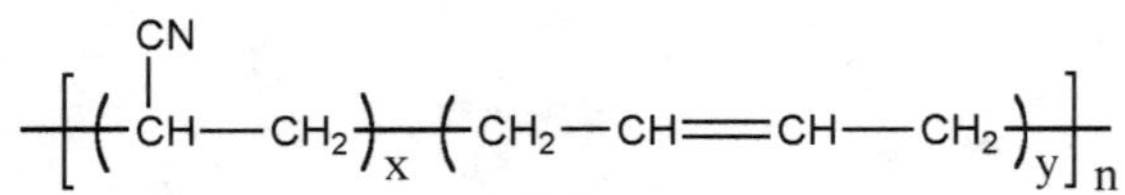
Neoprene (chloroprene)	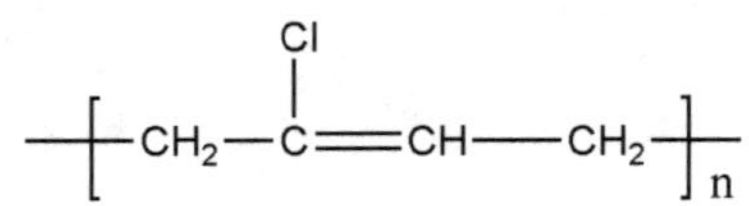
Neoprene/SBR (chloroprene; styrene-butadiene blend)	*See Neoprene; SBR*
Neoprene/Natural (chloroprene; isoprene blend)	*See Neoprene; Natural rubber*

Polyethylene (ethylene)	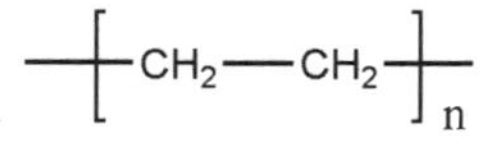
PVA (vinyl alcohol)	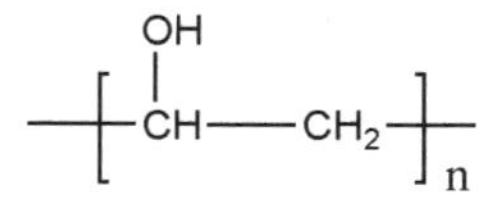
PVC (vinyl chloride)	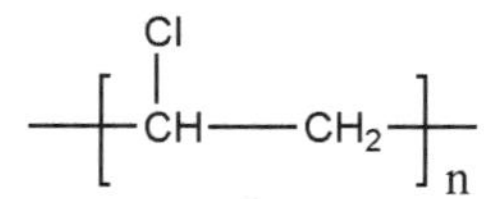
PVC/Nitrile (blend)	*See PVC; NBR*
Saran (85% vinylidene chloride/15% vinyl chloride copolymer)	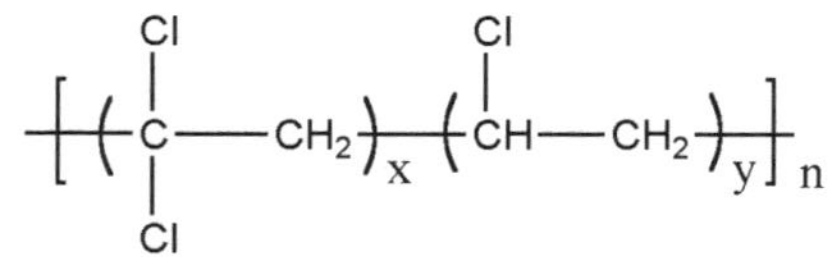
Saranex (laminate of polyethylene and saran)	*See structures elsewhere*

SBR (25% styrene/75% butadiene random copolymer)	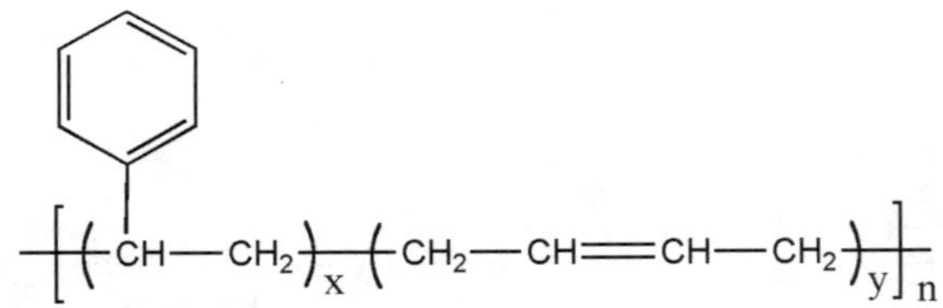
Silver shield, 4H (laminates of polyethylene and EVOH)	*See structures elsewhere*
Viton (hexafluoropropylene/vinylidene fluoride random copolymer)	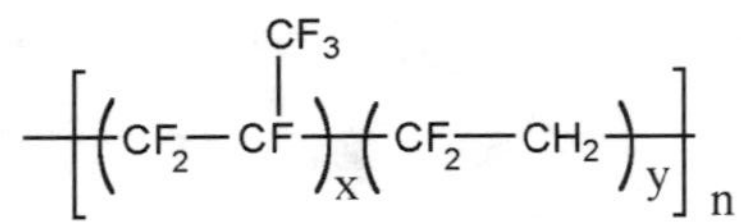
Vitrile (Viton; nitrile blend)	*See Viton; NBR*
Troionic (natural; neoprene; carboxylated nitrile blend)	*See Natural; Neoprene; NBR (carboxylated nitrile has carboxyl groups at the end of the polymer chains)*

Urethane (condensation product of a polyisocyanate and a polyol)	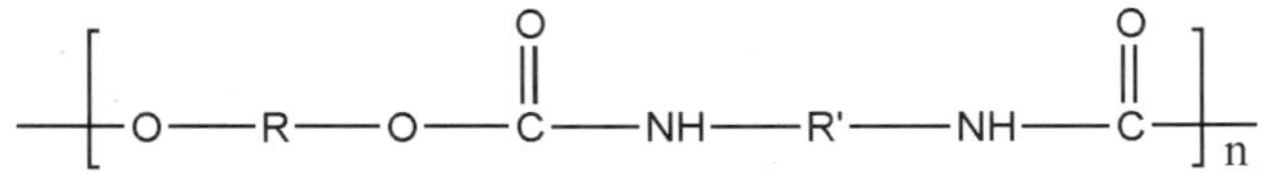
Commonly Coated Fabrics	
Cellulose or cotton (polysaccharide, or polyalcohol)	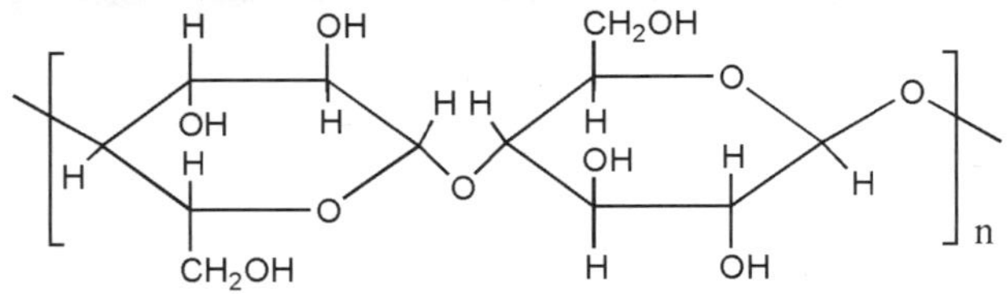
Disposaguard (cellulose reinforced with nylon, i.e., scrim reinforced)	*See cellulose*
Duraguard (non-woven polypropylene fibers)	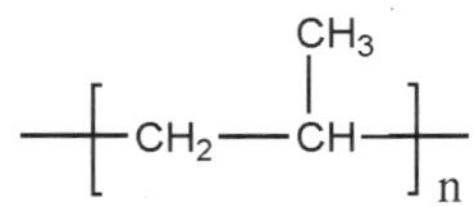
Fiberglass (fibers of glass, usually coated with silicone or silanes)	*May be added to many polymers (polyester, nylon, polypropylene)*

Gore-Tex (TFE, polyester or Nomex laminate)	*See structures elsewhere*
Nomex (aramid fiber, polyamide)	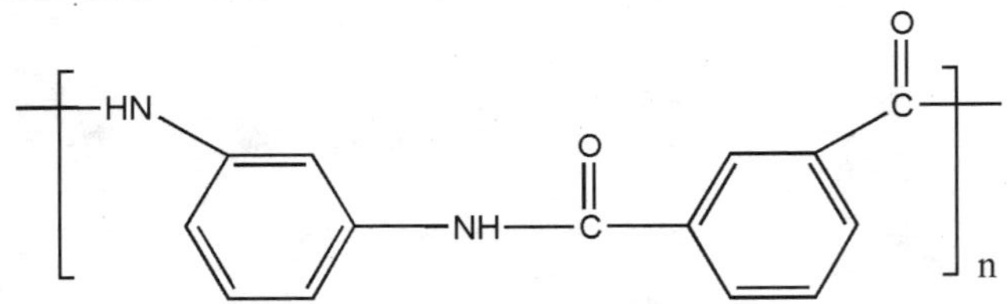
Nylon (polyamide, condensation product of diamine and dicarboxylic acid, e.g., nylon 66)	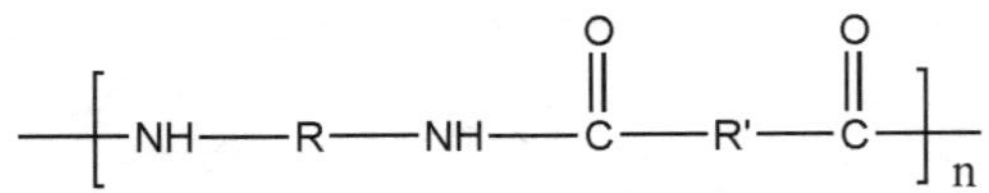
Polyester (condensation product of ethylene glycol and terepthalic acid e.g., Dacron and Sontara)	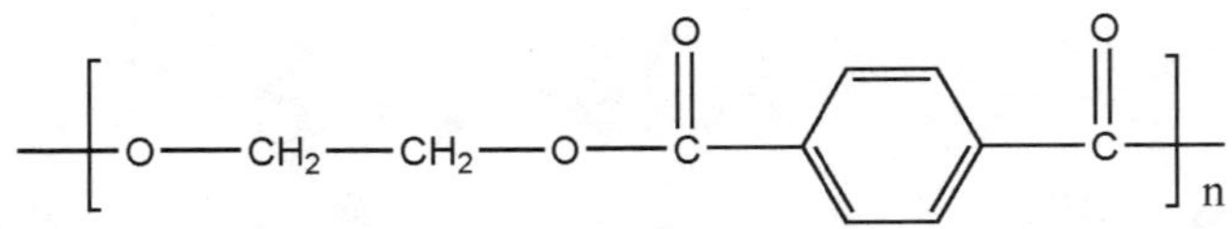
Rayon (regenerated cellulose fiber)	*See cellulose*

Safeguard (three layers of polypropylene)	*See polypropylene*
Tyvek (non-woven polyethylene fibers)	*See polyethylene*
Face Shield Polymers	
Acetate (cellulose acetate, acetate ester of cellulose)	*See cellulose*
Acrylic (methylmethacrylate)	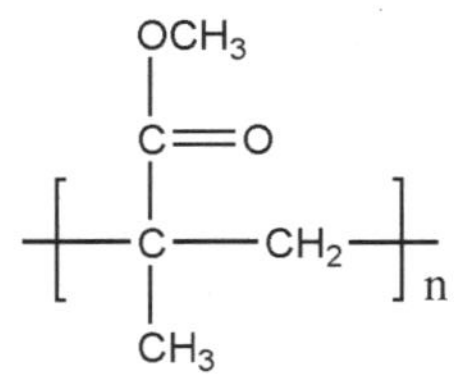
Cellulose propionate (propionic ester of cellulose)	*See cellulose*

CR-39 (diethylene glycol bis allylcarbonate)

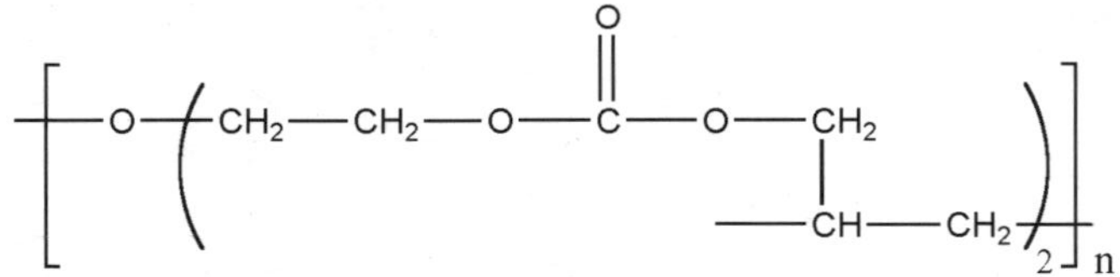

Polycarbonate (condensation product of bisphenol-A and phosgene)

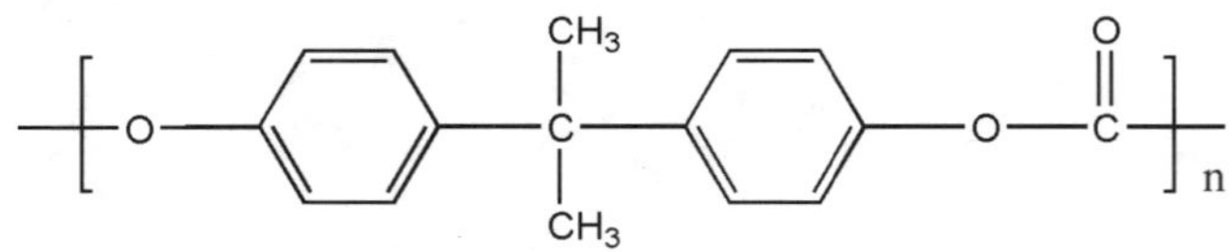

Polyester (condensation product of dicarboxilic acid and glycol; for example terephthalic acid and ethylene glycol)

See under "Commonly Coated Fabrics"

Polysulfone (condensation product of bis-phenol A and dichlorophenyl sulfone)

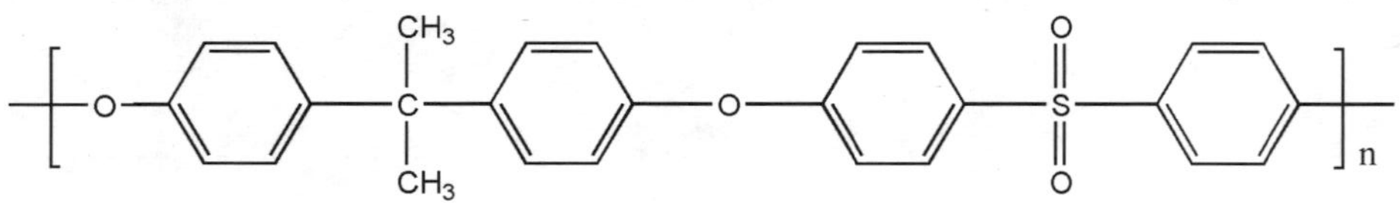

PVC (vinyl chloride)

See under "Flexible Films"

Surlyn (ethylene/ methacrylic acid, random copolymer)

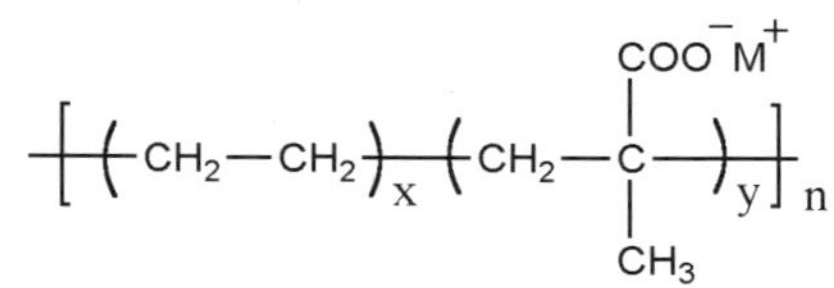

3

Solvent-Polymer Interactions

By Jimmy L. Perkins

Background

Although the major purpose of this chapter is to discuss solvent-polymer interactions as they relate to chemical protective clothing, the solutions of similar interaction problems must be mentioned first. These additional applications have longer histories in terms of research and use than protective clothing. For example, as one might expect from examining the vast array of plastic and elastomer products used in everyday life, there is a great interest in the possible effects that environmental contaminants may have upon a finished polymer product, such as how ozone can affect natural rubber products.[1]

Many polymer products are produced from a polymer resin dissolved in a solvent that later evaporates. In this chapter, "solvent" will be used to describe any non-polymer material of interest which contacts the polymer. The manner in which the solvent interacts with the polymer will largely determine the polymer drying time and the lasting quality of the product. Paints and coatings are perhaps the best examples of this application. Once the paint film has formed, the remaining solvent must diffuse through the polymer and evaporate to the outside. The manner in which this occurs can affect the quality of the final product.[2]

The automotive industry encounters situations that are somewhat similar to protective clothing problems. Gaskets and seals made of polymer compounds must have a resistance to the various solvents and chemicals with which they come in contact. For example, the polymer parts of carburetors must have solvent resistance to gasoline.[3]

In other applications of polymers, the promotion of solvent-polymer interactions is used in order to accomplish a desired effect. As such, polymer films may be used to separate a mixture of solvents. In a particular mixture, one component may permeate more rapidly through the film than the other components, separating that component from the mixture.[4]

No polymer is indefinitely resistant to a chemical, and consequently rates are very important in solvent-polymer interactions. The rate variable is used in applications involving the timed-release concept. If an active ingredient (such as a pesticide or drug) is divided into small particles and encapsulated in polymer spheres of varying diameters or varying polymer types, then the active ingredient will be released at varying times. In many similar situations, water is the solvent interacting with the polymer to release the active ingredient. Other applications involve the binding of the ingredient into the polymer matrix below the saturation point. At this point, the ingredient will diffuse through the polymer and evaporate to the environment at some determinable rate.[5]

The permeation process resulting from solvent-polymer interactions is used in several makes of passive dosimeters for sampling gases or vapors. By covering the dosimeter with a polymeric compound that is differentially permeable to a gas or vapor of interest, one may increase the selectivity of the device for that particular gas or vapor, and thus increase the passive dosimeter's accuracy.[6] Permeation-type dosimeters should not be confused with dosimeters that operate strictly on the principle of diffusion. Although some diffusion-type monitors have polymer membranes, their hole sizes are such that they do not interact with diffusing molecules (i.e., the organic vapor monitors of the 3M Co.).

In all the above examples, knowledge of solvent-polymer interactions is necessary to make useful predictions. As stated earlier, a given polymer is not indefinitely impermeable to a given chemical, although for certain solvents, the time necessary for detection of a chemical traversing the thickness of the polymer can be on the order of days or weeks. Many ketones will permeate the polymer Viton, a fluorocarbon, in a matter of minutes. However, propanol requires days before permeation can be detected.[7]

Given an understanding of the mechanisms involved in solvent-polymer interactions, the relationships between the appropriate solvent and polymer physical/chemical properties should allow some prediction of the interaction parameters, such as the rate of permeation. While the chemical and physical properties involved are many and varied, the most important drawback to predictive modeling is the fact that protective clothing products are never pure polymers. They contain additives that change the chemical composition of the products and make the prediction process much more complicated.

The main purpose of this chapter is to discuss the mechanisms by which solvents and polymers interact, and to discuss the previous attempts to predict or model these interactions. The previous modeling attempts show that qualitative predictions can be made about permeation rates and breakthrough times. This can be done using known physical/chemical data for the polymer and solvent. Perhaps most important, the accuracy of the predictions should improve, since the application of research to chemical protective clothing is relatively recent.

Solvent-Polymer Interaction Processes

The best way to visualize the interaction between solvents and polymers is to imagine that the polymer sheet is made up of individual polymer molecule chains, almost like a bowl of spaghetti or a tangled mass of earthworms. In some situations, the chains may be oriented in a more parallel fashion. Sometimes the chains are chemically (covalently) bonded at various points along their length. Through random motion, the chains open and close, creating "holes" between them. These random movements of the chains, coupled with the random kinetic movement of the solvent molecules, underlie the process of the solvent permeating through the polymer.

Usually, the solvent must first interact with the polymer chains before it can diffuse through the polymer. The magnitude of these interactions is dependent on the cohesive energy (the energy that holds liquid or solid molecules together) of the solvent and polymer. The cohesive energy for small molecules is related to vapor pressure—i.e., the more cohesive a solvent, the lower its vapor pressure.

Compounds that have small cohesive energies are gases at room temperature.[8] According to solubility theory, two molecules with similar cohesive energies will have maximum solution in one another. Consequently, if a solvent and polymer have similar cohesive energies, the likelihood that the solvent will interact with the polymer is greater.[9]

This solvent-polymer interaction is the "solution step" in the permeation process. As shown in Figure 1, solvent molecules come in contact with the polymer and are dissolved depending on the interaction between solvent and polymer. Imagine an initial "layer" of polymer molecules that the solvent would encounter. The number of solvent molecules entering that layer is dependent on the solubility of the solvent in the polymer (i.e., the similarity of the two compounds), the size of the solvent molecules, and the size of holes opening in the polymer layer.

In order to enter the second layer, solvent molecules must move forward through the first layer. Remember, however, that the random motion of the solvent molecules may move them backward or forward. As shown in Figure 1, a concentration gradient usually develops. The rate of molecules evaporating from the downstream membrane surface in Step 3 is dependent on diffusion (Step 2) and solution (Step 1).[10]

Other terms are used to describe interaction of the solvent and polymer in CPC applications including degradation, penetration, and permeation. Degradation is also a function of the cohesive energies (solubilities) of the solvent and polymer. During degradation, the polymer may disintegrate because of solvent action on the polymer itself or on plasticizers or other additives in the polymer. Generally, the greater the degradation caused by a solvent, the more readily the solvent will permeate the polymer. Swelling usually takes place prior to degradation, and crosslinked polymers will swell, but rarely degrade.

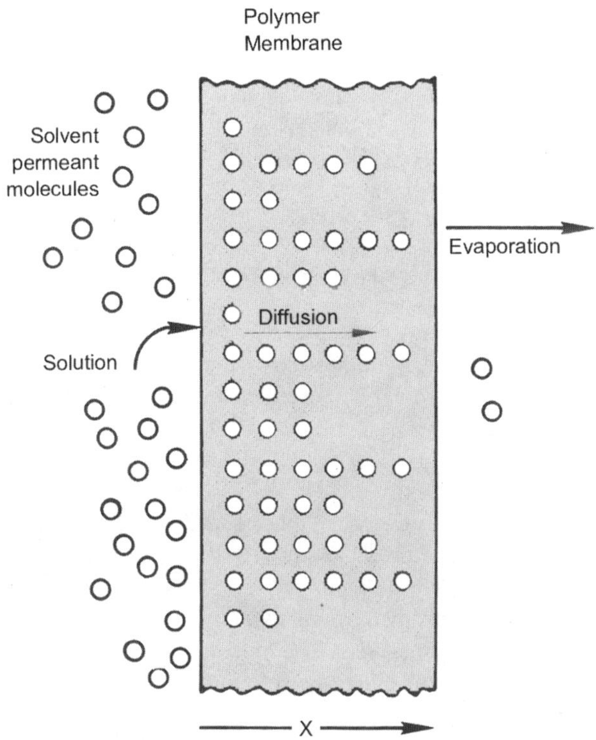

Figure 1. The permeation process.

Penetration is a physical process where bulk liquid solvent passes through the polymer membrane. This process is not dependent upon diffusion or solution, but instead relies on relatively large discontinuities that may occur in the polymer product because of construction processes. In effect, the solvent convectively flows through the pores with minimal interaction at the molecular level. Penetration is common in articles composed of paper, cotton, and other cellulose products, but can also occur through "pinholes," seams, and zippers in finished elastomers.

Permeation is the process in which a chemical passes through a polymer by means of solution and diffusion. In some cases, the solution step may be of little importance, as with N_2 passing through a thin sheet of polyethylene. However, for liquid solvents and CFC polymers, the importance of interaction or solution can vary greatly in magnitude. Since permeation is most dependent upon solution and diffusion, further detail about these processes is set forth in the following two sections.

The Solution Step

From solubility theory, two substances are soluble if, upon mixing, the free energy of the mixture is less than the sum of the free energies of the two pure substances. The free energy of mixing is defined as

$$\Delta G_m = \Delta H_m - T\Delta S_m \qquad \textbf{(1)}$$

where ΔG, is the free energy of mixing, ΔH_m is the enthalpy, T is temperature, and ΔS, is the entropy. The important point of this equation is that the enthalpy of mixing term (ΔH_m) must be less than the entropy term in order for the difference in free energy (ΔG_m) to be negative. The more negative the free energy difference, the better the solubility.

Solubility is dependent upon intermolecular forces. For example, if hydrogen bonds form between a polymer and solvent, the ΔH_m is small or negative, and solution (or negative ΔG_m) is assured. In effect, if these intermolecular forces are similar for two molecules, then solution may occur, since the enthalpy of mixing is quite small (or negative), causing the free energy to be negative.

The molecular forces that hold a liquid together are composed of three separate forces, as discussed in the Chapter 2 (see Figure 2 in that chapter on page 58). The relationship of the three forces is

$$C^2 = D^2 + P^2 + H^2 \qquad \textbf{(2)}$$

where C is the cohesive energy density, D is the dispersion parameter, P the polar parameter, and H the hydrogen bonding parameter.[11] The square root of the cohesive energy density is called the one dimensional, or total, solubility parameter (SP). The three energy components taken as a vector are often referred to as the three-dimensional solubility parameter (3-DSP). Since permeation is a function of solubility and diffusivity (see Equation [9], below), it follows that by knowing something about the mutual solubility of a solvent and a polymer, one can determine something about its permeability.

An important feature of solubility parameter theory is that the total SP of a mixture is the weighted sum of the SP for the components of the mixture. Hence, if a paint product requires a solvent with a certain 3-DSP and no such solvent exists, a mixture of solvents may be made which closely matches that 3-DSP. By the same token, the polymers used in automotive products, such as carburetor O-rings, can be chosen so they have the greatest amount of resistance to the chemicals with which they come in contact.

Simple Diffusion

There are many definitions for diffusion and many theories concerning the manner in which it takes place.[12] Put simply, diffusion is the random movement of molecules such that, given enough time, the distribution of molecules tends towards even concentration over space.

In contemplating the diffusion of molecules in a membrane, one must assume that the solution of the solvent and polymer occurs when a solvent molecule contacts the polymer surface. As the solvent diffuses through the polymer membrane, a concentration gradient is established that usually decreases linearly, from highest concentration at the contacted surface to lowest concentration at the opposite surface. The reason for the concentration gradient is that the unexposed surface from which the solvent evaporates is being continuously swept by a carrier gas or liquid. If this didn't happen, then no concentration gradient would occur, establishing instead a uniform and saturated distribution of solvent in the polymer. Of course, the workplace situation may be somewhere between these two extremes, since under a protective garment there may be varying degrees of air movement.

According to some diffusion theories, a molecule in a "hole" within the polymer will move from that hole due to translational energy.[13] The molecule will move to another hole, provided that other holes in its vicinity are greater than a certain size. (Obviously, the molecule cannot occupy a hole that is smaller than itself.) The hole size formed by the polymer chains is a function of the energy of the chains. If the system is heated, the hole sizes will increase and the rate at which the holes are formed will also increase, aiding or speeding up the diffusion process.

The mathematical equation that describes simple diffusion is Fick's First Law

$$J = -D\frac{dc}{dx} \qquad \textbf{(3)}$$

where J is the mass flux in mg/min-cm^2, D is the diffusion coefficient in cm^2/min, c is the solvent concentration in the membrane in mg/cm^3, and x is the distance in centimeters from the contacted membrane surface. The minus sign indicates diffusion occurs toward areas of lower concentration. Consequently, the flux rate or diffusion rate is dependent upon the concentration gradient (dc/dx), and the diffusion coefficient (D) is simply a constant for the system. At constant temperature, the concentration gradient reduces to the concentration on the exposed surface divided by the thickness of the membrane (C/L), assuming that the opposite surface is continuously swept free of chemical (i.e., the concentration is 0 at that surface). For a liquid solvent contacting the membrane surface, the concentration at that surface is equivalent to the solubility of the solvent in the polymer. For a gaseous permeant exposed to the membrane surface, the maximum concentration in the membrane surface is the free concentration of the gas or vapor.

The diffusion coefficient is a constant that relates mass flux to concentration gradient and applies exclusively to the specific polymer and solvent in question. In other words, the diffusion coefficient derived for the permeation of benzene through neoprene will not be applicable to benzene through polyvinylchloride, or benzene through air. Diffusion coefficients are highly dependent on temperature.[10]

For most solvents that are liquids at room temperature, the diffusion coefficient can vary with the solvent concentration at the membrane surface. This is due to the chemical interaction of solvent and polymer. Consequently, the diffusion coefficient for a liquid in contact with a membrane will not be the same as for a non-saturated vapor of the same liquid in contact with the membrane. Most CPC permeation tests, however, use a liquid; this could be especially important in situations where the protective clothing is intended to prevent contact with a vapor.

If Equation (3) is integrated, the more Familiar equation,

$$J = -D\frac{C_1 - C_2}{L} \qquad \textbf{(4)}$$

is derived, where the diffusion coefficient D is not a function of concentration. Here, C_1 and C_2 are the permeant concentrations upstream and downstream, respectively, and L is the membrane thickness. As stated earlier, C_2 is usually 0 and falls out of the equation. In the cases described above, where D is a function of concentration,

$$\bar{D} = -\frac{1}{L}\int_{C_2}^{C_1} D dC \qquad \textbf{(5)}$$

In this equation, the diffusion coefficient in the integral is related to concentration by some other empirical equation. Consequently, in cases where the diffusion coefficient is dependent upon concentration, $\bar{D}$ is used to indicate an average diffusion coefficient.

D or $\bar{D}$ can be determined by measuring the concentration in the exposed surface of the membrane and the flux rate. In the strictest sense, ideal diffusion (D) will take place only if the permeant does not interact physically with the polymer or polymer additives, such that the concentration gradient is altered. Any non-ideal diffusion behavior is called non-Fickian diffusion and probably occurs where polymer swelling is noticeable.

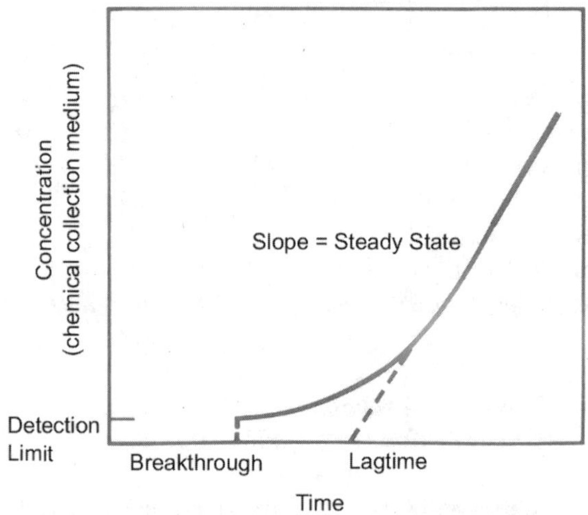

Figure 2. The lag time is the intersection point obtained by extending the steady-state permeation slope to the x-axis.

Another procedure for measuring diffusion coefficients is the lag time method. As shown in Figure 2, lag time is the intersection point obtained by extending the steady-state permeation line (slope) to the x-axis. From diffusion theory, the following equation relates diffusion coefficients to lag time:[12]

$$D = \frac{L_2}{6T} \qquad (6)$$

where L equals the thickness of the membrane and T equals the lag time.

Concentration in Equations (3), (4), or (5) is related to pressure by Henry's law,

$$C = Sp \qquad \textbf{(7)}$$

where C is concentration, p is partial pressure, and S, is the solubility coefficient. In the case of a solvent diffusing through a polymer that is continuously swept from the opposite side, C is the concentration of the solvent at the polymer surface, and p is its partial pressure. By substituting for C in Eq. (3), a new equation can be written:

$$J = DS\frac{dp}{dx} \qquad \textbf{(8)}$$

The coefficient DS is renamed P for permeability coefficient; thus

$$P = DS. \qquad \textbf{(9)}$$

The permeability coefficient is defined as a function of diffusion and solubility.

Molecular size and shape are known to affect diffusion rates. As might be expected from diffusion theory, the larger a molecule, the less likely the chance that a large enough hole will be formed to allow the molecule to diffuse. Consequently, diffusion coefficients are usually smaller as the molecule becomes larger in a homologous series.[14] Larger molecules also possess less kinetic energy, so the speed with which it is able to move to the next hole is less than that of a smaller molecule.

If an interaction occurs between polymer strands that are oriented effectively parallel to each other, then the possibility of hole formation of a certain size is reduced. This commonly occurs in polymers with a high degree of crystallinity, such as high-density polyethylene having a greater crystallinity than low-density polyethylene. Plasticizers, which increase the space between polymer strands and make the polymer more pliable, are necessary to "soften" these polymers for use in protective clothing. While increased crystallinity usually decreases the diffusion coefficient, the addition of a plasticizer increases the diffusion coefficient. Glassy polymers like PVC are below their glass-transition points at room temperature (*See Chapter 2*). Plasticizers must also be added to soften PVC for use in protective clothing.

Plasticizer considerations can be complex. For example, a chemical may have an initially high permeation rate, but at the same time it may interact with the plasticizer and leach it from the polymer. As more plasticizer is leached away, the permeation rate begins to decrease because the polymer strands become closer together. This is one example of how additives to polymers can cause non-Fickian diffusion, as mentioned above.[15]

Another polymer property that affects diffusion is crosslinking. Most accelerants used in crosslinking are organic, or crosslinks can occur directly between two polymer chains (covalent bonding). Crosslinking typically lengthens diffusion times because the polymer chains are pulled close together. However, if the crosslinking agent is a large organic molecule, the effect on diffusion can be more like that of a plasticizer. Also of importance is the idea that solvents can interact with the crosslinking agents. The net result may be non-Fickian diffusion.

The Evaporation Step

Permeation is often described as a repeating three-step process involving sorption of solvent molecules into holes forming between polymer chains, desorption of the solvent molecules, and then diffusion to the next available hole.[10] However, if the downward-gradient side of the polymer is constantly swept clear, the desorption or evaporation step likely plays a very small role in the rate determination of the overall permeability process.

Solubility is important both in terms of the solvent-polymer interaction during diffusion and the subsequent evaporation of the solvent from the opposite side of the polymer. A solvent or permeating substance with a very low vapor pressure will be slow to desorb from the opposite side.

However, such a solvent will also be slow in diffusing and, because of possible poor molecular interaction with the polymer, may also be poor in dissolving into the polymer. Thus, the evaporation step would also appear to be a function of solubility and diffusivity.

The Importance of Solubility and Diffusivity

Previous attempts to predict how well a solvent will permeate through a polymer have relied on the concepts of solubility and diffusivity. However, the diffusion and solubility concepts have rarely been combined for the purpose of predicting permeation through chemical protective clothing.

One might expect that a chemical permeating through a membrane would be affected most by its size relative to the holes that are forming in the membrane. Since the size of any hole formed is presumably a random process with respect to some statistical mean, smaller molecules will have a greater selection of hole sizes through which they can diffuse, and consequently will diffuse faster. By the same argument, for a homologous series of molecules, larger molecules should diffuse more slowly (assuming that molecular interactions or solubilities are equal among the different species). For example, propane (a small molecule) permeates fastest, and n-pentane (a large molecule) permeates slowest for a propane to n-pentane series against polyisobutylene.[14] Since the molecular interactions between these solvents and the polyisobutylene are presumably very similar because of chemical structure, then the major factor affecting permeation is the diffusion coefficient.

Molecular shape is also important. A branch on a molecule may impede the diffusion process more so than the main body of the molecule. Permeation experiments using a homologous series of phthalic acid esters reveal that there is an optimum size for the length of the ester molecule.[16] In the case where C_4 is the ester molecule, diffusion is fastest, but for appending groups smaller or larger than C_4, diffusion becomes slower. This is presumably because the smaller appendages do not sufficiently shield the C=O group, thereby promoting molecular interaction, and the longer branches effectively slow the diffusion process and add nothing to the shielding.

Diffusion and solubility are independent of one another in the sense that a highly soluble solvent may not necessarily have a rapid diffusion rate. By the same token, a molecule with very little solubility in a polymer may have a very rapid diffusion rate. Small diatomic gases or noble gases diffusing through polymers often do so quite rapidly, even though the gases apparently have no interaction with the polymer. The size of the noble gas becomes important with respect to diffusion.

When molecules are quite small, such as methanol, carbon disulfide, and acetone, permeation rates are unusually rapid when compared with rates predicted by solubility alone.[7] The molecules may be small enough that the physical interaction (i.e., solution) with the polymer is insignificant in the overall permeation rate. In these cases diffusion is the most important step, resulting in a much higher permeation rate than for a larger molecule where chemical interaction or good solution is probably necessary for the molecule to gain entrance to the polymer.

Consequently, solubility and diffusion, though not directly related, appear to be very much interwoven in the permeation process, and neither can be considered alone. This fact may be one reason why some previous attempts to predict permeation have been only partially successful. Most of these attempts have relied singularly on either diffusion coefficients or solubility parameters, rather than addressing the joint action of the two concepts.

Other Variables in the Permeation Processes

Two other variables must be considered in the permeation process: the thickness of the polymer and the temperature of the solvent/polymer system. The effect of temperature on the permeation process can be related by an Arrhenius equation,

$$P = P_0^{-E_p/RT} \qquad \textbf{(10)}$$

where E_p is the activation energy of permeation, P is the permeation coefficient, P_0, is the pre-exponential factor, R is the ideal gas constant, and T is the absolute temperature. Similar equations may be written for the diffusion coefficient and solubility coefficient. When natural logarithmic transforms of temperature data are plotted against permeation coefficients, a straight line with an intercept of P_0 and slope of $-E_p/RT$ is obtained.

Two examples in Tables 1 and 2 demonstrate this phenomenon. The temperatures shown in these tables are not uncharacteristic of those found in many manufacturing processes. Most laboratory permeation data are reported at 25°C. However, as can be seen in Table 2, a glove that was tested at 25°C in the laboratory may give vastly different results at 40°C. Note that the temperatures used here are test temperatures (solvent, polymer, and collection media.) Since many industrial processes are performed at higher than ambient temperatures, gloves coming in contact with solvents from those processes should be selected properly with respect to the actual temperatures of the solvents.

Thickness is also an important variable. The user of protective clothing should be aware that permeation rates and breakthrough times reported for a neoprene glove of a given thickness will not be applicable to a neoprene glove of another thickness. In the past, attempts have been made to standardize permeation data (such as the breakthrough time) by squaring it, or dividing it by the thickness of the glove or the square of the thickness of the glove. The latter relationship is derived from Equation (6), which shows that the lag time (see Figure 2) is related to the square of the thickness of the membrane. This equation holds only for Fickian diffusion, so in many cases its use is incorrect, and more importantly, there is no relationship between breakthrough time and lag time.

The use of the thickness-squared term (or any correction term), if appropriate, should yield a constant value when

Table 1:
Benzene Permeation through Neoprene.

Solvent T (°C)	Breakthrough Time (min)	Steady-State Flux (mg/cm2-min)
7	40	0.19
22	24	0.23
37	16	0.33

Source: Data from Ref. 17.

Table 2:
Methylene Chloride Permeation through Viton.

Solvent T (°C)	Breakthrough Time (min)	Steady-State Flux (mg/cm2-min)
25	61	0.12
30	47	0.16
35	35	0.20
40	29	0.21

Source: Unpublished data from author.

used for a series of breakthrough times for varying polymer thicknesses. Table 3 shows that this is not the case. The data in this table, derived from Ref. 15, show average breakthrough times and permeation rates for three thicknesses of neoprene latex gloves and three thicknesses of PVC gloves manufactured by Pioneer. For the neoprene gloves, 17 values of the breakthrough time and the permeation rate, derived for 17 different chemicals, were averaged over each of the three thicknesses. If division by thickness or thickness squared is an appropriate way to standardize for thickness differences, then the number derived from this quotient should be a constant over the range of thicknesses. As seen in Table 3, however, the number is not a constant. The same error can be seen for the PVC gloves, in which case only five sets of data (chemicals) were available for averaging.

Of the correction factors shown for permeation rate,

Table 3: Standardization of Permeation Parameters for Thickness.

Polymer	Thickness (mm)	BL[b]	B/L	B/L^2	PR[b]	PR x L	PR x L^2
PVC	0.13	4.3	33	234	1199	156	20
	0.20	9.05	45	226	616	123	25
	0.31	16.76	54	174	399	124	28
Neoprene	0.29	7	50	84	146	42	12
	0.47	17.2	37	78	82	39	18
	0.70	28.4	40	60	64	45	31

PR x L is theoretically correct, if Fickian diffusion is occurring. This gives fairly good agreement for the neoprene gloves, but yields more variability for the PVC gloves. The relationship between thickness and permeation parameters is not constant, owing to experimental error and non-Fickian behavior. Consequently, one should be very careful when applying thickness correction factors. Ideally, permeability coefficients would be the best parameter for comparing gloves of varying thicknesses (see Eqs. [8] and [9]), but they are rarely reported.

Attempts to Model Protective Clothing Permeation

Colletta et al. attempted to derive short-term diffusion coefficients (values for D that occur prior to steady state–see Equation [6]) for nine carcinogens in order to predict the permeation resistance of protective clothing.[17] Unfortunately, they concluded that, due to the complexities associated with permeation through "filled" elastomers, accurate diffusion coefficients could not be determined, and modeling was impractical.

However, solubility tests (in which a polymer was immersed in a solvent for an extended period of time) did show interesting predictive results. The weight gain of the polymer was an indicator of the solubility of the solvent in the polymer. When weight gain and solubility were high, short breakthrough times were usually observed; however, the inverse was not so consistently true, and no correlation studies were performed.

Weeks and McLeod did correlation studies of weight gain and breakthrough times.[18] A typical result of their work is shown in Figure 3 for data involving Arochlor 1254 (PCB). They did not report correlation coefficients. Other chemicals studied were 1,1,2-trichloroethane, 1,2-dichloroethane, and l,1,1-trichloroethane.

Holcombe studied the weight gain of Viton after its submersion in six different chemicals.[19] When the natural logarithm of these values was correlated with breakthrough time, a correlation coefficient of 0.9 was obtained.

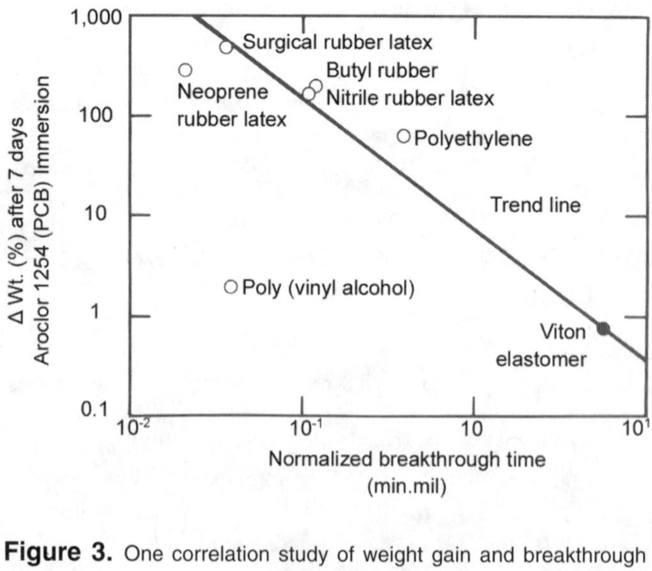

Figure 3. One correlation study of weight gain and breakthrough time.

Recalling that "like dissolves like" from the discussion of the 3-DSP, one might conclude that if a solvent and polymer have similar 3-DSP values, then the solvent should be highly soluble in the polymer. Henriksen used this concept in correlation studies with permeation parameters.[20] His preliminary work with permeation data from other researchers indicated that the natural logarithm of the steady-state permeation rate when correlated with polymer/solvent 3-DSP differences gave a correlation coefficient of (–)0.5 to approximately (–)0.7 The greater the difference between 3-DSP values for solvent and polymer, the lower the solubility, hence the negative correlation coefficients. Although these coefficients are not unity, they nevertheless indicate a dependence of permeation on solubility.

Bomberger et al. used permeability coefficients,[21] mainly derived from the work of Nelson et al.[15] and correlated those against 3-DSP differences of solvent and polymer pairs. This work yielded correlation coefficients of (–)0.5 to (–)0.98. The researchers thought that correlation coefficients could be improved if more data for low permeation rates were available. These data are lacking because hours, or in many cases days, are needed before steady-state permeation is reached. A good example is Viton and carbon tetrachloride, where steady-state rate was reached after approximately seven days and breakthrough occurred after three days.[7] Henriksen's study also had the same problem concerning the lack of long breakthrough-time data."

Perkins et al. also modeled permeation using the 3-DSP with the polymer Viton.[7] This work yielded a correlation coefficient of 0.69 for the ln values of breakthrough time, and a correlation of (–)0.65 for ln steady-state permeation rates when correlated against S-DSP differences of polymer and solvent. In this work, solvents with long breakthrough times were included, without a great improvement in correlation coefficients.

In the studies of Perkins et al.,[7,22] Bomberger et al.,[21] and Henriksen,[20] solubility was determined by calculating the S-DSP differences between the polymer and the solvent. If this difference is great, the solvent should not be very soluble in the polymer. The correlation coefficients obtained indicate the potential usefulness of the solubility parameter as a predictor of permeation. Indeed, Perkins et al. showed that by using confidence limits placed on the regression line, one can predict both the breakthrough time and the permeation rate. Even though the confidence limits were somewhat large, a 95% lower confidence limit on breakthrough time, and a 95% upper confidence limit on permeation rate, could be used to predict the probable value of those two parameters for an untested solvent. For example, if 3-DSP values for the solvent and polymer alone are known, a minimum breakthrough time of 340 minutes (95% confidence), and a maximum steady-state permeation rate of 4.3 mg/m^2s can be predicted for xylene and Viton.

Perkins et al. also used polymer thickness and solvent molecular volume in combination with 3-DSP differences to predict permeation parameters.[22] Molecular volume is thought to affect diffusion rate. In these studies of butyl, Viton, and nitrile rubbers:

1. Thickness was not significant in breakthrough time studies, but was significant in predicting steady-state permeation rate;
2. Molecular volume was significant in breakthrough time studies, but it was only significant for nitrile in permeation rate studies;
3. 3-DSP difference was by far the most significant variable;
4. When significant variables alone are considered, coefficients of determination (R^2) Were 0.38 to 0.55 for breakthrough time, and 0.57 to 0.64 for steady-state permeation rates.

All of the above results indicate that 3-DSP has utility, but only in semiquantitative applications. It may be used to predict which chemical will have the highest permeation rate for a polymer, or which polymer is most protective for a chemical.

In a good semiquantitative application of 3-DSP, Henriksen designed a new glove to protect against a chemical mixture.[20] In this case, the mixture consisted of various chemicals used in an epoxy resin process. Henriksen found that no single polymer protected against the hardener, the binder, or the various solvents used. Using 3-DSP data, he determined that a triple-laminate of polyethylene, polyvinyl alcohol–ethylene copolymer, and polyethylene, should provide protection against all the chemicals of concern for an acceptable period of time. A single-use glove was fabricated of the laminate and was demonstrated to be acceptable.

One drawback in the 3-DSP approach and in most of the theoretical considerations discussed here is that glove products are not pure polymers. They contain plasticizers, antioxidants, fillers, pigments, and other additives. The solubility theory relates only the mutual solubility of the polymer and the solvent, and therefore acts as a predictor of permeation based on that two-component system. However, if a plasticizer is present in a polymer that has a substantially different solubility parameter, then results may not be predictable. If a solvent is selected that is poorly soluble in the polymer but readily soluble in the plasticizer, the solvent will leach out the plasticizer, changing the polymer structure. In the case of PVC, which contains large amounts of plasticizer (frequently 50–60%), this effect is prominent, resulting in a very stiff unplasticized PVC membrane, which cracks readily and is difficult to use.

This effect can be seen even for elastomers containing smaller amounts of additive. For example, the polymer Viton contains an amine-derivative additive; although Viton is resistant to many chemicals, it will degrade readily when exposed to certain amines such as diethylenetetramine. Solubility theory predicts that the amines would have little solubility in pure Viton. However, they readily decompose the glove polymer through their interaction with the amine.[19]

Once an additive is removed from a polymer, thereby changing the polymer configuration, the diffusion step is also affected, since the hole size and the hole-formation rate will also be changed. With some CPC polymers that contain plasticizer, a rapid breakthrough may occur due to the diffusion of the chemical through holes created by the plasticizer. If the plasticizer is then leached out by the solvent, the holes will close. Hence the steady-state permeation rate in this instance is very low, while the breakthrough time is short. The polymer film is also hardened and cracked by the deplasticizing action.[18]

Since polymer recipes vary among manufacturers, one must calculate 3-DSP values for all glove products of a generic polymer. At present, values for 13 polymers have been published and it is hoped that all manufacturers will publish data for their gloves.[23]

Another problem in modeling permeation is that neat or pure compounds rarely occur in the workplace. Workers are usually exposed to mixtures of compounds, and gloves or protective clothing must be selected for those mixtures. Solubility parameters, as noted above, can be derived for mixtures, and this approach is used successfully in the selection of solvents for paint formulations.

However, the permeation of chemicals through glove polymers may be different. A solubility parameter calculated for a mixture of two chemicals may not be a good predictor of permeation parameters. In other words, chemical A and chemical B may give permeation parameters that are predictable by their relative solubilities in the polymer, but the solubility parameter for the mixture of the two may not predict the total permeation of the mixture. More research is needed.

The few permeation studies involving mixtures have so far yielded interesting results.[15,24,25] Although these results may be summarized, it should be emphasized that permeation data involving mixtures are few, and any conclusions drawn from them are tenuous. Nevertheless, for a mixture containing two very similar chemicals with respect to solubility parameters (such as, for instance, toluene and xylene) the permeation is approximately additive. If the mixture contains a chemical that readily permeates the polymer and another compound that does not readily permeate the polymer, the results are not additive (see Figure 4). For example, if only small amounts of a good permeant are added to large amounts of a poor permeant, the poor permeant may begin permeating

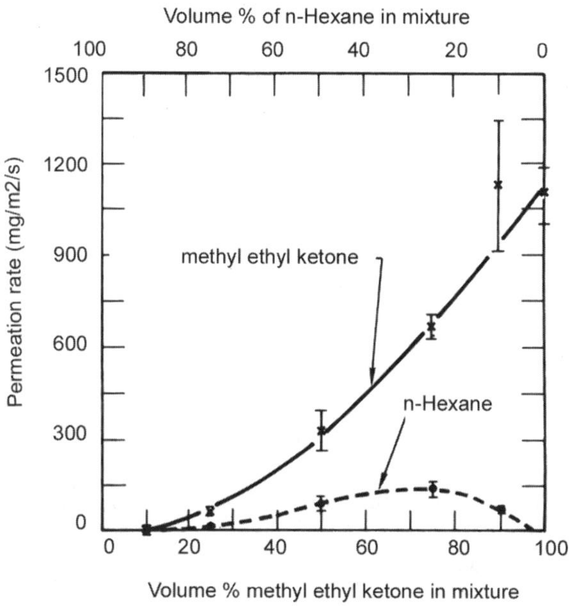

Figure 4. For each component in the mixture, steady-state permeation rate is plotted against mixture composition. As the MEK concentration in the binary mixture increases along the abscissa, the n-Hexane concentration decreases.

the polymer sooner than expected. In Figure 4, if permeation were additive, then one should obtain a line with zero slope for the sum of the permeation rates. In addition, a pure compound that does not break through in eight hours (permeation rate ~ 0) should not break through if the compound is mixed with another one. However, this is not the case. A good permeant may be changing the structure of the polymer internally to the extent that the calculated solubility parameter for that polymer is no longer applicable.

This process is similar to the manner by which a plasticizer affects a permeant, by opening up holes and allowing for greater diffusion. Such a permeant, once absorbed by the polymer, may also interact with the poor permeant. Under these changed conditions caused by the good permeant, the poor permeant now interacts with the "new" polymer structure and passes through the polymer. These interactions may not be quantitatively predictable with a 3-DSP approach. A better understanding of the process occurring with mixtures must be achieved and when more mixture data are generated, the 3-DSP approach can be modeled more thoroughly.

Expectations for the Future

Since recipes for polymers are often trade secrets, at best, only the nature of the additives and not the exact proportions may be known. This fact will continue to make modeling of permeation results difficult. One solution is to make the nature of polymer additives apparent with persistent research. Another solution is for manufacturers to be somewhat more lenient in allowing access to information, thus improving the ability of industrial hygienists to select protective clothing.

Industrial hygienists are trying harder to select the proper CPC to ensure workers' health and to reduce employers' liability. Consequently, vague guidelines for the selection of protective clothing are no longer useful. Industrial hygienists must have exact information, which for the most part, must come from manufacturers. Therefore, manufacturers must be certain that the information they disclose is not misleading. Certainly, from the hygienists' point of view, manufacturers should willingly issue as much information about their protective clothing products as possible, even at the risk of losing some commercial advantage.

Industrial hygienists, however, must often deal with imprecision. Steady-state permeation rate and breakthrough-time data derived from permeation tests are inaccurate. In a review of permeation data, Bomberger et al. state that permeation parameters may vary as much as one order of magnitude when reported by different researchers for the same generic polymer-solvent combinations.[21] The many potential reasons for this discrepancy probably relate to differences in the permeation test cells used, glove lot differences, and detection limits. Permeation testing applied to protective clothing has begun only recently (see Chapter 5). As time progresses, one can expect that the precision of test results should improve, along with our attempts to model those test results. In the meantime, users of this information must realize its imprecise nature and must be willing to err on the safe side when choosing protective clothing.

The user of permeation predictive tools must also realize that at present these can only predict the best polymer for a given application, rather than accurately quantify the permeation parameters. Therefore, the industrial hygienist should probably not select a particular piece of protective clothing because its breakthrough time is 60 minutes, for instance, and the job task will last only 50 minutes. Preferably, the industrial hygienist should select the best protective clothing for the application relative to other clothing types, and have knowledge of the permeation parameters.At the same time, the hygienist should also realize that inaccuracy exists in those parameters, whether experimentally determined or predicted by a model.

With so many new developments in the areas of permeation testing and protective-clothing selection, it is not surprising that use of permeation test data is still at a very elementary level. One can look forward in the future to agreement in permeation test results among researchers, to increased knowledge of permeation theory among practicing industrial hygienists, and to improved methods for predicting permeation parameters. These factors should allow the industrial hygienist to make more rapid and more accurate decisions and should also remove the need for many permeation tests on the large number of chemicals and polymers now present in the workplace.

References

1. Stern, A.C., H.C. Wohlers, R.W. Boubel, and W.P. Lowry, *Fundamentals of Air Pollution*, 3rd ed., Academic Press, New York (1994).
2. Hansen, C.M., "Solubility in the Coatings in Industry," Farg Och Lach 4:69-77 (1971).
3. Beerbower, A., and J.R. Dickey, "Advanced Methods for Predicting Elastomer/Fluids" in *Interactions*, ASLE Transactions 12:1-20 (1969).
4. Rodgers, C.E., V. Stannett, and M. Szwarc, "The Sorption, Diffusion, and Permeation of Organic Vapors in Polyethylene," J. of Poly. Sci. 45:61-82 (1960).
5. Cardarelli, N., Controlled Release Pesticides Formulations, CRC Press, Cleveland, OH (1976).
6. Reiszner, K.D., and P.W. West, "The Collection and Determination of Sulfur Dioxide Incorporating Permeation and West-Gaeke Procedures," Envir. Sci. Tech. 7:523-526 (1973).
7. Perkins, J.L., A.B. Holcombe, M.C. Ridge, M.K. Wang, and W.E. Nonidez, "Skin Protection, Viton, and Solubility Parameters," Am. Ind. Hyg. Assoc. J. 47:03 (1986).
8. Burrell, H., "Solubility Parameter Values," in *Polymer Handbook*, 4th ed., J. Branderup Immergut, and E.A. Grulke, Eds., John Wiley & Sons, New York (1999).
9. Hildebrand, J., and R. Scott, The Solubility of Non-Electrolytes, 3rd ed., Reinhold, New York (1949).
10. Yasuda, H., and V. Stannett, "Permeability Coefficients" in *Polymer Handbook*, 4th ed., J. Branderup, E. Immergut, and E.A. Grulke, Eds., John Wiley & Sons, New York (1999).
11. Hansen, C.M., The Universality of the Solubility Parameter, IEC Product Research and Development, 8:2-11 (1968).
12. Crank, J., and G.S. Park, "Methods of Measurement" in Diffusion in Polymers, J. Crank and G.S. Park, Eds., Academic Press, New York (1968).

13. Kumins, C.A., and T.K. Kwei, "Free Volume and Other Theories," in Diffusion in Polymers, Academic Press, New York (1968).

14. Prager, S., and F.A. Long, "Diffusion of Hydrocarbons in Polyisobutylene," Am. Chem. Sec. J. 73:4072-4075 (1951).

15. Nelson, G.O., B.Y. Lum, G.J. Carlson, C.M. Wong, and J.S. Johnson, "Glove Permeation by Organic Solvents, Am. Ind. Hyg. Assoc. J. 42:217(1981).

16. Ueberreiter, K., "The Solution Process," in *Diffusion in Polymers*, J. Crank and G.S. Park, Eds., Academic Press, New York (1968).

17. Colletta, G.C., A.C. Schwope, I.J. Arons, J.W. King, and A. Sivok, *Development of Performance Criteria for Protective Clothing Used Against Carcinogenic Liquids*, U.S. Department of Health, Education and Welfare, National Institute of Occupational Safety and Health, Pub. No. 79-106, Cincinnati, OH (1978).

18. Weeks, R.W., and M.J. McLeod, "Permeation of Protective Garment Material by Liquid Halogenated Ethanes and a Polychlorinated Biphenyl," U.S. Department of Health, Education and Welfare, National Institute of Occupational Safety and Health, Pub. No. 81-110, Cincinnati, OH (1981).

19. Holcombe, A.B., Use of Solubility Parameters to Predict Glove Polymer Permeation by Industrial Chemicals, Masters Project, University of Alabama at Birmingham, School of Public Health (1983).

20. Henriksen, H.R., Selection of Materials for Protective Gloves, Polymer Membranes to Protect Against Epoxy Products, Danish National Labor Inspection Service, Lyngby,Copenhagen DK-2800 (1982).

21. Bomberger, D.C., S.K. Brauman, and R.T. Podoll, Studies to Support PMN Review: Effectiveness of Protective Gloves, US Environmental Protection Agency, Office of Pesticides and Toxic Substances, Technical Directive 68, Washington, DC (1984).

22. Perkins, J.L., M.C. Ridge, and W.E. Nonidez, "Predicting Permeation Properties of Butyl, Viton, and Nitrile Rubbers," presented at the 2nd International Symposium on the Performance of Protective Clothing, American Society for Testing and Materials Committee F23, Philadelphia, PA (1987).

23. Perkins, J.L., and A. Tippitt, "Use of the Three-Dimensional Solubility Parameter to Predict Glove Permeation," Am. Ind. Hyg. Assoc. J. 46:455 (1985).

24. Mickleson, R.L., M.M. Roder, and S.P. Berardinelli, "Permeation of Chemical Protective Clothing by Three Binary Solvent Mixtures," Am. Ind. Hyg Assoc. J. 47:236(1986).

25. Spence, M.W., "Chemical Permeation Through Protective Clothing Material: An Evaluation of Several Critical Variables," presented at American Industrial Hygiene Association Conference, Portland, OR (1981).

4

Types of Chemical Protective Clothing

By Jeffrey O. Stull

Introduction

The evaluation of chemical protective clothing (CPC) designs, design features, performance, and applications require an understanding of the types of chemical protective clothing available for protection. The types of chemical protective clothing available in the marketplace, and the choices available to the end user, have changed dramatically over the past two decades and continue to become more diverse. Chemical protective clothing exists in a variety of designs, materials, and methods of construction, and each has its own advantages and disadvantages for specific protection applications. End users should have an understanding of the different types of chemical protective clothing and their features in order to make appropriate selections. It is important to realize that CPC which appears to be similarly designed, may offer significantly different levels of performance. Thus, chemical protective clothing performance must be carefully scrutinized in addition to design and features. Furthermore, CPC must be properly sized to provide adequate protection. Improperly sized or ill fitting chemical protective clothing may reduce or eliminate the protective qualities.

This chapter provides a thorough description of the different types of chemical protective clothing, in terms of the general designs, design features, materials, methods of fabrication, performance features, and sizing. A detailed description of full body protection and standards is provided in the Chapter 6.

Classification of Chemical Protective Clothing

Chemical protective clothing may be classified by its design, performance, and intended service life. These three characteristics will permit the end user to understand the type of CPC being considered or used, as well as indicating its potential limitations.[1] The following sections describe how CPC can be "type" classified in these three ways.

Classification by Design

Classification of CPC by its design usually reflects how the item is configured or the part of the body area or systems that it protects. For example, footwear by design provides protection to the wearer's feet. For the purposes of this chapter, the types of CPC can be generally categorized as:

- Full-body garments,
- Partial-body garments,
- Gloves,
- Footwear, and
- Face and eye protection.

Table 1 shows the different CPC designs associated with different areas of body protection.

Classification of chemical protective clothing by design may also provide an indication of specific design features that differentiate CPC items of the same type. For example, totally-encapsulating chemical protective suits are configured with significant design differences when compared to splash suits. Some designs of chemical protective clothing may offer varying protection against hazards in different parts of the CPC item. A coated glove may offer chemical protection to the fingers and palm area, but the uncoated portions of the glove on the back and wrist areas will offer no chemical protection. Nevertheless, it is important to realize that even apparent CPC coverage of a specific body area, in and of itself, does not guarantee protection of that body area.

Table 1:
CPC Design Types by Body Area

Body Area(s)	Type
Entire body	Totally encapsulating suit 2-piece suit (hooded jacket with visor with pants or overalls)
Torso, head, arms, and legs (excluding hands, feet, and face)	Hooded coveralls 2-piece "splash" suit (hooded jacket and pants or overalls)
Torso, arms and legs	Coveralls 2-piece "splash" suit (jacket and pants or overalls) Smock
Top torso and arms	Coat or jacket Lab coat
Bottom torso and legs	Pants
Torso (front) and arms	Sleeved apron
Torso (front)	Apron
Head and face	Hood with visor
Head	Hood
Face and eyes	Faceshield Ventilated goggles Non-ventilated goggles Full facepiece of respirator
Hands	Gloves Mittens
Foot	Boots Overboots or overshoes Booties Boot or shoe covers

A further means of distinguishing chemical protective clothing by design is to indicate the materials used in its construction. Materials will possess different characteristics that impact both the performance and wearability of the CPC item. Rubber materials behave much differently than plastic materials. Owing to their elasticity, rubber materials are more likely to offer form fitting designs but may weigh more than comparable plastic materials, which tend to offer better chemical resistance. Since material choices vary with the type of item, a more detailed discussion is provided in the respective section on the available and characteristics of materials for different types of chemical protective clothing.

Classification by Performance

Classification of chemical protective clothing by performance indicates the actual level of performance to be provided by the item of CPC. This may include a general area of performance or a more specific area of performance. While two items of CPC might be considered chemical protective clothing, one item may provide an effective barrier to liquids, but not to vapors while the other item provides an effective barrier to both liquids and vapors. Classification of CPC by performance is best demonstrated by actual testing or evaluation of the chemical protective clothing with a standard test that relates to the type of desired protection (see Chapter 5). These tests are used as a demonstration of protection against the anticipated hazards and often become the basis of claims by the manufacturer for their products. However, intended or manufacturer-claimed performance does not always match actual performance. Furthermore, performance claims should be uniformly applied to all parts of the CPC item, i.e., the seam should offer the same performance as the material, and the performance classification should be limited by the weakest element of the CPC design.

The specific classification of chemical protective clothing will be related to the types of protection offered against chemical hazards. Table 2 describes a hierarchy of chemical barrier performance.

In general, chemical protective clothing that protects against gases and vapors will also protect against liquids and particulates, and CPC effective against liquids will also prevent penetration of particulates. However, there are exceptions. Some chemical protective clothing based on adsorptive materials (such as those used in air-purifying respirator cartridges) may prevent penetration of gases and chemical vapors, but not when exposed to liquids if the CPC is splashed by or immersed in liquid chemical.

Table 2:
Hierarchy of Chemical Barrier Performance

Type of Chemical Barrier	Protection Offered*
Permeation-resistant	Prevents or limits any contact with chemicals in the form of gases/vapors, liquids, or particulates (solid)
Vapor penetration-resistant	Prevents wearer contact with atmospheric vapors or gases
Liquid penetration-resistant**	Prevents wearer contact with liquids
Particulate penetration-resistant	Prevents wearer contact with particles

* Protection offered is specific to chemical or chemical characteristics.

** Liquid penetration resistance is sometimes subdivided into two categories where chemical protective clothing may be resistant to liquid penetration under pressure as might be associated with spraying liquid (e.g., from a burst pipe) versus CPC that may limit penetration under conditions of a light splash or mist.

Other performance features may be related to the non-chemical hazards in the workplace, the durability of the CPC in different use environments, or the impact of the CPC on the wearer. The respective performance of CPC against these properties can similarly be ranked or rated, but may create tradeoffs between desired characteristics.

Classification by Service Life

The classification of chemical protective clothing by expected service life is based on the useful life of the CPC item. Thus, service life reflects the longevity of the product in relation to the user's expectations. The service life of chemical protective clothing generally fits into three classes:

- Disposable after a single use—Products that are relatively inexpensive, which cannot be adequately, cleaned, reserviced, or maintained after use; or it is easier to dispose of and replace the CPC, rather than provide care or maintenance.

- Limited use—Some cleaning, care, and maintenance is possible, but the CPC may not be reusable under hard physical conditions; or the CPC is eventually degraded by use and maintenance processes.
- Reusable—Can be readily cleaned and maintained, and still continue to provide acceptable performance.

Unfortunately, chemical protective clothing manufacturers do not always specify the service life of their products and may not indicate conditions for retirement of CPC. Some chemical protective clothing may also have limited "shelf life" (i.e., time in storage before use) because of material degradation that can take place in storage due to heat, ozone, or material self-degradation. Of course, any item of chemical protective clothing can be rendered unusable if irreversibly contaminated or damaged in use, but these distinctions are often addressed in the CPC's design, design features, choice of materials, and performance.

Chemical protective clothing service life is a function of three factors:

1. Durability–How CPC maintains its performance with use,
2. Ease of serviceability–The user's ability to care for, maintain, and repair CPC so that it remains functional for further use, and
3. Life cycle cost–The total costs for purchase, using, and maintaining an item of CPC.

Durability – Chemical protective clothing durability is demonstrated by the length of time that the CPC item provides acceptable performance, given the range of use conditions, care, and maintenance. Unacceptable performance may be evident through physical changes in the chemical protective clothing item such as:

- Rips, tears, or separation of materials and seams,
- Thin spots or cracks in coated materials or protruding fibers,
- Unexplained material discoloration, and

- Diminished functionality of CPC component parts.

Unacceptable performance may also not be readily evident unless products are carefully examined by product manufacturers or subject to destructive testing.[2] Estimates of product durability can be made through product testing for simulating product wear but most often are derived from field experience involving actual product use. Some products are expected to lose some performance in certain property areas, so the end user must examine the acceptability of any drops in performance related to protection.

Ease of Reservicing – Some types of chemical protective clothing can be serviced or repaired to extend service life. Servicing and repair is considered part of a regular care and maintenance program to allow CPC to meet its expected service life. NASA has an elaborate capability for cleaning, decontaminating, repairing, and testing Propellant Handlers' Ensembles used for protecting launch site personnel during hypergolic fuel operations.[3] Other types of chemical protective clothing may not be easily repaired or cannot be repaired without manufacturer assistance.

The most significant aspect of reservicing chemical protective clothing is the ease of decontamination. This is an important issue because the end user must have confidence that the CPC item is contamination-free. One of the reasons for the increasing popularity of disposable clothing is because its use obviates the difficult decision of determining if the clothing is clean enough for reuse. However, in many work environments, the levels of contamination may be low or the type of contamination may be readily removed by standard decontamination practices. Otherwise, destructive testing of clothing may be needed to determine if the selected decontamination process adequately removed the chemical contaminants.[4]

If an item of CPC cannot be reserviced to an acceptable level of performance, then it cannot be reused. In addition, if the cost to repair or maintain the chemical protective item represents a significant proportion of the item's original cost, then the CPC item will probably not be reused.

Life Cycle Cost – The life cycle of chemical protective clothing includes all aspects of its selection, use, care, and maintenance until its ultimate disposal. Therefore, the life cycle cost of chemical protective clothing is the sum of all costs associated with an item of CPC. To compare different products on the same basis, life cycle cost is usually represented as the cost per use for a CPC item.

The following costs should be considered in determining the life cycle cost:

- Purchase cost,
- Labor cost for selection/procurement of CPC,
- Labor cost for inspecting CPC,
- Labor and facility costs for storing CPC,
- Labor and materials costs for cleaning, decontaminating, maintaining, and repairing CPC, and
- Labor and fees for retirement and disposal of used CPC.

The total life cycle cost is determined by adding the separate costs involved in the chemical protective clothing life cycle and dividing by the number of CPC items and by the number of uses per item. A detailed method for estimating chemical protective clothing life cycle cost is provided in Schwope and Renard.[5] Table 3 provides an example of a life cycle cost analysis between two different types of totally-encapsulating chemical protective suits.

Overall Type Classification of CPC

The overall representation of chemical protective clothing should be made with respect to the three classification systems:

- Design classification is needed to establish which parts of the body will be protected.
- Performance classification will indicate the type of protection that will be offered by the CPC.

Table 3:
Example of Life Cycle Cost Estimates for Two Chemical Protective Suits

Cost	Product #1: Disposable*	Product #2: Reusable	Comments
Purchase	$1,200	$3,600	Reusable product 3x price of disposable suit
Decontamination	$50	$250 (five uses)	Includes labor and consumables for both types
Maintenance	$50	$300 (five uses)	Each suit must be pressure tested upon receipt
Storage	$25	$25	Proportional to bulk – same for both types
Disposal	$25	$25	Proportional to weight – same for both types
Subtotal	$1,350	$4,200	
Number of uses	1	5	
Cost per use	$1,350	$840	

*Disposable suit is provided with overcover to provide the same physical ruggedness as the reusable suit.

- Service life classification will establish the expected longevity of the CPC.

The following specific examples illustrate the application of this type classification approach:

- The coated glove shown in Figure 1, while covering the hand and wrist, provides liquid penetration protection only to the majority of the hand, but not the wrist since the coating does not extend into the wrist area. For certain chemicals, the glove may also provide permeation resistance in the coated area, but if vapors are associated with the exposure, the glove will not completely protect the hand. The glove is considered reusable only if used for protection against chemicals that do not degrade or significantly permeate the glove coating.

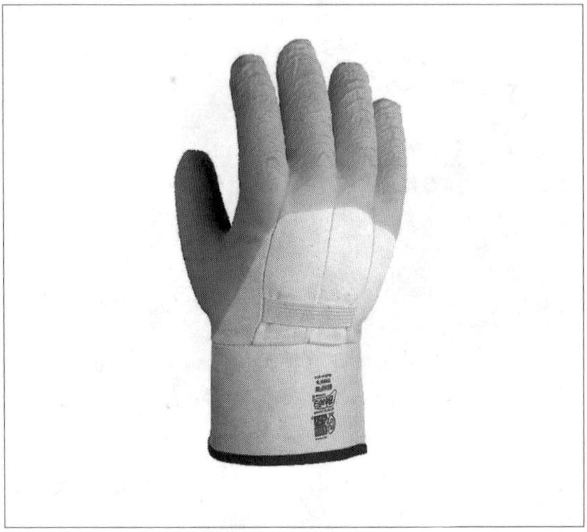

Figure 1. Example of a coated glove providing partial protection of the hand. (courtesy of Best Manufacturing)

- Figure 2 shows a rubber overboot, which completely covers the wearer's feet and ankles. Since the boot is constructed of a continuous rubber material, it will offer liquid penetration resistance performance and will limit chemical permeation for some chemicals, depending on the performance of the material against specific chemicals. Like in the preceding glove example, the boot is considered reusable if properly decontaminated and the residual chemical is removed.
- The chemical protective clothing item pictured in Figure 3 is a hooded coverall constructed of a coated non-woven material that provides liquid splash protection to the wearer's torso, arms, legs, and head (excluding the face). The coverall is considered disposable because the material has low durability, cannot be easily decontaminated if splashed and has a relative low cost for a single use. If this CPC item is combined with gloves, boots, and full facepiece respirator, then protection is afforded to the wearer's hands, feet, and face. However, the interfaces of between the clothing

Figure 2. Example of a rubber overboot providing barrier protection of wearer's feet. (courtesy of Norcross Safety Products, LLC)

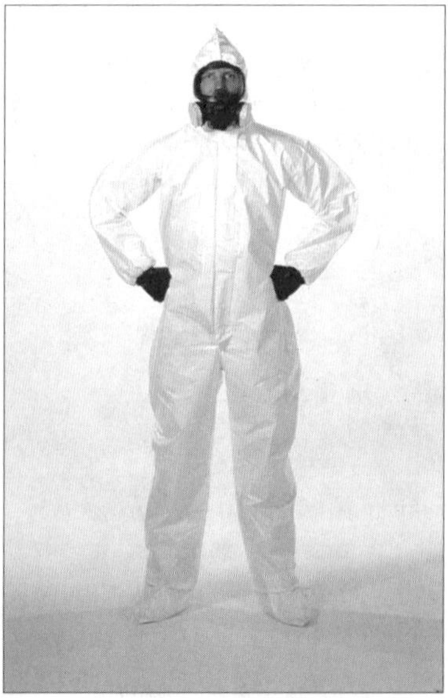

Figure 3. Example of a disposable coverall providing full-body protection against liquid splashes. (courtesy of DuPont)

items (e.g., garment sleeve end to glove) and the integrity of clothing seams and closure will affect the level of overall performance of the ensemble.

Further understanding the type classification of chemical protective clothing requires examining the specific designs, design features, materials, and other attributes associated with specific CPC types. For the purpose of this chapter, this discussion is broken into the following areas:

- Body protection (full- or partial-body garments covering the torso, arms, legs, and head, or portions thereof),
- Hand protection,
- Foot protection, and
- Face and eye protection.

Body Protection

Protection of the body from chemicals may be provided from either full-body garments or partial body garments. Full-body garments are designed to provide protection to the wearer's upper and lower torso, arms, and legs. Full-body garments may also provide protection to the wearer's hands, feet, and head when auxiliary CPC is integrated with the garment to form a suit. Full-body garments may be single- or multi-piece clothing items, including:

- Full-body suits,
- Jacket and trouser combinations,
- Jacket and overall combinations, and
- Coveralls.

Partial-body garments provide protection to only a limited area of the wearer's body including the upper torso, lower torso, arms, legs, neck, or head. Examples of partial body protective garments are:

- Hoods, head covers, and bouffants for head/or face protection,
- Aprons, smocks, lab coats for front or upper torso

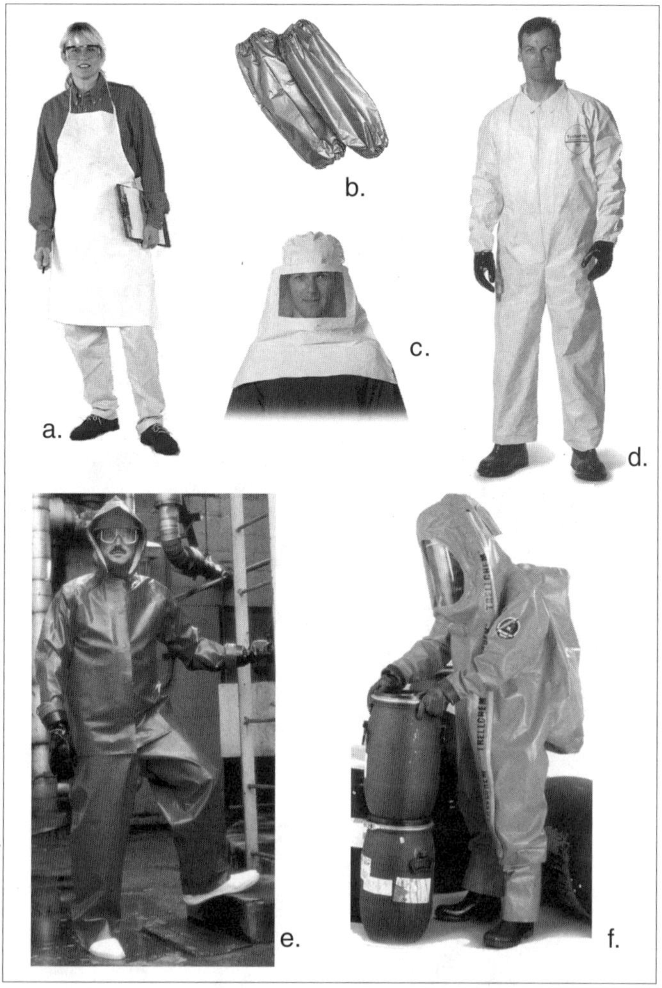

Figure 4. Types of chemical protective garments:

(a) Bib-style apron (courtesy of Kimberly Clark)

(b) Sleeve protector (courtesy of North Safety Products)

(c) Hood with visor (courtesy of DuPont)

(d) Disposable coverall with collar, front zipper, and elasticized wrists and cuffs (courtesy of DuPont)

(e) Disposable 2-piece splash suit with hooded jacket and bib pants

(f) Totally-encapsulating chemical protective suit with attached gloves and booties (worn inside boots) (courtesy of Trelleborg)

protection,

- Shirts for upper torso and arm protection,
- Pants for lower torso and leg protection,
- Sleeve protectors for arm protection, and
- Chaps or leggings for leg protection (rarely used for chemical protection).

Several examples of protective garments are shown in Figure 4. Table 4 indicates some of the characteristics and features associated with each of these types of chemical protective garments.

The extent of body protection varies with the garment design. Many garment designs do not provide uniform protection for all areas of the body covered by the garment. The specific materials and design features associated with the garment design will also influence protection.

Garment Materials

Materials used in chemical protective clothing include seven basic types:

1. Textiles,
2. Unsupported rubber or plastics,
3. Microporous film fabrics,
4. Adsorbent-based fabrics,
5. Coated fabrics,
6. Plastic laminates, and
7. Combination or specialized materials.

Textile Materials – Ordinary textile materials are generally not considered suitable chemical protective fabrics for protection against chemicals. However, special non-coated textile materials are used for a variety of applications involving particulates and light liquid spray from relatively non-hazardous chemicals. Though woven textiles are not often found in chemical protective clothing, tightly woven, repellent-treated

Table 4:
Design Characteristics and Features Associated with Chemical Protective Garments

General Garment Type	Specific Garment Type	Characteristics and Features
Full-body garments	Full-body suits	• One-piece garments which may offer a variety of entry options depending on the type and placement of the closure • Generally known as totally encapsulating suits, that encapsulate the wearer and other protective equipment such as the respirator (some design may permit a respirator facepiece to integrate with a suit hood) • Includes other CPC attached to the suit such as gloves and footwear
	Jacket and pants combinations	• Mimic normal wearing apparel • Head protection provided by hood, usually without a visor • Generally provide some overlap of the waist portion of trousers with or without a collar, hood, or wrist protection • Generally use front closure • Pants usually rely on a zipper, snaps, or other front fly closure or drawstring/elastic waist • May have openings with closures at the foot end to allow entry while wearing footwear
	Jacket and overall combinations	• Similar to jacket and pants combination • Distinguished by a higher "bib" style pants (overalls) and permit a shorter jacket • Some jacket combinations designed as pullover with hood • Overalls often use straps or gusset with snaps for adjust garment on lower torso • Bib overalls usually provided with suspenders
	Coveralls	• One-piece garments, usually with a front closure, and have options for attachment of hoods, type of sleeve end (open or elastic), type of pant cuff end (open, elastic, or bootie)

Table 4 (continued)
Design Characteristics and Features Associated with Chemical Protective Garments

General Garment Type	Specific Garment Type	Characteristics and Features
Partial-body garments	Hoods	• Cover the wearers head and include either a face/eye opening or may be provided with an integrated faceshield or visor • Usually pullover design • Length affects integration with upper torso garments • Type of visor and size may vary (if present) • Size and type of face opening accommodates respirator or eye/face protection • Face opening closure options (elastic, pliable/stretchable material, drawstrings, ties, snaps, or hook and loop closures) • Bulk of the hood on the wearer's crown may affect the fit of helmets and respirator straps
	Head covers and bouffants	• Provide protection to the wearer's upper head, but are most often used to contain or cover the wearer's hair • Bouffant style head covers are secured to the wear's head by elastic around the periphery • Head covers may also be used as helmet covers as an aid to preventing contamination
	Aprons	• Consist simply of a flat piece of material contoured to the front of the body for providing lower or lower/upper torso protection • Lower part extends around the wearer's sides and ties in the back • Aprons covering the upper torso are designated 'bib' aprons with a strap around the wearer's neck that holds up the 'bib' • Generally available in more than one size, designated by unisex rectangular dimensions, and are adjusted by the tie straps at the top or sides
	Lab coats	• Provide torso, arm, and upper leg protection • Of varying lengths and are generally offered in alphabetic sizing • Use front closures usually with snaps or buttons • May have open neck area or collar

Table 4 (continued)
Design Characteristics and Features Associated with Chemical Protective Garments

General Garment Type	Specific Garment Type	Characteristics and Features
Partial body garments	Smocks	• Provide front torso, arm, and upper leg protection • Of varying lengths and are generally offered in alphabetic sizing • Use back closures with ties • May have open neck area or collar
	Shirts	• Provide protection to upper torso and arms (long-sleeved) • Mimic regular wearing apparel with differences in collar and type of closure
	Pants	• Provide protection to lower torso and legs • Mimic regular wearing apparel with differences in waist (fly or elastic)
	Sleeve protectors	• Provide protection from the wearer's hand to the shoulder area • Usually secured to the wearer's arm by several means by elastic ends
	Chaps	• Partial pants which are open at the sides/back and are intended to provide wearer front leg protection (rarely used for chemical protection) • May incorporate belt or straps and be flared at bottom to cover top of wearer's footwear
	Leggings	• Protect the wearer's lower leg (generally knee height) • Have elastic top or adjustable tops with snaps or tie closures • May have a flare for protecting the top of wearer's footwear

fabrics can provide some low minimum protection against liquid exposure.[6] More common are nonwoven fabrics that have demonstrated barrier performance against particles and repellency of liquids. Two predominate examples of nonwoven fabrics are flashspun polyethylene (Tyvek®) and spunbond/meltbown/spunbond (SMS) polypropylene (Kleenguard®). These textiles are used because of their relatively low cost, but also because the materials provide a structure of microfibers that filter out dry particulates and many water-based liquids.

Unsupported Rubber and Plastic Materials – Normally, chemical protective clothing materials include supporting textile fabrics to provide strength. However, there are some CPC materials that do not include a fabric substrate. The rubber material or plastic is thick enough to provide sufficient strength for clothing use. Examples of polymers used in these materials are polyvinyl chloride and chlorinated polyethylene. Because the materials are continuous, they offer a barrier to liquids and can be used in the construction of CPC intended for protection against liquids and gases.

Microporous Film-Based Materials – A relatively new class of CPC materials uses microporous films. As the name implies, microporous films have millions of microscopic pores per square inch of the film structure. In most cases, the pores are irregularly shaped with tortuous paths through the film. These pores are small enough to prevent the passage of most liquids, but still allow vapors and gases to pass through the material (see Figure 5). This material feature makes the film "breathable" and is considered desirable for specifically allowing moisture vapor to transfer through chemical protective clothing to lessen the effect of wearer heat stress. The microporous films are generally glued or laminated to woven or nonwoven fabrics for physical support. Owing to their physical structure, these fabrics provide barrier performance against liquids but not gases. The type of liquids held out by microporous materials will depend on the surface tension, as lower surface tension liquids penetrate easier than higher surface tension liquids.

Adsorbent-Based Materials – One class of CPC materials use adsorbents added to textile layers for providing

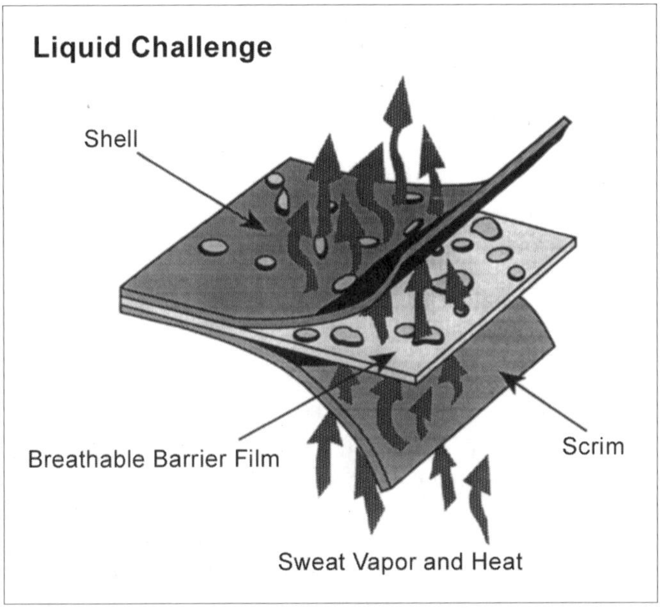

Figure 5. Configuration of microporous material for providing liquid barrier performance and breathability.

chemical protection against hazardous vapors or aerosols. These engineered materials include adsorbents, such as activated charcoal or other sorbent materials. The principle of material operation is similar to cartridges used in air-purifying respirators. The inner and outer layers of these fabrics are treated with liquid repellent to limit liquid contact that can saturate the adsorbent layer. For the most part, these materials have been used in the military applications for protection against chemical warfare agents. In general, these materials can be optimized for adsorbing different chemical classes, but generally are most effective for large molecule chemicals.[7]

Coated Fabrics – Up until the mid 1980s, the majority of chemical protective clothing used coated fabrics. Woven fabrics such as cotton, nylon, and polyester are coated with a polymer such as butyl rubber, neoprene, polyvinyl chloride, or polyurethane to provide a continuous coating over the fabric substrate. Coating thickness is important for barrier performance. Coatings that are too thin may be prone to pinholes and other defects. For some materials, coating may be applied to both sides to create a thicker and more-chemically resistant material. Materials using this construction tend to be heavy, but relatively rugged and are used in a variety of chemical protective clothing. Today, the most common clothing using coated fabrics are splash suits, but some higher-end totally-encapsulating suits may be made out of coated fabrics. Much of the same reusable splash clothing is similar to rainwear.

Plastic Laminates – The majority of chemical protective clothing today uses plastic laminates. These materials combine various plastic polymer films with nonwoven substrate fabrics resulting in a chemical-resistant, lightweight, relatively inexpensive material. The majority of substrates are based on polypropylene and films are based on polyethylene and polyvinyl alcohol polymers, but more sophisticated films may employ a variety of different plastic films in a single laminate. Like coated fabrics, the plastic layer may be applied to one or both sides of the fabric substrate. Materials with plastic film on both sides tend to provide greater chemical resistance but at the expense of increased stiffness.

Combination or Specialized Materials – This category includes chemical protective clothing materials that do not fit into the classes above. These fabrics generally attempt to combine the best attributes of each class. For example, coated fabrics that have been laminated with a plastic film on one side, provide the flexible features associated with rubber-based fabrics and the high levels of chemical resistance. Microporous films can be combined with adsorbent-based materials to provide a film with liquid chemical resistance that adsorbs large molecular weight chemical vapors. The development of one unique special material was based on applying Teflon to woven or nonwoven fiberglass substrate fabrics.

Table 5 provides some characteristics and examples of these materials.

Table 5:
Types of Chemical Protective Garment Materials

Material class	Characteristics	Examples	Type of protection
Textiles	Primarily nonwoven	Flashspun polyethylene (Tyvek) SMS Polypropylene (Kleenguard)	Dry particulate Light spray from water-based chemicals
Unsupported rubber or plastics	Thick rubber or plastic film	Polyvinyl chloride, PVC Chlorinated polyethylene, CPE (Chemturion)	Liquid splash, liquid or gas/vapor permeation
Microporous-film based	Polymer film with microscopic pores laminated to fabric	PTFE (Goretex) Polypropylene (NexGen, Kleenguard Ultra)	Liquid splash
Adsorbent-based	Material incorporating sorbent layer	Fabric/carbon/fabric (Lifetex)	Gas/vapor penetration
Coated fabrics	Woven fabric coated with rubber or plastic on one or both sides	PVC/Nylon Polyurethane Nylon Neoprene/Nomex Butyl/Nylon/Nylon	Liquid splash, liquid or gas/vapor permeation
Plastic laminates	Plastic film laminated to one or both side of nonwoven fabric	Polyethylene/Tyvek* Hazard-Gard I, II Tychem SL, BR, TK CPF I, II, III, IV Responder, Responder Plus	Liquid splash, liquid or gas/vapor permeation
Combination or specialized	Combines one or more of above material technologies or uses unique materials	Neoprene/Nylon /Neoprene/plastic laminate VPS) Teflon/fiberglass/Teflon (Challenge)	Liquid splash, liquid or gas/vapor permeation

* Fabric is actually coated with polyethylene

Garment Design Features

In addition to material choices, features affecting the design of garments include:

- The type and location of seams,
- The type, length, and location of the closure system(s),
- The design of interface areas with other chemical protective clothing or equipment, and
- The types, function, and location of hardware.

Seams – Seams play a critical role in protective clothing because they directly affect the integrity of the CPC in providing protection against specific chemical challenges. General types of seams used in chemical protective clothing include:

- Sewn seams,
- Glued seams, and
- Sealed.

Seams may also be bound or reinforced with other material pieces including additional fabric or tape, or even covered with a top coating of a polymer. A number of seam constructions are illustrated in Figure 6.

The type of seam and how it's applied is usually affected by the material used in the garment's construction, and the intended integrity or performance of the chemical protective clothing.

- Sewing can be applied to any textile-based material.
- Gluing is usually performed for coated fabrics and unsupported rubber materials.
- Sealing is applied to materials where the surface can be melted to create a bond with the surface of another material. This seaming approach is used for plastic laminates, thermoplastics, and other film-based materials. Sealing can be accomplished by using heat, ultrasonic radiation, or other means.

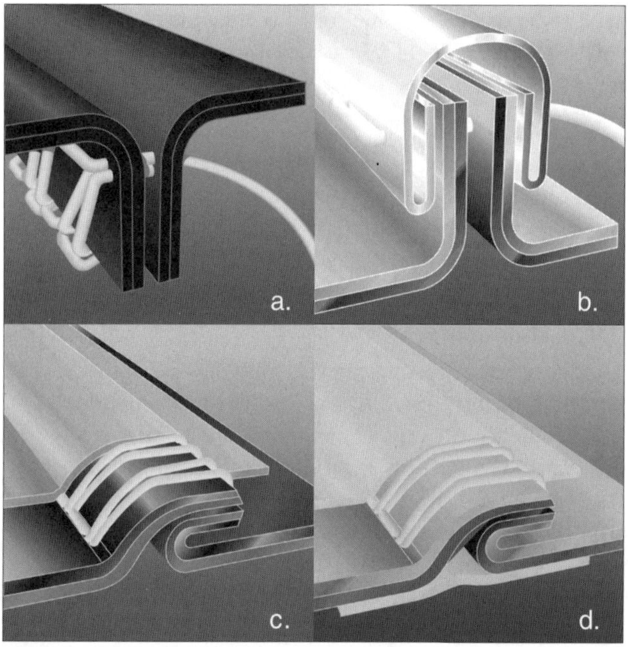

Figure 6. Configuration of chemical protective clothing seam types: (courtesy of DuPont)

(a) Serged seam

(b) Bound seam

(c) Sewn seam with tape seal on one side

(d) Sewn seam with tape seal on both sides

Different seam constructions can be used in protective clothing. A serged seam construction (Figure 6a) is popular for many styles of dry particulate-protective clothing because of its simplicity. However, lap or fell seams are also used in the fabrication of some CPC.

Many CPC seams are sewn to provided needed garment strength against the stresses of wear and use. However, sewing by itself produces stitching holes that can provide a pathway for chemical penetration. Therefore, some seams will be bound (see Figure 6b) for limiting these pathways for chemical penetration. Further protection of a sewn seam can occur when strips of tape directly cover the holes. The tape can be applied by either glues or by sealing the extra material over the seam (Figure 6c). In some cases, seams are taped on both sides to provide increased chemical resistance (Figure 6d).

The location of seams is also important. Not only do a large number of seams increase the garment manufacturing cost, more seams can create potential exposure to the wearer in the event of seam failure. Therefore, some clothing designs are created to limit the number of total seams and seams that are on the front of the garment where most chemical exposures occur.

Closures – Closures are typically the "weak link" in the chemical protective clothing barrier. Closures are necessary to allow people to don and doff CPC, but should also not lower the integrity provided by the clothing. The simplest closures are zippers or a series of snaps. Because of the open construction of these closures and their needed placement often on the front of the garment, pathways for chemical penetration can be created. For this reason, many chemical protective clothing designs use storm flaps to cover the zipper or snaps. An extra storm flap on the interior side of the zipper can provide additional liquid protection. On many disposable or limited-use CPC, adhesive strips may be incorporated into the storm flap to keep it in place. Reusable CPC sometimes uses hook and look closure tape to secure storm flaps, or as the principal closure system.

Closure systems offering increased integrity against chemical penetration include:

- Liquid-repellent zippers,
- Two-track extruded closures, and
- Special pressure-sealing zippers.

Liquid-repellent zippers are conventional zippers that use rubber or plastic-coated tape instead of woven cloth on the sides of the zipper. These zippers have a special chain (teeth) that is coated to limit liquid penetration.

Two-track closures involve two extruded pieces of plastic that fit together to provide a seal much like many plastic sandwich bags. These closures offer good integrity against liquids (and vapors) but can be difficult to seal over a long length as might be required in a full-body garment. Some two-track closures use a zipper-like pull to seal the length shut but this design still leaves part of the end open. In addition, the types of plastics that work well for these closures are also limited and may not have chemical resistance compatible with the garment material.

The pressure-sealing zipper uses a compressible rubber or plastic material with a zipper chain to push the two sides of the closure material together for creating a seal. These zippers tend to be bulky and expensive but are generally required when total encapsulation of the wearer is required.

As with seams, the location of the closure will be important. In some cases, the conventional design will require the closure to be in the front of the garment. Some items such as totally-encapsulating suits can have closures on the side or rear of the clothing item.

Interface Areas – The interfaces between the chemical protective clothing item and other CPC or equipment can vary with the design. Principle CPC interfaces include:

- Amount of overlap for multi-piece garments,
- Upper torso garment sleeve to glove,
- Lower torso pant cuff to footwear,
- Upper torso garment collar to hood, and
- Hood to respirator.

If different pieces of chemical protective clothing are worn, the items should provide sufficient overlap to cover the wearer's skin particularly when the wearer is reaching or bending over. This is particularly an issue for jacket and pant sets of CPC.

The sleeve end for garment to glove interfaces include a number of designs:

- For garments where there are no integrity issues, the sleeve end may be open. For a reusable garment, the sleeve end is hemmed; a disposable garment sleeve may be unfinished. If liquid leakage in the sleeve to glove area is a concern, duct tape is used but not recommended for providing integrity against liquid penetration.

- Elasticized sleeve ends provide somewhat better conformance of the glove with the sleeve. This design feature is particularly common for disposable and limited use clothing. However, chemical penetration may still occur at the interface. The application of tape provides only a questionable improvement of integrity for chemical liquids or vapors.
- For suits requiring a higher level of protection, the gloves are attached to the garment sleeve end either permanently, or in a manner in which they can be detached. These designs often include the incorporation of a hard ring into the sleeve end. An additional section of material at the sleeve end may be formed over the glove to act as a "splash" guard. The separate chapter on full-body protection includes descriptions of the different approaches used in attaching gloves to full-body garment sleeves.

Similar approaches are used for the pant cuff bottom in the interface area with footwear:

- A straight cuff is used. The cuff is generally pulled over the boot to keep liquid from entering the top of the boot or footwear. Like gloves, tape is sometimes used to keep the cuff in place.
- An elasticized cuff is used on some garment leg ends. This helps the pant leg stay in place on top of the boots. Tape may also be used, but only as a temporary measure for keeping the pant leg over the footwear.
- Some lower torso garment designs may use a covered zipper at the bottom of the leg so that the wearer can more easily insert his or her foot with a boot or shoe on it through the pant leg.
- The garment material is fashioned into a bootie that is attached to the bottom of the garment leg and is worn

inside the footwear. Garment booties are then worn inside an outer boot. This design may also include a splash guard mounted at the bottom of the garment legs that pull over the top of the boots.

Hoods can involve multiple interface areas. The bottom of the hood may be required to interface with the top of an upper torso garment. Most often, the length and flare of the hood control this interface. The bottom of the hood also may have slits to accommodate the wearer's shoulders. Without a visor, the hood will create the interface with the wearer's respirator facepiece or eye/face protection. This interface provides limited protection unless some device or aid to close the hood face opening exists around the device. For this reason, hoods often incorporate drawstrings or elastic, or may have some other feature to cinch the hood opening around the respirator facepiece. Hoods may also need to accommodate hard hats or other head protection and hearing protection.

Figure 7 shows a progression of design features that intended to improve the integrity of coveralls.

Sizing

There are few uniform sizing practices for the design of chemical protective garments. The availability of sizing often depends on the specific type of garment and the relative volume of garments sold by the manufacturer. Sizing may be based on individual measurements for custom sizing, numerical sizing for regular wearing apparel (chest or waist size), or alphabetic sizing (e.g., small, medium, and large). Partial body garments are likely to be offered in fewer sizing choices. Sleeve protectors may vary in three lengths, while two sizes of aprons will be available, and only one size of hood provided. It is important to realize that garment sizing must often take into account that the garment will be worn over regular work clothing and may need to accommodate different kinds of equipment, such as respirators, hard hats, and other devices.

Some sizing systems for garments usually use two or more wearer dimensions, such as height and weight, or height

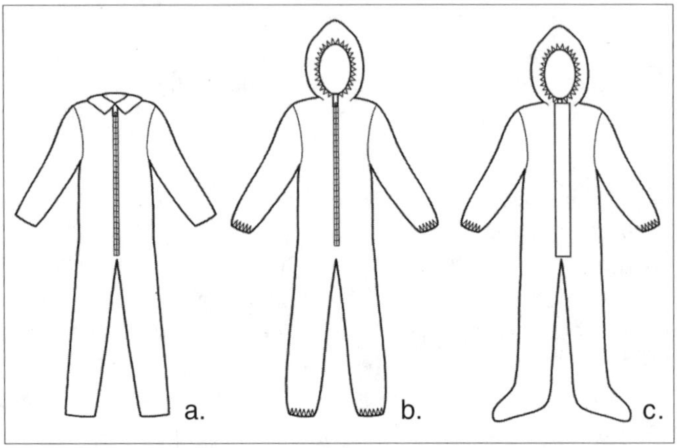

Figure 7. Progression of disposable coverall design with increasing barrier protection:

(a) Coverall with collar, front closure, cut sleeves and pant cuffs.

(b) Coverall with hood, front closure, elasticized sleeves and pant cuffs.

(c) Coverall with hood, front closure with storm flap, elasticized sleeves and booties

and chest circumference. This practice allows the wearer to determine the correct size by using personal physical data. An example of such a sizing system, based on ANSI/ISEA 101-1996, *American National Standard for Limited-Use and Disposable Coveralls – Size and Labeling Requirements,*[8] is shown in Figures 8 and 9.

Unfortunately, sizing between manufacturers is often inconsistent. ANSI/ISEA 101-1996 provides a system of sizing for disposable or limited use coveralls but has not found widespread acceptance among manufacturers. In addition, the sizing of protective garments often does not address the needs of women and special worker populations.[9]

Performance Characteristics

Different methods of evaluation are used to determine the performance of chemical protective garments and their materials and components. These include many of the test

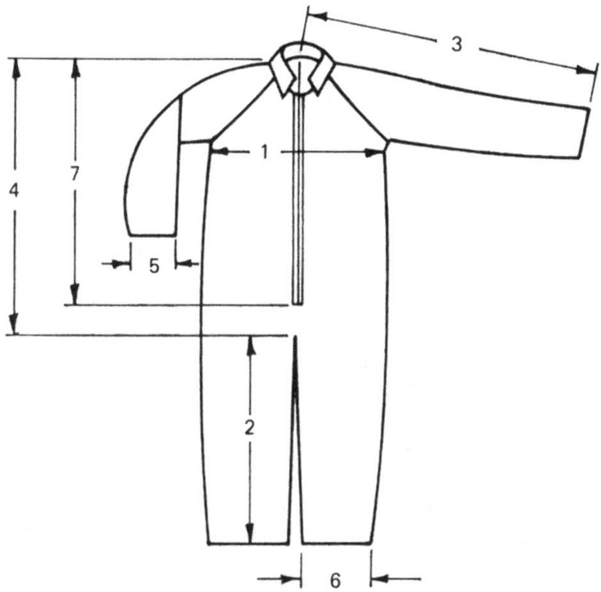

Figure 8. Coverall measurement locations from ANSI/ISEA 101, American National Standard for Limited Use and Disposable Coveralls—Size and Labeling requirements (courtesy of the International Safety Equipment Association)

methods described in *Chemical Protective Clothing: Test Methods and Standards*. Table 6 provides a list of key performance properties for chemical protective garments. These properties are organized by the part of the garment type/part and performance area.

Outside of NFPA standards on full-body chemical protective suits for hazardous materials emergencies, there are no standards specifying the minimum performance of chemical protective garments.

Hand Protection

Gloves are the most common form of handwear used to provide protection against chemical hazards to the wearer's

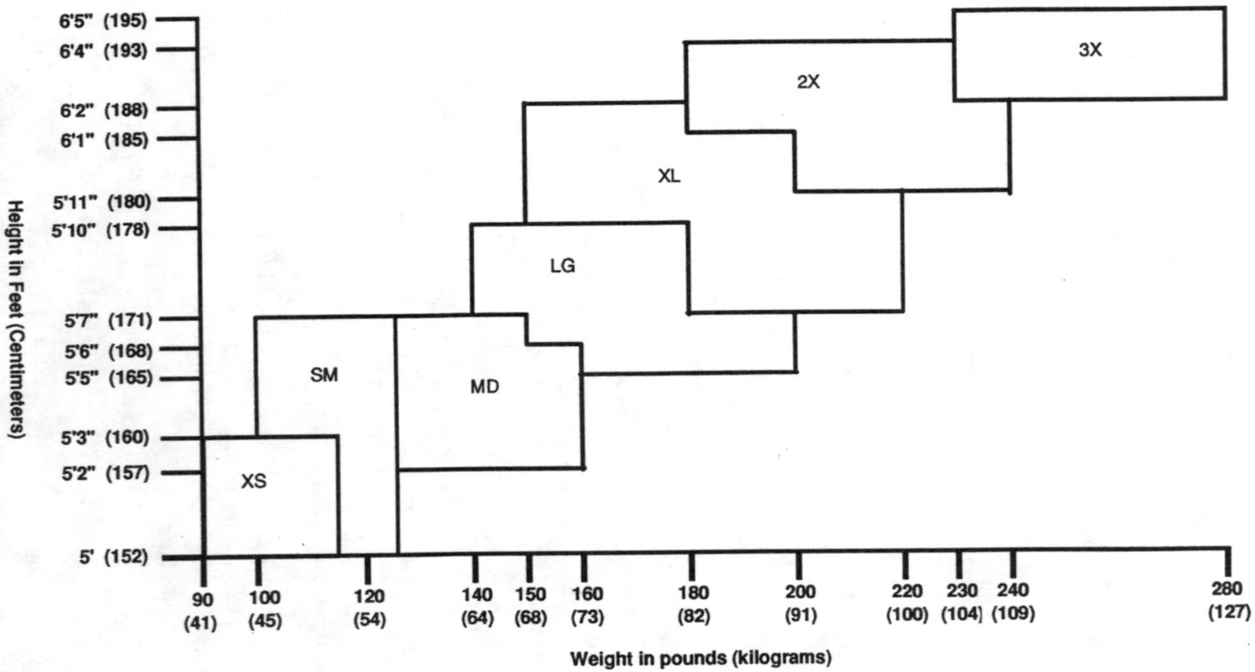

Figure 9. Sizing chart from ANSI/ISEA 101, American National Standard for Limited Use and Disposable Coveralls—Size and Labeling requirements (courtesy of the International Safety Equipment Association)

hands and wrists. Mittens are rarely used for chemical protection and finger cots are generally used for avoiding contamination of the fingertips in non-hazardous settings. The majority of gloves used for chemical protection use a film or coating that covers the entire hand, but some gloves that have partial barriers may be used in less hazardous applications.

Glove Types

Chemical protective gloves can be classified into three basic types:

- Unsupported gloves (including flock-lined gloves),
- Supported gloves, and
- Flat film gloves.

Figure 10 shows representative examples of these three glove types.

Unsupported Gloves – Unsupported gloves are made of a cured polymer film. Of the different glove types, unsupported gloves offer a higher degree of dexterity and come in a greater selection of polymers, lengths, and thickness. For flock-lined gloves, the inside surface of the glove is coated with adhesive and covered with shredded processed cotton or synthetic fibers to improve extended wearing comfort and to facilitate donning and doffing. The thickness of an unsupported glove is generally expressed as the single-wall polymer thickness to the nearest thousandth of an inch (known as "mils"), measured in the work area of the glove, avoiding any embossed pattern or raised grip surface where possible.

Supported Gloves – Supported gloves or coated work gloves are fabric liners that have been coated with a polymer. Compared to the other types of gloves, supported gloves can offer increased thermal, abrasion, and tear resistance at the sacrifice of dexterity. Supported gloves are manufactured in various glove construction techniques using several polymers.

Flat Film Gloves – Since most glove rubber and plastic

Table 6:
Key Performance Properties for Chemical Protective Garments

Type or part of garment	Performance area	Performance property	Suggested standard method	Evaluation output*
Full-body garments	Integrity	Gas/vapor integrity	ASTM F 1052	Ending pressure ↑
		Liquid integrity	ASTM F 1359	Evidence of leakage ↑
	Human factors	Fit and function	ASTM F 1154	Ease of completing tasks ↑
Garment materials	Chemical resistance	Permeation resistance	ASTM F 739	Breakthrough time ↑ Permeation rate ↓
		Liquid penetration resistance	ASTM F 903	Evidence of penetration ↓
		Hydrostatic resistance	AATCC 127	Penetration pressure ↑
	Physical strength	Tensile strength	ASTM D 5034	Breaking strength ↑
		Tear resistance (trapezoid)	ASTM D 1117	Tear strength ↑
		Burst strength (mullen)	ASTM D 751 (A)	Burst strength ↑
		Burst strength (ball)	ASTM D 751 (B)	Burst strength ↑
		Puncture propagation tear resistance	ASTM D 2582	Tear resistance ↑
	Durability	Abrasion resistance	ASTM D 3389 ASTM D 4157	Number of cycles when damage occurs ↑
		Flex fatigue	ASTM F 392	Number of cycles when damage occurs ↑

Table 6 (continued)
Key Performance Properties for Chemical Protective Garments

Type or part of garment	Performance area	Performance property	Suggested standard method	Evaluation output*
	Flammability	Flame spread	16 CFR Part 1610	Flame spread time ↑
		Ease of ignition; burning characteristics	ASTM F 1358	Afterflame time ↓ Burn distance ↓
Garment seams	Chemical resistance	Permeation resistance	ASTM F 739	Breakthrough time ↑ Permeation rate ↓
		Liquid penetration resistance	ASTM F 903	Evidence of penetration ↓
		Hydrostatic resistance	AATCC 127	Penetration pressure ↑
	Strength	Seam breaking strength	ASTM D 751	Seam strength ↑

* Upward arrow represents that maximum value is desirable; downward arrow represents that minimum value is desirable.

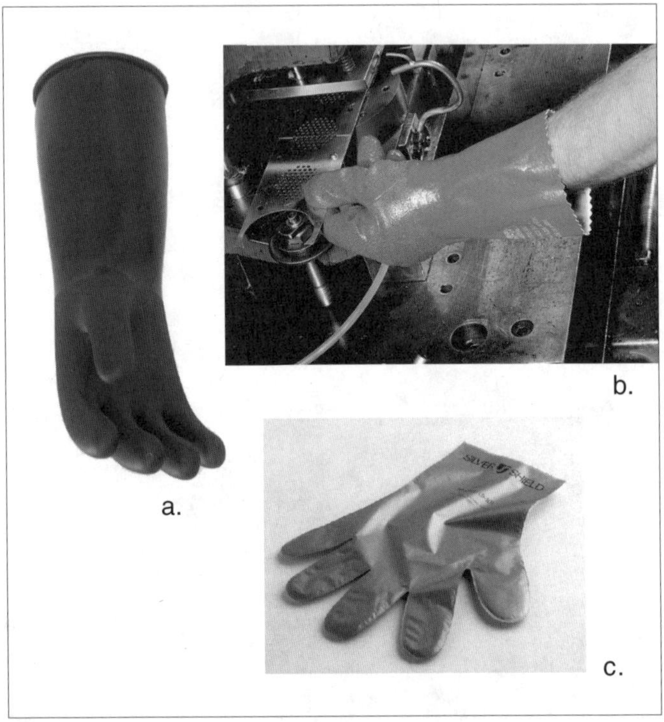

Figure 10. Types of chemical protective gloves:
(a) Unsupported butyl rubber glove (courtesy of Guardian Manufacturing Co.)
(b) Supported Nitrile glove (courtesy of North Safety Products)
(c) Flat film glove (courtesy of North Safety Products)

polymers have limited chemical resistance compared to garment materials, a difference type of glove was created to provide broader chemical resistance. Flat plastic films, similar to the plastic laminates used in garments but without the nonwoven substrate, are fashioned into gloves by sealing processes. While significant improvements in chemical resistance are attained, the resulting two-dimensional gloves generally sacrifice dexterity and physical hazard resistance.

Other Glove Types – Impregnated gloves, a fourth type commonly used to protect against cuts and abrasions, are not considered liquid-proof, and are therefore not usually considered to be chemical-resistant. Liquid penetration into these gloves is caused by the manufacturing process since the gloves are sewn from pieces of coated fabric.

Glove Construction

Chemical protective gloves are usually manufactured by the dip molding or dipping process. In this process, forms or molds made of porcelain or metal in the approximate size and shape of a hand are immersed in a "rubber" solution (see Figure 11). When the form is removed from the liquid, a coating of rubber is deposited, which is further processed into a finished glove. Both unsupported and supported gloves are made this way except that for a supported glove, a fabric liner is placed over the form prior to dipping.

The liner is constructed of either knit or woven fabrics. The more common styles of liner construction include:

- *Two-piece* – The glove liner is made from two pieces of material sewn in the general form of a glove.
- *Clute cut* – The front of the glove liner is made from one piece while the back is sewn from a number of parts. This style provides a formed glove. The design provides a seamless palm and is considered to offer a comfortable fit.
- *Gunn cut* – The back of the glove liner is made from one piece. The second and third fingers are set in with a seam across the bottom of each. The seams in the finger areas extend two-thirds of the way around each finger, eliminating exposed seams in the wear areas.

The liquid "rubber" solution can be either a latex system, a solvent-based solution, or in some cases, a plastisol. The latex technique is the most common process for the manufacture of chemical protective gloves. In a latex system, rubber particles are suspended as colloid in an aqueous dispersion. The latex system is compounded by conventional wet-mixing techniques with additives to enhance the desired physical properties. The non-water-soluble ingredients are added as dispersions or emulsions. Glove films are coagu-

Figure 11. Manufacture of gloves by dipping process.

lated, normally using chemical means, and the thickness of the film is controlled by the specific gravity of the coagulant, the dwell time in the latex, and the latex solids content and/or viscosity. In most cases, lightweight and medium-weight unsupported gloves can be made in a single dip. This process is safer to use in the manufacture of gloves and less expensive than the solvent process.

In the solvent process, dry rubber is mixed or milled with compounding chemicals into a homogeneous mixture and dissolved in the appropriate solvents. The forms are immersed in this rubber solution, then removed, allowing the solvent to evaporate. This leaves a thin film of dried rubber. Repeated dips are used to build up the desired thickness before the gloves are cured. This type of multiple dipping results in longer processing times, and uses volatile and flammable solvents instead of water, thus making this manufacture more costly and more hazardous than the latex process. Multiple-dip gloves, however, have certain superior chemical-resistance properties over latex gloves made from the same polymer.

In a plastisol system, a polymer such as polyvinyl chloride is dispersed in a liquid plasticizer (e.g., dioctyl phthalate [DOP]). Forms are dipped, withdrawn, and allowed to drain. Glove thickness is controlled by the plastisol viscosity, form temperature, and drain time. The gloves are then "fused" at a high temperature and removed from the forms.

Flat film gloves are made by cutting hand silhouettes from the plastic film and sealing the periphery with a hot wire device. Some newer industry designs use a three-piece pattern to create an opposable thumb for improvement of hand dexterity.

Glove Materials

Glove Polymers – Glove polymers include natural rubber or latex, synthetic rubber, and plastics. Common glove polymers include:

- Natural rubber
- Synthetic rubber
- Chloroprene (Neoprene)
- Nitrile rubber
- Butyl rubber
- Fluorocarbon rubber (Viton®)
- Plastics
- Polyethylene (PE)
- Polyvinyl alcohol (PVA)
- Polyvinyl chloride (PVC)
- Ethylene vinyl alcohol (EVOH)

Tables 7 and 8 provide a general comparison of polymer properties and chemical resistance, respectively. Further descriptions of the common glove polymers are provided below:

- ***Natural rubber*** is found in nature in over 200 plants, but its most common source for gloves is the Hevea Brasiliensis tree. Natural rubber dispersed in water is known as latex. Natural rubber has a very high elasticity compared to other glove materials, excellent cut and tear resistance, and outstanding grip and temperature resistance. While it's flexible and durable over a wide range of temperature (0° to 300°F), it has poor flame resistance. In general, natural rubber withstands water, alcohols, and some ketones, but has poor chemical resistance against most hydrocarbon and organic solvents. It is the most common material for both unsupported and supported gloves.

- ***Chloroprene***, known by its DuPont tradename of Neoprene, was the first commercial synthetic rubber. It has good abrasion and cut resistance and resists degradation due to aging, sunlight, ozone, oxidation, and weather. Neoprene is also flame resistant and demonstrates heat stability up to 200°F. It provides excellent resistance to a broad range of chemicals including acids, alcohols, fats, caustics, refrigerants, ketones, detergents, and fertilizers. Neoprene is used in unsupported and supported gloves.
- ***Nitrile rubber*** is a copolymer of acrylonitrile and butadiene. Nitrile rubber offers excellent physical hazards resistance to punctures, cuts, snags, and abrasion. While it is not flame-resistant it demonstrates functional performance in temperatures ranging from 25°–300°F. Nitrile rubber has considerable resistance to oils, fuels, and certain organic solvents, and is used as an unsupported glove polymer, as well as a coating for dipped gloves.
- ***Butyl rubber*** is a copolymer of isobutene (97% to 99.5%) and isoprene (0.5% to 3%). Butyl rubber has good resistance to light and ozone, low air permeability, excellent flexing properties and heat resistance, good flexibility at low temperature, good tensile strength and tear resistance. It does exhibit poor chemical resistance to petroleum oils and gasoline but excellent resistance to corrosive chemicals, vegetable oils, phosphate ester

Table 7:
General Comparison of Different Glove Polymer Physical Properties and Relative Cost

Glove Polymer	Abrasion Resistance	Cut Resistance	Puncture Resistance	Tear Resistance	Flexibility	Heat Resistance	Flame Resistance	Ozone Resistance	Relative Cost
Natural rubber	E	E	E	E	E	F	F	P	Medium
Neoprene	E	E	E	G	G	G	E	E	Medium
Nitrile rubber	E	E	G	G	E	G	G	F	Medium
Butyl rubber	F	G	G	G	G	E	P	E	High
Viton	G	G	G	G	G	G	F	E	Very High
Polyethylene	P	F	F	P	P	P	P	E	Low
Polyvinyl alcohol	F	F	F	G	P	G	P	E	Very High
Polyvinyl chloride	G	P	G	G	F	P	F	E	Low
Ethylene vinyl alcohol	P	P	F	P	F	P	P	E	Medium

Ratings: E - excellent; G - good; F- fair; P - poor

Note: Ratings are subject to variation depending on formulation thickness, and the type of fabric support

Source: Primarily from: Guidelines for the Selection of Chemical Protective Clothing, 3rd Edition, American Conference of Governmental Industrial Hygienists, Cincinnati, Ohio, 1987.

Table 8: General Comparison of Different Glove Polymer Chemical Resistance

Glove Polymer	Alcohols	Aldehydes	Amines	Esters	Ethers	Halogen Compounds	Hydro-carbons	Inorganic acids	Inorganic bases	Ketones	Nitriles
Natural rubber	*	N	N	N	N	N	N	*	R	N	N
Neoprene	R	N	*	N	*	N	N	*	R	N	*
Nitrile rubber	R	N	*	*	*	N	*	*	R	*	N
Butyl rubber	R	R	*	R	*	N	N	*	R	*	R
Viton	R	*	R	N	*	R	R	R	R	N	R
Polyethylene	*	*	*	*	*	N	*	R	R	N	R
Polyvinyl alcohol	*	*	*	R	*	*	*	N	N	*	R
Polyvinyl chloride	*	N	N	N	*	N	N	*	*	N	N
Ethylene vinyl alcohol†	*	R	R	R	*	*	R	N	N	*	R

† Usually combined with polyethylene in flat film gloves

Ratings: R - recommended; N - not recommended; * - mixed performance for chemical class

Note: Ratings vary with glove polymer thickness and are provided only as general assessment of polymer chemical resistance. When available, use specific data for the selected glove and chemical combination.

Source: Information adapted and consolidated from *Guidelines for the Selection of Chemical Protective Clothing*, 3rd Edition, American Conference of Governmental Industrial Hygienists, Cincinnati, Ohio, 1987.

oils, and some ketones. It is currently provided only in unsupported gloves.

- ***Fluorocarbon rubber***, known under the tradename Viton® is a copolymer of vinylidene fluoride and hexafluoropropylene. It has relatively poor physical hazard resistance, and is thermally stable to 450°F. Viton® offers good resistance against a wide variety of chemicals, especially chlorinated and aromatic solvents. However, the relative high expense of fluorocarbon rubber relegates its use to unsupported gloves.

- ***Polyethylene (PE)*** is a polymer of ethylene and is one of the most common plastic polymers in the world. While typically used in packaging, polyethylene is either used in thin flat film gloves for food handling or as one of the polymer layers in chemical-resistant flat film gloves. Compared to other plastic and rubber materials, polyethylene shows very little effect for chemical exposure. However, PE tends to create a relatively weak film with low abrasion, cut and puncture resistance. PE also melts at low temperatures and is thermally stable to only 180°F.

- ***Polyvinyl Alcohol (PVA)*** is a polymer of vinyl alcohol. PVA offers good resistance to punctures, cuts, snags, and abrasion. This material provides good resistance to organic chemicals such as aliphatic and aromatic hydrocarbons, chlorinated solvents, esters, and most ketones, but is not resistant to water, water-based solutions (like acids and bases), or light alcohols. PVA is generally provided as a coated glove by a limited number of manufacturers.

- ***Polyvinyl Chloride (PVC)*** is a synthetic thermoplastic polymer of vinyl chloride. PVC offers good abrasion resistance but may be susceptible to punctures, cuts, and snags. While it is flexible, it does not provide the tactile sensitivity associated with most rubber products. PVC melts at approximately 180°F. PVC is effective against water and most aqueous solutions, detergents, and diluted bases and acids. It only has limited chemical resistance to organic solvents. PVC is one of the more common coatings for coated work gloves. It is also used in very thin unsupported gloves.

- ***Ethylene Vinyl Alcohol (EVOH)*** is a copolymer of ethylene and vinyl alcohol. EVOH is one of the most chemical-resistant polymers available. It is generally used as one of the polymers in flat film gloves and consequently is only available as flat sheets of material. EVOH is highly resistant to a wide range of hazardous materials, but possesses low resistance to physical damage. EVOH is only stable to 160°F and melts at 375°F.

Several of these polymers are also combined to provide improvements in glove performance. A notable example for combining polymers occurs in flat film gloves where a three-layer film construction of polyethylene, ethylene vinyl alcohol, and polyethylene (PE/EVOH/PE) permits superior chemical resistance when compared to conventional polymers or polyethylene and ethylene vinyl alcohol by itself. For supported and unsupported gloves, the combination of polymers can occur either through blending two or polymers or by alternating the dipping of the glove form. For example, there are several gloves that have Neoprene over natural rubber.

Lining Materials – Knit fabrics, such as interlock and jersey, generally provide better comfort and fewer seams in glove liner construction. Interlock knits generally provide more flexibility and ease of laundering while jersey knits provide more cushioning and insulation. Woven fabrics, such as cotton flannel and rayon often require more pieces in the liner glove pattern but may offer relatively more strength. Cotton linings provide good absorption of perspiration and are generally considered to offer relatively good hand comfort compared to rayon and synthetic fibers.

One issue related to supported gloves is the integrity of the coating. Supported gloves tend to provide less chemical resistance than unsupported gloves because of the increased potential for defects in the rubber or plastic film resulting from the non-uniform fabric liner surface. The polymer itself will fill voids in the areas between fibers to create a thicker coating between fibers, but provide a thin coating on the top of fibers.

Manufacturers generally compensate for these variances in thickness by providing a heavier polymer thickness on supported gloves.

Glove Design Features

Principal glove design features include the cuff design and style of surface finish or grip designs.

Cuff Designs – Gloves are finished in a variety of cuff designs. The type of cuff design will depend on the construction style for the glove as well as the intended purpose. The most common cuff designs for unsupported gloves include:

- ***Rolled*** – A cuff style where the glove end is rolled and held in place with an adhesive originally designed to prevent glove tear.
- ***Pinked or serrated*** – A cuff style where the glove end is trimmed but using a cutting blade that forms a series of V-shaped cuts along the edge of the glove end.
- ***Straight*** – A cuff style where the glove end is cut across the opening

Supported cuff designs, shown in Figure 12, include:

- ***Gauntlet*** – An extension of the glove body intended to offer protection to the lower forearm. Gauntlets generally flare out from the forearm to permit movement.
- ***Safety cuff*** – A cuff style where a wide band using 2 to 4 inches of a relatively stiff material that is attached to the end of the glove. Safety cuffs are intended to provide greater wrist protection and cooler wearing, as well as allow the glove to be quickly removed if caught in equipment.
- ***Slip-on*** – A cuff style with the glove end having no cuff material attached.
- ***Wristlet (knit wrist)*** – A cuff style using a soft, highly elastic tubular knit material. Wristlets provide close fitting of the glove around the wearer's wrist and can function or prevent debris from entering the glove.

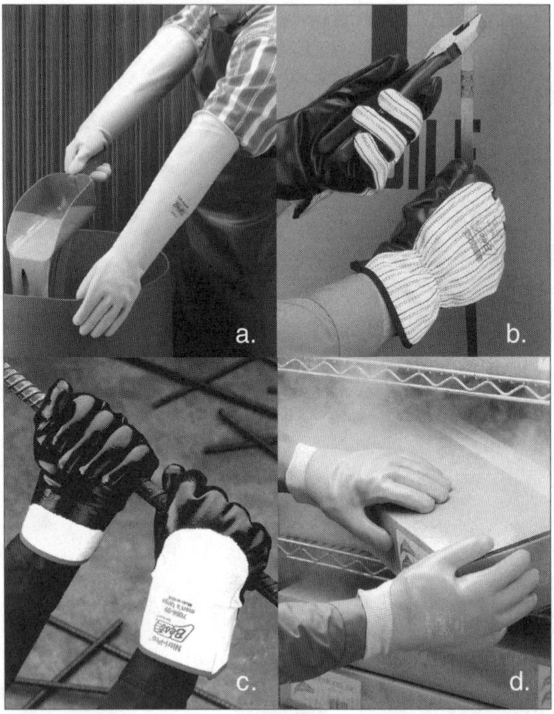

Figure 12. Cuff designs for supported gloves: (courtesy of Best Manufacturing)

(a) Gauntlet

(b) Safety cuff

(c) Slip-on cuff

(d) Knit wrist

Plastic laminate gloves only use straight cut cuffs.

Grip Designs – The grip design is the style of finish on the exterior glove surface. Some grip designs provide increased friction against dry or wet surfaces. Grip designs vary with the type of glove and intended application. Finish styles for unsupported gloves are affected by the form used in glove manufacturing and range from smooth to various types of textured or raised surfaces that are named differently by the respective manufacturers. Raised finishes may have uniform patterns or coarse patterns for increased glove surface roughness. Finish styles for supported gloves include smooth, rough, or crinkle finishes. Plastic laminate gloves

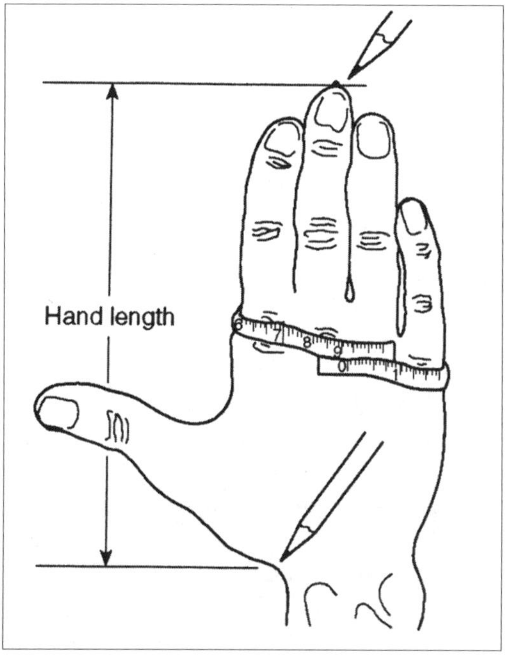

Figure 13. Measurement of hands for glove sizing.

generally only have a smooth finish consistent with the outer plastic layer.

Glove Sizing

Glove sizing is specified based on hand circumference and length. Glove sizing is indicated using either numerical hand sizes or qualitative size ratings, such as small, medium and large. Numerical hand sizes use the circumference of the hand at the widest area to specify glove size in inches (see Figure 13). Glove length is measured from thc tip of the fingers. For some styles of gloves, the length will increase with the glove size. For other gloves, the length may be fixed or an option for each particular glove style. Table 9 shows the relative hand and arm protection provided by different ranges of glove length.

Glove fit is not consistent among different manufactur-

Table 9:
Typical Glove Length and Protection

Protection	Length
Hand protection only	Up to 12 in.
Midarm protection	13-15 in.
Elbow length protection	16-18 in.
Shoulder length protection	30-32 in.

ers, even for the same basic glove material and style. Sizing variability occurs because manufacturers use different forms for dipped gloves. Not only do hand dimensions vary, but also the basic shape of the hand can change. Moreover, the availability for different glove sizes will depend on the glove style and relative sales volume.

Glove Performance Characteristics and Issues

Specific properties are associated with glove performance. Some of these properties are associated with all types of chemical protective clothing, while others tend to be specific to gloves. Table 10 provides a listing of the key glove performance properties.

Recommended classification of glove performance to some of these properties is provided in ANSI/ISEA 105, *American National Standard for Hand Protection Selection Criteria.*[10] This standard requires testing of gloves, using specific test methods for selected performance properties and classification of the respective results to a specific performance level. For each performance property, there are a different number of performance levels for differentiating glove performance.

It is important to point out that the glove performance can vary significantly from the reported generic polymer performance. This is because the specific glove polymer

Table 10:
Key Performance Properties for Chemical Protective Garments

Part of glove	Performance area	Performance property	Suggested standard method	Evaluation output*
Whole glove	Integrity	Liquid integrity	ASTM D 5151	Evidence of leakage ↓
	Human factors	Dexterity	ASTM F 2010	Percent increase with gloves ↓
		Fine dexterity/tactility	EN 420	Diameter pin picked up ↓
Glove materials	Chemical resistance	Permeation resistance	ASTM F 739	Breakthrough time ↑
				Permeation rate ↓
		Liquid penetration resistance	ASTM F 903	Evidence of penetration ↓
		Degradation resistance	ANSI 105	Pct. change in tensile strength ↓
	Physical strength and durability	Abrasion resistance	ASTM D 3389	Number of cycles at which damage occurs ↑
		Cut resistance	ASTM F 1790	Blade travel distance ↑
		Puncture resistance	ASTM F 1342	Puncture force ↑
		Tear resistance	ASTM D 2261	Tear resistance ↑
	Flammability	Flame spread	16 CFR Part 1610	Flame spread time ↑
		Ease of ignition;	ASTM F 1358	Afterflame time ↓
		Burning behavior		Burn distance ↓
Glove seams	Chemical resistance	Permeation resistance	ASTM F 739	Breakthrough time ↑
				Permeation rate ↓
		Liquid penetration resistance	ASTM F 903	Evidence of penetration ↓
		Hydrostatic resistance	AATCC 127	Penetration pressure ↑

* Upward arrow represents that maximum value is desirable; downward arrow represents that minimum value is desirable.

formulations and dipping processes can vary extensively between manufacturers. For example, Mickelson and Hall reported significant differences in permeation breakthrough times for neoprene and nitrile gloves for several chemicals.[11]

In some supported glove styles, the coating may not complete cover the entirety of the hand. Generally, glove palms, fingers, and portions of the back, are coated to provide chemical resistance to the majority of the hand. However, the non-coated portions of the glove will provide no barrier protection and can actually facilitate chemical exposure if liquid wicks onto the non-coated portions of the liner and migrates into the glove interior. This phenomenon can also occur for supported gloves provided with a knit wristlet or other glove end that is not coated. For this reason, careful consideration must be given to the use of coated gloves where the coating is not continuous.

Foot Protection

Chemical protective footwear is used to protect the wearer's feet from chemical exposure and related hazards. Depending on the footwear height, additional protection may be afforded to the wearer's ankles and lower and upper legs. The principal approaches to chemical protective footwear include:

- Boots – footwear that covers at least the wearer's ankle providing both barrier and physical protection to the foot and lower leg.
- Overboots or overshoes – footwear that is worn over conventional footwear providing either barrier or physical protection or both to the foot.
- Booties – either an extension of the garment leg (for full-body coveralls and suits) or a separate garment-like item worn directly over the foot for barrier protection; generally considered a partial body garment as much an item of footwear
- Boot or shoe covers – footwear items similar to overboots or overshoes, but made of lightweight materials similar to booties.

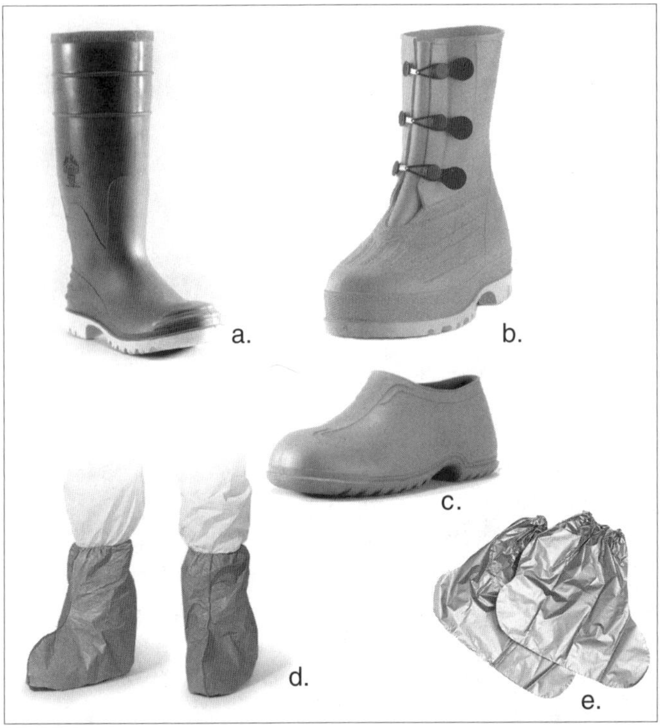

Figure 14. Types of chemical protective footwear:
(a) Boot (courtesy of ONGUARD Industries)
(b) Overboot (courtesy of Tingley Rubber)
(c) Overshoe (courtesy of ONGUARD Industries)
(d) Bootie on end of chemical protective garment (courtesy of DuPont)
(e) Boot cover (courtesy of North Safety Products)

Figure 14 shows several types of chemical protective footwear.

Like the other items of chemical protective clothing, the choice of appropriate footwear must take into account the other hazards present, requirements for use with other CPC, and preferences of the end user. The same issues that apply to design, performance and service life are also important to footwear.

Footwear Design Features and Materials

Boots – Conventional footwear has several distinct features that include:

- An upper material,
- The type of outer sole,
- The type of closure (if present),
- Linings,
- Interior supports (metatarsal or arch supports, insoles), and
- Protective hardware (toecaps, metatarsal plates, puncture resistant devices or midsole plates, and ladder shanks).

Figure 15 shows the different terms associated with the areas and features of a standard rubber boot. Unlike other items of CPC, a chemical protective boot has several areas of changing thickness and materials that must all function together to provide barrier protection as well as resist other hazards common to footwear (i.e., impact from dropping objects, puncture through the sole, and adequate traction on slippery surfaces).

Boot Upper – The upper will be either an unsupported rubber/plastic or coated fabric. The upper material includes the shaft, quarter, and vamp footwear areas extending down to the sole. Boot linings may consist of woven, nonwoven textiles, coated barriers, or foam materials. For chemical protective footwear, these materials are intended primarily for comfort much in the same way that flocking and liners in gloves perform.

Boot Sole – The boot sole consists of several parts. The bottom of the boot or outer sole (often simply referred to as the sole) is likely to be compounded rubber and must include a tread and wear surface to provide traction. Several different wear surfaces are offered by each manufacturer to offer slip resistance performance. Most boots include a heel to aid in walking over uneven surfaces. Puncture-resistant

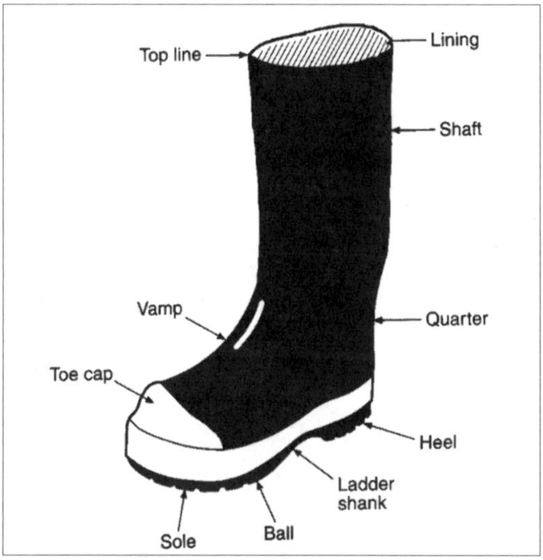

Figure 15. Diagram of chemical protective boot design.

footwear will generally include a metal plate above the outer sole to deflect nails from penetrating the sole into the user's foot. Shanks are long metal pieces extending the length of the arch in the footwear sole and are designed to keep the bottom of the footwear from bending when stepping onto ladder rungs. The insole provides cushion for wearer comfort and may be removable for cleaning and replacement.

Other Boot Features – Chemical protective boots may have other types of features present. For impact resistance, boots may include metal hardware or reinforced materials on the toe and metatarsal (vamp) areas. Footwear may be of the step-in type, or include a closure system on the front of the footwear that permits the footwear to widen for donning and create better ankle support. The latter type of footwear uses a "gusset" in the front of the boot that is secured by some form of closure devices. Closure devices may include buckles, snaps or straps.

Boot Materials – Over the past several years, the variety of materials used in chemical protective boots has increased. Many boots are still based on polyvinyl chloride

(PVC) and neoprene rubber. However, a number of other polymers and polymer blends are being used that are based on butyl rubber, nitrile rubber, polyurethane, and PVC. The chemical protective footwear industry has developed several material proprietary formulations with permeation resistance against a broad range of chemicals.

Boot Manufacturing Methods – Chemical protective boots are made in one of two different ways; injection molding and making boots by hand on a last. The part of the footwear mold in the shape of a foot is called a last. The last may be covered with a fabric sock to create the liner and help position hardware in the footwear.

A two-stage injection molding process is generally used for chemical protective boots that incorporate physical protection. The two stages permit different upper and sole materials. Thus, in this process, the boot manufacturer is able to produce a lightweight upper boot with an optimized longwearing sole having good traction. The two-stage process also results in a better-fitting boot with improved chemical resistance. Hardware such as the toecap and midsole plate are placed in the mold and encapsulated during pre-injection stage for the footwear upper. The application of different materials in this process is dependent on the ability to injection mold the material. Consequently, injection-molding processes are performed with polyvinyl chloride, polyurethane, nitrile rubber and several blends of these materials.

Hand-made boots are fabricated by placing individual hardware onto the last using adhesives. The boot is then heated to cure (vulcanize) the rubber and form the final boot. These types of boots are made in stages with a large number of component parts. This manufacturing approach, while more labor-intensive, typically permits a greater range of sizes since only a last is used for each size and width, as opposed to the complete molds that are used in injection molding.

Overboots and overshoes – Overboots or overshoes are intended to be worn over regular footwear to provide barrier protection to the wearer's feet. These products differ from regular chemical protective boots in that they lack the physical protection features such as impact and compression resistance, and the sole puncture resistance associated with regular boots. As a consequence, overboots and overshoes are made using a single-stage injection molding process. These products have the same appearance as soft-sided rubber rain boots. Overboots are generally provided with a gusset that expands to allow the wear to insert their shoe or boot into the overboot. The gusset area is closed with a snap or buckle. Overshoes are short enough to slip over the wearer's shoes.

Using the one-stage process, the sole of the boot is made from the same polymer formulation as the upper. This means the wear and traction characteristics of the sole will not be optimum, especially when compared to chemical protective boots. Polymers used in overboots and overshoes include:

- Butyl rubber,
- Natural rubber,
- Neoprene rubber,
- Nitrile rubber, and
- Polyvinyl chloride.

Booties and boot/shoe covers – Less durable chemical-resistant footwear is available in the form of booties, boot covers, and shoes. Booties are sock like extensions that are worn directly over the feet. In many cases, booties are built into the pants of garments, particularly coveralls and totally-encapsulating suits. In these cases, the booties are often constructed of the same material as the garment. Boot and shoe covers are worn over existing footwear but differ from overboots and overshoes since they are made from relatively inexpensive disposable garment materials. The materials used in their construction result in footwear items that lack the ruggedness of boots or overboots/overshoes. In fact, boot and shoe covers are intended to be disposable but may include a special material on the wear surface for improved traction and abrasion resistance.

Footwear Sizing

Most chemical protective boots use the same footwear sizing system as employed for conventional footwear. This system is based on the Brannock measuring scale, in which foot length and width are the two key dimensions used for choosing footwear. This method requires the individual to measure his or her foot and then to select the corresponding labeled footwear size (e.g., 7D or 9EE). It also involves designation of sizes by the manufacturer for their products so that appropriate fit with the corresponding dimensions is provided. However, one manufacturer's size may not fit as well as the same size from a different manufacturer because:

- Manufacturers address fit differently and may use different allowances for ease (the difference between footwear measurements and wearer dimensions).
- Individual feet are shaped differently, and a two dimensional system does not fully capture the characteristics of feet to ensure fit.

Availability of chemical protective boot sizes is based on the materials of construction, manufacturing methods, design features, and relative production volume. In particular, rubber boot makers are more constrained by the individual lasts or molds used for making footwear and consequently are less likely to provide a full range of sizes unless warranted by the demand for the specific footwear. The most common practice is to offer chemical protective boots in full sizes and in one width.

Less sizing options are provided for other types of chemical protective footwear. Overboots and overshoes are typically offered in limited sizes (small, medium, large, etc.). Since there is less variation in the design of regular work shoes, overshoe sizes are represented as corresponding to the size of shoe over which they can be worn. Booties sizing will be proportional to the size of the clothing to which they are attached. Since booties must accommodate a large range of foot sizes, booties will tend to be oversized for most individuals. Boot and shoe covers may be available in only one size and will include an elastic top to snug around the top of the footwear or wearer's leg.

Footwear Performance Characteristics

Some of the same tests that apply to other forms of chemical protective clothing (garments and gloves) also apply to footwear. However, since footwear uses some unique features and components, other special testing applies to footwear. Table 11 summarizes the key performance properties for footwear indicating the suggested test methods and types of results reported by these tests.

Requirements for occupational footwear are governed by ANSI Z41, *American National Standard for Personal Protective Footwear*.[12] The OSHA regulations in Title 29 of the Code of Federal Regulations Part 1910.137,[13] require that footwear comply with the 1991 edition of that standard. ANSI Z41 establishes toe impact and compression resistance criteria for three classes of performance. The standard also sets requirements for metatarsal, conductive, electrical insulative, sole puncture-resistant, and static-dissipative footwear.

There are standards that specifically address footwear performance and industrial protection. Footwear criteria are provided in the NFPA standards for chemical emergency response.

Eye and Face Protection

Many chemicals in the workplace can cause significant eye damage, including permanent loss of vision, if they are splashed into an eye. Facial scarring and disfigurement can also result from direct contact with corrosive chemicals. Approaches for providing face and eye protection against hazardous chemicals include the use of:

- Goggles,
- Faceshields,
- Visors, incorporated into separate hoods, coveralls, or suits, and
- Respirator full facepieces.

Figure 16 shows examples of these approaches. In

Table 11:
Key Performance Properties for Chemical Protective Footwear

Part of footwear (if present)	Performance area	Performance property	Suggested standard method	Evaluation output*
Whole boot	Integrity	Liquid integrity after flexing	FIA 1209	Evidence of leakage ↓
Footwear upper	Chemical resistance	Permeation resistance	ASTM F 739	Breakthrough time ↓
				Permeation rate ↑
		Liquid penetration resistance	ASTM F 903	Evidence of penetration ↓
		Degradation resistance	ASTM D 471	Pct. weight change ↓
	Physical strength and durability	Abrasion resistance	ASTM D 3389	Number of cycles at which damage occurs ↑
		Cut resistance	ASTM F 1790	Blade travel distance ↑
		Puncture resistance	ASTM F 1342	Puncture force ↑
		Tear resistance	ASTM D 2261	Tear resistance ↑
	Flammability	Flame spread	16 CFR Part 1610	Flame spread time ↑
		Ease of ignition;	ASTM F 1358	Afterflame time ↓
		Burning behavior		Burn distance ↓
Footwear sole	Physical hazard resistance	Puncture resistance	ANSI Z41	Puncture force ↑
		Abrasion resistance	ASTM D 1630	Change in thickness ↓
	Traction	Slip resistance	ASTM F 489	Coefficient of friction ↑
Footwear toe	Physical hazard resistance	Impact and compression resistance	ANSI Z41	Resulting clearance ↑
Footwear seams	Chemical resistance	Permeation resistance	ASTM F 739	Breakthrough time ↑
				Permeation rate ↓
		Liquid penetration resistance	ASTM F 903	Evidence of penetration ↓

* Upward arrow represents that maximum value is desirable; downward arrow represents that minimum value is desirable.

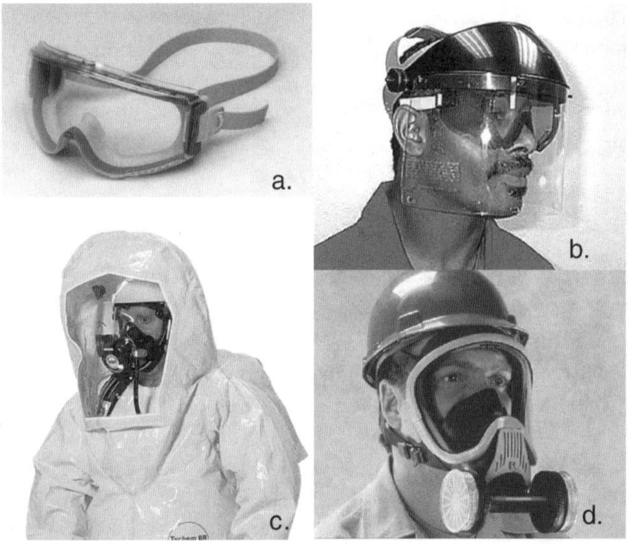

Figure 16. Types of face and eye protection approaches for chemical exposure:

(a) Goggles

(b) Faceshield

(c) Visor in chemical protective suit (courtesy of DuPont)

(d) Full-face respirator (courtesy of MSA Company)

addition to chemical protection, eye and facewear may be needed to provide protection to the wearer's eyes and face from other hazards including impact, optical or light radiation (intense light) and glare.

ANSI Z87.1, *Practice for Occupational and Educational Eye and Face Protection*[14] defines a number of general types of eye and face protectors, including safety glasses, faceshield, goggles, and welding helmets. A selection chart in the standard indicates the specific types of eye and face protectors that are recommended for chemical exposures. For chemical splash, the standard specifies the use of goggles with indirect or no ventilation, and faceshields. Where ventilated goggles are used, the ventilation area of the goggles should be well protected from splash entry. For irritating chemical mists, only goggles with no ventilation are recommended. Safety glasses, while used in many applications should not be relied on for chemical protection. Likewise, faceshields are not primary protection and should be only worn over protectors offering primary eye protection (i.e., goggles).

Goggles

The most frequently used types of eye protector for chemical protection are goggles. Goggles consist of lens holders, lenses, and a band to secure the goggles to the wearer's head. Goggles are designed to fit snugly to the face or eyes, providing a seal that prevents liquid chemical exposure to the eyes in the event of a splash. There are two principal designs of goggles – eye cup goggles that only cover the eye sockets and include a bridge piece of plastic to connect the two eye cups, and cover goggles, which protect the bridge of the nose and may accommodate eyeglasses. Cover goggles are by far the more popular of the two designs. Cover goggles are further differentiated by the type of ventilation provided:

- Goggles with direct ventilation – openings are provided on the top or sides of the goggles to prevent fogging of the lenses.
- Goggles with indirect ventilation – openings in the goggles are also provided but are partially covered splash guards to prevent liquid entry.
- Goggles with no ventilation – goggles have no openings.

The lens holder is usually made from a soft pliable polymer, such as PVC, using injection-molding techniques. Lenses are also made of plastic material, but must have sufficient optical properties to prevent minimal distortion and haze to affect the wearer's vision. Goggles lenses may also be impact-resistant, shaded for protection against non-ionizing radiation, or photochromic (darkens when exposed to and fades when removed from ultraviolet light or sunlight). Many goggles permit the replacement of lenses, when scratched. The bands used with goggles are either adjustable, elastic, or both.

The type and rigidity of the materials in goggles will affect their sizing or adjustment. While many flexible goggles are offered in only one size, some products are available in several sizes, usually represented as small, medium, and large. Rigid goggles often require a cushioning material at the edges, which fit onto the face, and are more likely to have adjustable straps.

Faceshields

The faceshield is another common form of eye and face protector used for industrial chemical protection. Unfortunately, most users do not understand that this type of CPC is designed to prevent only direct splash exposures to the face, and does not provide complete eye protection. According to both ANSI Z87.1, safety glasses or goggles must also be worn under the faceshield for primary eye protection. This concern is made evident when a splash, stream, or jet of chemical can hit the worker's chest, splash up under the faceshield, directly into the wearer's eyes.

Faceshields consist of a frame or headgear that support a window which is curved to surround and cover the wearer's face. The headgear assembly may be provided with or without a crown (top of the head) protector. Typically, the frame can be adjusted to fit the circumference of the wearer's head. Other features of the faceshield include the material, thickness, size, and shape of the window for protection from a variety of different hazards. Commonly used window materials are plastic or glass. ANSI Z87.1 requires that faceshield lens materials be thick enough to prevent damage from impact. No specific requirements are provided for the size or areas of coverage for faceshield, except as defined by the test requirements for impact resistance in ANSI Z87.1. The length of the faceshield can vary, with few faceshield designs providing protection to the wearer's chin. Some faceshields may permit the replacement or substitution of the window. Faceshield windows may also be attached to protective helmets in place of their own headgear.

Garment Visors

Visors are generally incorporated into separate hoods, coveralls, or suits to offer chemical barrier protection for the heat and face area. In general, the visor is constructed from a material that provides clear undistorted vision as well as chemical resistance. While having some physical integrity, visors generally do not offer the same physical impact resistance provided by faceshield and other primary face protection.

Garment visors vary in their size, stiffness, materials, and method of integration with the garment.

Visor Size – The size of the visor affects the wearer's peripheral vision. In some garment designs, such as totally-encapsulating suits, visor size has a significant effect on the wearer's field of vision because the suit visor does not move with the wearer's head. Consequently, more recent designs of these garments provide extra-large visors. This is less of a problem for hoods, which will rotate when the wearer turns his or her head.

Visor Stiffness – Visors can be flexible or rigid. Reusable suits and garments tend to use stiffer visors since overall weight may be less of consideration while many disposable garments incorporate relatively flexible visors to maintain low weight.

Visor Materials – The principal materials used in visor construction are polyvinyl chloride, polycarbonate, and polymethacrylate. Since the visor material must provide optical qualities, only transparent materials can be chosen as visors. When greater chemical resistance is required, the visor is made of a composite material that includes fluorinated ethylene propylene (FEP) laminated to PVC. The FEP layer is kept thin because the material becomes opaque at large thicknesses, while the PVC provides support and scratch resistance.

Visor Integration with Garment – The joint between the visor and the garment is important for maintaining the integrity of the garment item. In some cases, the visor material is directly sealed to the garment material. In other cases, the visor may be perforated and sewn in place, and then the seam is covered with tape. Some more robust designs for reusable clothing may use a gasket material to seal a rigid visor into a frame.

Respirator Facepieces

The facepieces of respirators also are relied upon for face and eye protection from chemical exposure. Full facepieces that cover the entire face generally offer the primary face and eye protection to many workers against chemicals, for both respiratory and dermal exposures. These facepieces are available for air-purifying and air-supplying respirators.

Performance Properties

The type of performance applied to face and eye protection items depends on the type of item and its intended protection. Goggles and faceshields are subject to the requirements provided in ANSI Z87.1. The criteria in that standard focuses on the physical protection and optical qualities of those devices. Faceshields are evaluated for impact resistance, physical penetration resistance, prismatic power, haze, transmittance, flammability resistance, corrosion resistance, and cleanability. While the committee has been working on a splash test to evaluate protector performance in preventing liquid contact, no such test or criteria are available.

Similarly, the National Institute for Occupational Safety and Health (NIOSH) has established performance criteria for facepieces that are part of respirators covered in regulations under Title 42 Code of Federal Regulations Part 84.[15] Nevertheless, no liquid penetration or chemical permeation criteria are currently applied to these respirator components. Instead, NIOSH requirements rely on whole respirator tests for inward leakage through quantitative fit testing.

The visors used in suits, including the seams for integrating the visor in the garment, should be evaluated for the same properties as garments. These properties are as important as optical properties (such as visor clarity, light transmittance, and haze) that are specific to the visor. However, industry practice has typically not included chemical resistance of visors. At an incident in Benicia, California in the mid-1980's, the failure of the visor in a totally encapsulating suit illustrated the potential life-threatening consequences of a visor failure. A leaking tank car of anhydrous dimethylamine required the response of a hazardous materials team for incident evaluation and mitigation. Though the suit was recommended as compatible with the leaking chemical, the visor of one responder's suit broke open during the incident. Fortunately, the self-contained breathing apparatus (SCBA) inside the suit protected man's respiratory tract, but his unprotected skin was exposed to the dimethylamine. An analysis of the incident by the National Transportation Safety Board (NTSB) found that the polycarbonate visor material was unsuitable for dimethylamine exposure, though the suit itself was recommended. This prompted the NTSB to recommend the development of standards for protective clothing for hazardous chemical emergencies with requirements applying to all parts of the ensemble.[16]

Summary

A wide variety of clothing and equipment is available in the workplace for chemical protection. These items fall into categories for body protection (full and partial body garments), hand protection (gloves), foot protection (boots, overboots/overshoes, booties, and boot/shoe covers), and eye/face protection (goggles, faceshields, garment visors, and respirator facepieces). The items can also be "type" classified by their general design, performance, and service life. Specific details related to the choice of materials, design features, construction methods, and performance features further serve to differentiate among the type of chemical protective clothing (Additional information is provided in Chapter 6.)

Informed health and safety professionals can improve their selection of chemical protective clothing by understanding the different attributes of CPC types, particularly the advantages, disadvantages, and limitations associated with each type of item. This information is essential to the selection of appropriate chemical protective clothing.

References

1. Stull, J.O., Selecting Chemical Protective Clothing, *Occup. Health & Safety*, (December 1995), 20-24.

2. Bray, A., and J.O. Stull, A New Nondestructive Inspection Method to Determine Fatigue in Chemical Protective Suit and Shelter Materials, *Performance of Protective Clothing: Fifth Volume, ASTM STP 1237* (J. S. Johnson and S. Z. Mansdorf, eds.), American Society for Testing and Materials, Philadelphia, PA, (1996), 281-295.

3. Dudzinski, D., Kennedy Space Center Maintenance Program for Propellant Handlers Ensembles, *Performance of Protective Clothing: Second Symposium, ASTM STP 989* (S. Z. Mansdorf, R. G. Sager, and A. P. Nielson, Eds.), American Society for Testing and Materials, Philadelphia, PA, (1988), 492-500.

4. Garland, C., and A. Torrence, Protective Clothing Materials: Chemical Contamination and Decontamination Concerns and Possible Solutions, *Performance of Protective Clothing: Second Symposium, ASTM STP 989* (S. Z. Mansdorf, R. G. Sager, and A. P. Nielson, Eds.), American Society for Testing and Materials, Philadelphia, PA, (1988), 368-375.

5. Schwope, A.D., and E.P. Renard, Estimation of the Cost of Using Chemical Protective Clothing, *Performance of Protective Clothing: Fourth Volume, ASTM STP 1133*, James P. McBriarty and Norman W. Henry, eds., ASTM, Philadelphia, PA, (1992), 972-981.

6. Fraser, A.J., and V.B. Keeble, Factors Influencing Design of Protective Clothing for Pesticide Application, *Performance of Protective Clothing: Second Symposium, ASTM STP 989* (S. Z. Mansdorf, R. G. Sager, and A. P. Nielson, Eds.), American Society for Testing and Materials, Philadelphia, PA, (1988), 565-572.

7. Baars, D.M., D.B. Eagles and J.A. Emond, Test Method for Evaluating Adsorptive Fabrics, *Performance of Protective Clothing, ASTM STP 900* (R. L. Barker and G.

C. Coletta, Eds.), American Society for Testing and Materials, Philadelphia, PA, (1986), 39-50.

8. ANSI, ISEA 101, *American National Standard for Limited Use and Disposable Coveralls – Size and Labeling Requirements*, Industrial Safety Equipment Association, Arlington, Virginia (1996).

9. ANSI, ISEA 105, *American National Standard for Hand Protection Selection Criteria*, Industrial Safety Equipment Association, Arlington, Virginia (2000).

10. Mickelson, R.L., and R.C. Hall, A Breakthrough Time Comparison of Nitrile and Neoprene Glove Materials Produces by Different Glove Manufacturers, *American Industrial Hygiene Association Journal,* 48, 941 (1987).

11. Keeble, V.B., M.B. Prevatt, and S.A. Mellian, An Evaluation of Fit of Protective Coveralls Manufactured to a Proposed Revision of ANSI/ISEA 101, *Performance of Protective Clothing: Fourth Volume, ASTM STP 1133* (J. P. McBriarity and N. W. Henry, Eds.), American Society for Testing and Materials, Philadelphia, PA, (1992), 675-690.

12. ANSI, Z41, *American National Standard for Personal Protection – Protective Footwear*, National Safety Council, Itasca, IL (1999).

13. OSHA, *Foot Protection*, Title 29 Code of Federal Regulations, Part 1910.137, U.S. Government Printing Office, Washington, DC (1999).

14. ANSI, Z87.1-1989, *American National Standard for Occupational and Educational Eye and Face Protection*, American Society of Safety Engineers, Des Plaines, IL (1989).

15. NIOSH, Title 42 Code of Federal Regulations, Part 84, U.S. Government Printing Office, Washington, D.C. (1999).

16. NTSB, J. Burnett, A.J. Investigation Report: Chemical Accident, Benicia, CA, National Transportation Safety Board, Washington, D.C. (1984).

Test Methods & Standards

5

By Norman W. Henry III and Jeffrey O. Stull

Introduction

Chemical protective clothing, including gloves, garments, footwear, or full ensembles, can be selected properly only when performance data indicate that their resistance to chemicals lasts for the duration of anticipated, worst-case chemical exposures. Such dependable resistance mandates the need for rigorously tested protective clothing. This clothing must also demonstrate acceptable integrity for overall protection and provide sufficient strength, durability, and physical hazard resistance. In addition, chemical protective clothing must be functional, allowing the wearer to safely perform the required tasks in an acceptable level of comfort.

Test methods serve several purposes in evaluating chemical protective clothing:

1. Test methods provide a basis for evaluating the effectiveness of clothing and material in providing protection against a specific hazard.
2. Test methods allow comparison of the relative performance among similar or different products.
3. Test methods aid in the development of new products by setting consistent procedures for evaluating performance.

For the most part, test methods should mimic workplace conditions as closely as possible. For example, in conducting chemical resistance testing, temperature and other appropriate environmental parameters (i.e., humidity) should be duplicated; the actual chemicals of interest should be used in proper form and concentrations; and most importantly, exposure by the chemicals to protective clothing materials should be on the normal outside surface only.

One-sided exposure is an important point when testing chemical protective clothing. In actual use, spills, splashes, and leaks will wet or coat an item's outside surface. The chemicals rarely, if ever, come in contact with edges or inside surfaces. (Actually, if edges or inside surfaces were to be contacted, the clothing has then failed in its mission of providing protection.) Moreover, many chemical protective clothing items are constructed with different layers to achieve the desired properties. To truly simulate end-use conditions, test method designs should accommodate this fact and isolate all such inner areas.

In some cases, test methods may not simulate exactly the conditions of exposure. Certain tests, such as those used to measure the strength of materials, may appear to poorly replicate effects on the material as it is used in actual practice. Nevertheless, these methods provide an effective evaluation of the material property and usually rank material performance consistent with observations of clothing qualities in use.

While test methods produce a result, which by itself is useful, specifications provide a basis for determining the acceptability of a test result. Many specifications provide minimum performance criteria that set specific pass/fail levels for specific tests on specific products. While there are limited specifications and other full product standards for chemical protective clothing, there is a desire among the end-user community for these types of standards to simplify selection decisions.

This chapter reviews the current status of efforts to develop test methods and standards that are appropriate for chemical protective clothing.

The History of Standards Development

Early Practices

In the protective clothing business, both manufacturers and marketing people proclaim the advantages of their newly developed clothing materials for use in the workplace environment. Formerly, however, their claims for protection were often not substantiated by test methods that demonstrated performance of a particular material against hazardous chemical substances.

Up through the early 1980s, most protective clothing manufacturers and users depended on simple immersion tests as primary sources for performance information on chemical resistance. Because such tests permit edge and inside-surface contact, data had been generated and published that could be inaccurate and misleading. Immersion tests were particularly inappropriate for evaluating multilayered composites.

Furthermore, protective clothing manufacturers often marketed their products on the basis of the results of crude test methods designed to measure physical hazard resistance. Consequently, protective clothing items would be selected on their ability to withstand abrasions, cuts, and punctures. Many of these tests were based on evaluation of the material as it might be used in other applications such as in gaskets or o-rings, such that the material in the form of protective clothing was never evaluated.

Much of the information developed from this early immersion and physical-property testing still persists in currently available literature; because of this, many occupational health and safety professionals are still selecting protective clothing items that do not represent an optimum match between exposures and materials.

A number of manufacturers and a few users developed their own test methods and data-reporting schemes based on convenient criteria for simple apparatus design and testing conditions for in-house applications and, perhaps, external marketing efforts. Often, these criteria bore little resemblance to the users' actual work environments. Minimum emphasis was placed on standardization, which makes comparison of data from one source to another difficult, if not impossible. This also made the selection process confusing for users, especially for those without laboratory facilities of their own. Also, the identification of performance goals for new product development had been, at best, inexact for the manufacturers.

Adrian L. Linch contributed to one of the earliest publications describing a comprehensive test method for evaluating the performance of protective clothing.[1,2] He described a permeation test method for measuring the breakthrough time and diffusion rate (permeability) of protective clothing materials exposed to hazardous chemicals. This was the first time that a chemical permeation resistance was recognized as an important criterion for measuring the degree of protection that clothing can provide. It was also the first reference to a standard permeability test cup procedure for determining chemical resistance.[1] Results of some of Linch's studies led to the development of a "Chem-Proof" air suit used for complete isolation from the work environment.

The lack of routine or comparable test methods for the evaluation of protective clothing made the selection process imprecise. Some early surveys indicated that one out of every four workers in the United States is exposed to some form of skin irritant, and that 1 percent of these workers developed skin disorders from such exposures. Even today, occupational diseases account for two-thirds of all job-related diseases, and 7 out of 10 industrial claims paid by insurance companies are for temporary disability resulting from dermatitis.

Early Standards Development

Liability issues, as well as the swirl of confusion surrounding measurement of protective clothing performance, prompted an aggressive movement toward standardizing test methodology and reporting. Since the late 1970s, manufacturers, users, and governmental agencies have been working together to address broad issues relating to protective clothing performance. In the United States, the principal forum for these activities has been the American Society for Testing and Materials (ASTM) Committee F23 on Protective Clothing. Within the ASTM framework, this committee endeavored to develop test methods, practices, guides, classifications, and specifications for protective clothing. In Europe, the comparable group is Technical Committee 162 on Protective Clothing. There is also Subcommittee 13 of Technical Committee 94 within the International Standard Organization, which has responsibility for developing protective clothing standards.

United States governmental organizations, such as the National Institute of Occupational Safety and Health (NIOSH) sponsored a program to investigate the performance of protective materials against selected carcinogenic chemicals. This program originated as a result of the Occupational Safety and Health Administration (OSHA) recommendation that protective clothing materials be "impermeable" to carcinogenic liquids. Since OSHA did not provide guidelines for assessing the barrier properties of protective clothing materials, NIOSH sponsored a study on the development of performance criteria for protective clothing used against carcinogenic liquids. A technical report on this subject was written and published in October 1978 by NIOSH.[3] Results of this work included the following:

1. A description of the form and extent of worker needs for protective clothing.
2. A recommended criterion for clothing resistance to permeation and an overview of several other functional requirements.
3. Recommended test methods for evaluating the performance of clothing materials.
4. Data showing the resistance of currently used clothing materials to permeation by selected carcinogens.

In 1986, NIOSH listed dermatitis as one of the 10 leading work-related diseases in the United States and sponsored a national symposium to plan a strategy for the prevention of work-related dermatological conditions.[3] This helped to establish greater interest in the standardization of protective clothing by general industry and other governmental organizations.

In a related set of circumstances, the National Transportation Safety Board (NTSB) investigated an accident involving the release of anhydrous dimethylamine from a rail car in Benecia, California during August 1983. During the incident, personnel wearing totally encapsulating suits responding to the spill had their facepieces craze and crack when exposed to the chemical. This was despite the use of clothing that purportedly was recommended for the dimethylamine. As a result of this accident, the NTSB recommended that the Environmental Protection Agency, the Federal Emergency Management Agency, and Occupational Safety and Health Administration, and the U.S. Coast Guard work closely with organizations such as ASTM's F23 committee. These recommendations also led to the development of a committee on Hazardous Chemical Protective Clothing within the National Fire Protection Association specifically for establishing standards on first-responder chemical protective suits.

A New Generation of Standards

Since the late 1970s, a wide range of standards has been developed that relate either directly or indirectly to chemical protective clothing. Many of these include test methods, but there are an increasing number of specifications available to the industry for the specification of chemical protective clothing. A list of standards by performance property or application is provided at the end of this chapter in Appendix A. The main property areas for test methods and standards include:

- Chemical resistance
- Strength
- Physical hazard resistance
- Durability
- Product function.

Product specifications cover a full range of protective clothing types and applications. A description of these test methods and other standards is provided in the following sections.

Chemical Resistance Testing

Chemical resistance test approaches for protective clothing may be segregated into small-scale, material-based tests and full item evaluations. Material test approaches can be classified into three types that describe how chemicals may interact with materials:

1. degradation resistance,
2. penetration resistance, and
3. permeation resistance.

A number of different procedures exist for the measurement of each phenomenon depending on the type of chemical challenge and the level of sophistication for performing the tests. Of the material testing approaches, both penetration and permeation resistance testing allow assessment for the barrier qualities of a protective clothing material, whereas degradation resistance does not. Penetration testing may involve chemical particles, liquids, or vapors (gases). This chapter specifically addresses liquid and vapor challenges since at the time of this writing, agreed-upon penetration test methods for particle penetration had not been developed.

The individual procedures available for measuring chemical degradation, penetration, and permeation resistance are described in the sections below.

Degradation Resistance

Degradation is defined by ASTM's F-23 Committee as the "change in a material's physical properties as the result of chemical exposure." Physical properties may include material weight, dimensions, tensile strength, hardness, or any characteristic that relates to a material's performance when used in a particular application. As such, the test is used to determine the effects of specific chemicals on materials. In some cases chemical effects may be dramatic showing clear incompatibility of the material with the chemical. Figure 1 shows a specimen of a protective clothing material before and after its exposure to a selected chemical, illustrating a severe case of material chemical degradation. In other cases, chemical degradation effects may be very subtle.

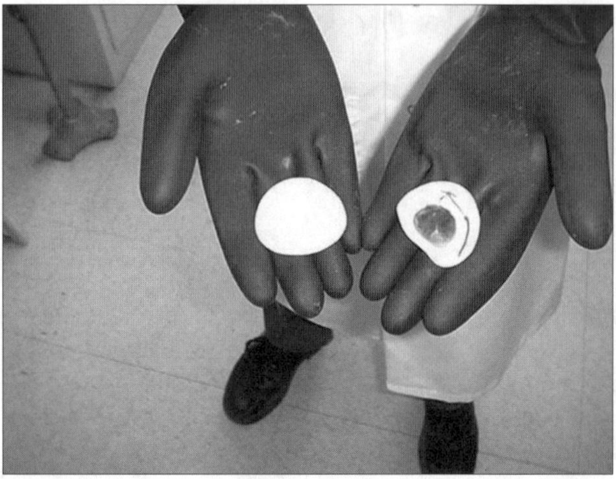

Figure 1. A polyvinyl chloride coated nylon before and after exposure to dimethylformamide. (Photo courtesy of Intertek Testing Services, Cortland, New York.)

Various groups have examined different approaches for measuring the chemical degradation resistance of barrier materials, but no single generalized test method has been developed by consensus organizations within the United States, Europe, or internationally that can apply to all protective clothing materials.[4] Nevertheless, a variety of techniques are commonly used for rubber and plastic materials within different barrier material industries. These procedures and their utility in evaluating chemical barrier materials are discussed below.

Specific Testing Approaches

While not specific to any particular product, a few test methods have been developed for evaluating chemical resistance of different materials:

- ASTM D 471, *Test Method for Rubber Property—Effect of Liquids*
- ASTM D 543, *Test Method for Resistance of Plastics to Chemical Reagents*
- ASTM F1407, *Test Method for Resistance of Chemical*

Table 1:
Comparison of Chemical Degradation Test Methods

Test Method	Type of Contact	Contact Period	Determination	Sample Handling
ASTM D 471	Both immersion and one-sided	22 to 760 hours	Weight, volume, or other physical property change	acetone rinsing followed by blotting
ASTM D 543	Immersion only	Up to 7 days	Weight, volume, or other physical property change	solvent rinsing depending on chemical followed by blotting
ASTM F 1407	One-sided	Unspecified	Weight change	blotting
ANSI/ ISEA 105	One-sided	1 hour	Change in puncture resistance	none, sample remains in special test cell for puncture resistance testing

Protective Materials to Liquid Permeation—Permeation Cup Method, developed by ASTM's F23 Committee also allows for determining chemical degradation of protective clothing materials; however, this method is primarily intended to provide a simple technique for measuring chemical permeation of protective clothing.

In late 1999, the Industrial Safety Equipment Association (ISEA) developed a new degradation resistance test method aimed at gloves. This test method became the first standard to address degradation resistance specific to chemical protective clothing.

These techniques have been grouped by the type of approach used in the following sections. Table 1 summarizes the key characteristics, differences, and applications of each approach for measuring the chemical degradation of protective clothing materials.

Degradation Tests Using Immersion-Based Techniques. ASTM D 471 and D 543 establish standardized procedures for measuring specific properties of material specimens before and after immersion in the selected liquid(s) for a specified period of time at a particular test temperature. Test results are reported as the percentage change in the property of interest. ASTM D 471 provides techniques for comparing the effects of selected chemicals on rubber or rubber-like materials, and is also intended for use with coated fabrics. ASTM D 543 covers the testing of plastic materials, including cast and laminated products, and sheet materials for resistance to chemical reagents. In each test, a minimum of three specimens are used whose shape and size are dependent on the form of the material being evaluated and the tests to be performed. An appropriately sized vessel, usually glass, is used for immersing the material specimens in the selected chemical(s). Testing with volatile chemicals typically requires either replenishment of liquid or a reflux chamber above the vessel to prevent evaporation.

The two sets of procedures prescribe a variety of exposure conditions and recommended physical properties. In general, the test methods can be applied to any type of liquid chemical challenge. ASTM D 471 cites a number of ASTM oils, reference fuels, service fluids, and reagent grade water. ASTM D 543 lists 50 different standard reagents that include representative inorganic and organic chemicals. In ASTM D 471, 17 different test temperatures ranging from –75° C to 250° C and 5 different immersion periods (22 to 760 hours) are recommended. ASTM D 543 suggests a 7-day exposure at either 50 or 70° C. For determining chemical degradation resistance, ASTM D 471 specifies procedures for measuring changes in mass, volume, tensile strength, elongation, and hardness for rubber materials, and breaking strength, burst strength, tear strength, and coating adhesion for coated fabrics. Measurements for material specimen mass and dimensions are recommended in ASTM D 543, while other properties may be selected that are appropriate for the material's application.

Both test methods indicate that the selected exposure conditions and physical properties measured should be representative of the material's use. For protective apparel material testing, this will usually mean specifying significantly shorter test periods and ambient temperature exposures. Since the methods are intended for comparing materials against similar chemical challenges, no criteria are given for determining acceptable performance.

Degradation Tests Using One-Sided Exposure Techniques. Section 12 of ASTM D 471 provides a procedure for evaluating the effects of chemicals when the exposure is one-sided. This technique is particularly useful for the evaluation of protective clothing materials, particularly those involving coated fabrics, laminates, and any nonhomogenous material. In this procedure, the material specimen is clamped into a test cell that allows liquid chemical contact on its normal external (outer) surface. Usually, changes in mass are measured for this testing approach since the size of the material specimens is limited.

During the mid-1980s, ASTM F23 Committee developed a procedure, which while intended for measuring chemical permeation resistance, also serves as a useful technique for evaluating chemical degradation resistance of protective clothing materials.[5,6] ASTM F 1407 employs a lightweight test cell in which the material specimen is clamped between a Teflon-coated metal cup filled with the selected chemical and a metal ring (flange). The entire cup assembly is inverted and allowed to rest on protruding metal pins that hold the test cell off the table surface. In mode of permeation testing, the weight of the entire assembly is monitored; however, for use as degradation test, the test cell serves as a convenient means for evaluating changes in material mass and thickness. Visible observations are also recorded as part of the testing protocol. Figure 2 shows a photograph of the permeation test cup specified by this method.

The Industrial Safety Equipment Association undertook the development of a degradation resistance test method for providing a means to evaluate and classify glove performance in a new standard specific for hand protection. The test is contained in ANSI/ISEA 105, *Criteria for Hand Protection Selection Criteria.* The test is based on the principle of one-

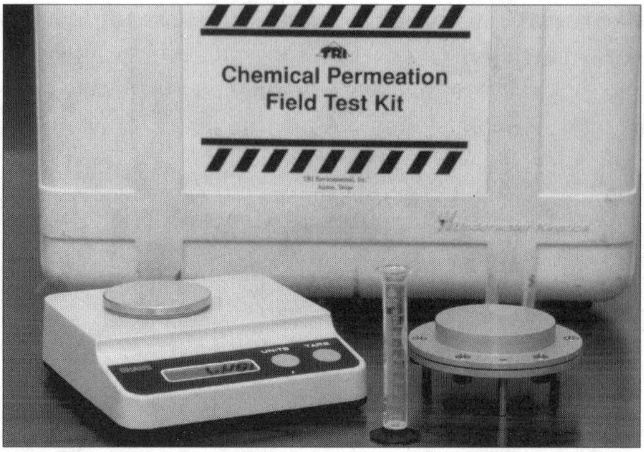

Figure 2. The permeation test cup and field test kit conforming to ASTM F1407 may also be used for measuring barrier material chemical degradation resistance. (Courtesy of TRI/Environmental, Inc.)

sided exposure, employing a small flask containing the test chemical. The protective glove specimen is sealed to the top of the flask and the flask is inverted to allow the test chemical to contact the specimen for a period of 1 hour. The flask is then positioned upright and mounted in a tensile testing machine outfitted with a compression cell and puncture probe. This test apparatus is used to measure the force required to puncture the specimen. The puncture resistance of exposed samples is then compared to the puncture resistance of pristine specimens. The percentage change in puncture resistance can thus be used as a measure of glove material degradation resistance to specific chemicals.

Each of the test methods described thus far for surface contact is provided for the testing of liquids. Protective clothing specimen chemical degradation can also be evaluated for chemical gas or vapor exposures. During one study for screening protective suit chemical resistance,[7] a special test cell was designed for evaluating material performance against gases. This cell was configured to provide a leak-free seal with the material specimen and to allow the flow of the gas into and out of the test cell. Degradation resistance data for three protective clothing materials are shown in Table 2 for both liquids and gases.

Table 2:
One-Sided Immersion Degradation Resistance Data for Selected Materials and Chemicals

	Viton/Chlorobutyl Laminate			Chlorinated Polyethylene			FEP/Surlyn Laminate		
Chemical	Percent Weight Change	Percent Elong.[a] Change	Visual Obs.	Percent Weight Change	Percent Elong. Change	Visual Obs.	Percent Weight Change	Percent Elong. Change	Visual Obs.
Acetaldehyde	10	0		24	11		0	-5	
Acrylonitrile	9	0	delam.	35	failed		-1	0	
Benzene	2	0		60	failed		0	0	
Chloroform	4	0		72	failed		0	0	
Dichloropropane	3	0		120	failed		-2	0	
Ethyl Acrylate	17	0	curled	160	failed		0	0	
Ethylene Oxide	2	0		13	11		0	0	
Hydrogen Fluoride	4	0	discol.	2	11	discol.	4	0	
Nitric Acid	9	0	discol.	8	-6		-1	0	

[a]Abbreviations: FEP - Fluorinated Ethylene Propylene, elong. - tensile elongation. obs. - observations, delam. - delaminated, discol. -discolore

[b]Percent elongation based on elongation measured using ASTM D412 for exposed and unexposed samples, 'failed' results indicate materials not tested due to weight changes over 25%

Source: Adapted from reference 8.

Application of Test Data

Chemical degradation by itself cannot fully demonstrate product barrier performance against chemicals. This form of chemical resistance testing does not ascertain the barrier properties of materials. While a material that shows substantial effects when exposed to a chemical can be ruled out as a protective membrane, it remains uncertain whether materials that show no observable or measurable effects provide a barrier against the test chemical. For this reason, chemical degradation data are typically used as a screening technique to eliminate a material from consideration for further chemical resistance testing (i.e., penetration or permeation resistance).[8,9]

To better understand the application of chemical degradation resistance data, it is instructive to know how these data are now being used with the protective clothing industry and how they could be used in a comprehensive material evaluation program.

Current Protective Clothing Industry Practices. Within the chemical protective clothing-related apparel industries, a great deal of chemical degradation resistance data have been generated and are presented in various product literature. Unfortunately, most presentations of material chemical degradation resistance are based on qualitative ratings such as "excellent," "good," "fair," "poor," and "not recommended." Ratings of this type provide little information to the end user particularly when the basis of the ratings is not explained or cannot be related to a particular application.

Manufacturers that use degradation data often base their ratings on either observed visual changes or weight gain. In a few cases, tensile or breaking strength differences are also used to qualify material chemical degradation. Ratings are then based on arbitrarily set levels of degradation. One manufacturer uses the following scale for its degradation rating system:

Rating	***Percent Weight Change***
Excellent	0–10%
Good	11–20%
Fair	21–30%
Poor	31–50%
Not Recommended	over 50%

This scale and probably all other rating systems used in the market have little bearing on the selection and use of protective clothing materials. Their presentation simply helps the end user to compare chemical effects on different materials. Until one standard test method is consistently used in the industry for measuring chemical degradation resistance of protective clothing materials, chemical effect information cannot be compared from one source to the next because different test approaches and criteria are used.[9]

The majority of chemical degradation resistance data are reported in the glove industry. This is because most gloves are made from elastomeric materials. As a class of materials, elastomers, when compared to plastics, show greater affinity for chemical adsorption and swelling.[10] Therefore, elastomeric materials are generally more susceptible to measurable chemical effects. This is particularly true today, because the majority of garment materials are composed of different plastic layers that have few observable degradation effects.

The creation of a new degradation resistance test for gloves will help standardize manufacturer reporting of degradation data. Not only does ANSI/ISEA 105 establish uniform test procedures, but it also sets specific levels of measured degradation resistance for classifying glove performance. This classification scheme, shown below, will provide a basis for end users to better compare glove products.

Performance Level	***Percent Change in Puncture Resistance***
0	over 80%
1	< 80%
2	< 60%
3	< 40%
4	< 20%

Recommended Use of Degradation Testing. Chemical degradation resistance testing is best used to qualify material candidates for subsequent barrier forms of chemical resistance testing. Materials that show significant signs of chemical degradation as determined by relevant criteria can be eliminated from further consideration in the required application. The premise for using degradation testing in this fashion is to configure the test-and-choose criteria that reflect how the material will be used and what barrier performance should be demonstrated. Establishment of a degradation testing protocol should include the following decisions:

- How long should the exposure be conducted?
- Should the exposure be one-sided or by complete immersion?
- What material properties should be measured?
- How should material specimens be handled following exposure?
- What criteria should be used for accepting or rejecting a material for a given chemical?

In general, the length of chemical exposure should be as long as the maximum duration for which chemical contact can occur or for the same period of time being used for the barrier test procedure selected. In some cases, longer exposure periods are used because the longer chemical contact can accentuate degradation effects that may be difficult to observe or measure.

Immersion offers the easiest approach for measuring chemical degradation resistance but may not be suitable for some materials. As described for ASTM D 471 and D 543 test methods, immersion-based degradation testing is best applied to homogeneous materials. Nevertheless, the ease of placing material specimens into a container filled with chemical can also be applied to more complex material to determine if separation of layers or substrate occurs as the result of the chemical exposure. The principal argument against immersion-based testing is that many products are typically exposed on their external surfaces only. This means that the substrate and different layers inside the material matrix are not likely to be exposed to chemicals unless the external surface is breached in some manner. For this reason, one-sided exposures are considered more realistic for the evaluation of protective clothing chemical degradation. On the other hand, one-sided degradation testing is more difficult to perform, usually limits the size of the material specimen for subsequent physical property testing, and may not provide any more information about the materials' performance following chemical exposure.

Physical properties should be selected to measure a material characteristic of interest. For the use of degradation testing as a means for screening material chemical resistance, observable changes in specimen appearance and weight gain are suggested. While disintegration of a test specimen is an easily recorded event, other differences in the visual appearance of test specimens may not be easily discerned. When chemicals are darkly colored, it may be impossible to carefully examine the material. Furthermore, it is difficult to achieve consistent material performance determinations using visual observations as the sole basis for rating material chemical degradation resistance. Nevertheless, the use of visual observations offers the easiest approach for conducting degradation testing with a minimum of specimen handling. When used, operator comments should be confined to certain observations. Examples include:

- discoloration (if the material is not darkly colored),
- curling,
- swelling,
- delamination (for multilayer or fabric supported materials), and
- disintegration.

Depending on both the nature of the chemical and material being evaluated, other "standardized" observation categories can also be created.

Measurement of weight change provides a simple, quantitative approach for assessing material degradation resistance. Material specimens may either gain or lose weight depending on how the chemical interacts with the material. Weight gain is caused by adsorption or absorption of a chemical on or in the material specimen and is often evidenced by swelling. Specimen weight loss is primarily due to some disintegration of the material or removal of particular components, such as plasticizers and other additives. In some cases, both phenomena can occur and result in very little weight change even though significant material effects have occurred. Weight change is always indicated as the percent change based on the sample's original weight.

Chemical degradation resistance testing involving the measurement of weight change is often affected by the techniques used for specimen handling. This handling can remove the chemical from the material's surface. In addition, removal of the material specimen from the chemical allows evaporation of volatile chemicals both off the surface and from inside the material matrix. The design of a degradation testing procedure must involve uniform specimen handling techniques to avoid systematic errors. Each test method defines slightly different procedures:

- ASTM D 471 specifies quickly dipping the specimen in acetone, blotting with a lint-free blotter paper, and then placing it in a tarred, stoppered weighing bottle.
- ASTM D 543 requires similar handling but indicates that the rinsing should be done with water, acetone, or not at all depending on the exposure chemical.
- ASTM F 1407 uses blotting only (without any sample rinsing).
- ANSI/ISEA 105 does not involve sample handling since the specimen is not removed from the exposure vessel prior to puncture resistance testing.

The choice of specimen handling procedures should be based on the nature of the chemical being tested and may need to be different for different chemicals, particularly when both low vapor pressure, nonvolatile, and high vapor pressure, volatile chemicals are tested. The former class of chemicals should require some blotting, with the time between removal from the chemical and weighing not as critical. Handling of specimens that have been exposed to high vapor pressure, volatile chemicals will most likely dictate no specimen rinse and a uniform period between specimen removal and weighing.

Development of screening criteria should be based on experience with the material(s) being evaluated and the barrier test chosen for assessment of chemical resistance. For example, small weight changes in some materials, such as Teflon or polyethylene laminates, may provide a clue about rapid permeation. But for other types of materials, these changes may not be significant for their permeation resistance. Whether the material specimen includes substrate fabric can also be a factor in setting acceptance and rejection criteria. Some fabrics can readily absorb test chemicals, particularly when immersion-based testing is used. If degradation testing is performed for screening materials prior to penetration resistance testing, then only severe material changes need be considered, which could lead to penetration of liquids through the material of interest.

The last consideration for using chemical degradation resistance testing is economic. The costs of these tests for degradation screening combined with subsequent barrier tests should not exceed the cost of barrier testing alone with all material and chemical combinations.

The other use for chemical degradation resistance testing is to identify potential modes of a product's failure. This form of degradation testing is most suitable to nonplanar components such as seams, closures, gaskets, or any other accessories used on the product. The chemical performance of these items can be evaluated by degradation tests that combine chemical exposure with an appropriate physical property test. Some examples include:

- Evaluation of rigid-formed materials for environmental stress cracking (see ASTM D 1975);
- Assessment of transparent material clarity following chemical exposure using light transmittal and haze testing (see ASTM D 1003);
- Measurement of changes in seam or closure strength (see ASTM D 751 for rubber materials or coated fabrics, D 1683 for textiles and F 88 for plastic materials);
- Operability of zippers (see ASTM D 2062); and
- Hardness of product gaskets or interface materials (see ASTM D 2240).

Overall, chemical degradation resistance testing can be a useful means for evaluating chemical barrier materials. However, it is important that the limitations of this testing be recognized and that decisions for barrier product selection and use never be based solely on chemical degradation resistance data.

Penetration Resistance

In the U.S. protective clothing industry, the ASTM F-23 Committee has defined penetration as "the flow of chemicals through closures, porous materials, seams, and pinholes and other imperfection in a protective clothing material on a non-molecular level." This definition is intended to accommodate both liquids and gases, but all U.S., European, and international test methods focus on liquid penetration. Liquid suspended in air as aerosols and solid particles can also penetrated protective clothing materials, but the discussion of penetration resistance in this chapter relates to liquids exclusively because particle-based methods are still under development.

Much of the liquid penetration resistance testing pertains to water as the challenge. The ASTM D-11 Committee on Rubber and Rubber Products (which includes coated fabrics) defines water repellency, waterproofness, and water resistance for coated fabrics as follows:

- Water repellency—the property of being resistant to wetting by water;
- Waterproofness—the property of impenetrability by liquid water; and
- Water resistance—the property of retarding both penetration and wetting by liquid water.

Liquid repellency and penetration resistance are related since wettability of the fabric affects the ability of the liquid to penetrate.

For porous fabrics, a liquid of surface tension, $\tilde{a}$, will penetrate given sufficient applied pressure, p, when its pores are of diameter, D, according the relationship known as Darcy's Law:

$$D = k\frac{4\gamma\cos\Theta}{p} \qquad (1)$$

where: Θ = contact angle of liquid with the material, and

k = shape factor for the material pores.

For nonporous fabrics, particularly coated fabrics or laminate materials, liquid penetration may still take place as the result of degradation. Given a sufficient period of contact, chemicals may cause deterioration of the barrier film to allow pathways for liquid to penetrate. In this sense, penetration testing allows both an assessment of material barrier performance to liquid chemicals and chemical degradation resistance.

There are two fundamentally different approaches used in liquid penetration resistance test methodologies.

1. Runoff-based methods, and
2. Hydrostatic-based methods.

Runoff-Based Tests

Runoff-based techniques involve contact of the liquid chemicals with the material by the force of gravity over a specified distance. The driving force for penetration is the weight of the liquid and the length of contact with the material specimen. Usually, the material specimen is supported at an incline, allowing the chemical to run off, hence the name for this class of penetration tests. Three features characterize runoff-based tests:

1. Impact of the liquid from a stationary source onto a material specimen,
2. Orientation of the material specimen at an incline with respect to the point of liquid contact, and
3. Use of a blotter material underneath the material specimen to absorb penetrating liquid.

Runoff-based tests differ in the distance separating the liquid source from the point of contact with the material specimen, the type of nozzle through which liquid is delivered, the amount of liquid and the rate at which it is delivered, the angle of the incline, and the type of test measurements made.

Runoff-Based Test Methods. There are a number of liquid penetration tests that are based on runoff techniques. These include:

American Association of Textile Chemist and Colorist (AATCC) Test Methods
AATCC 42, Water Resistance: Impact Penetration Test
AATCC 118, Oil Repellency: Hydrocarbon Resistance Test

European Committee on Standardization (CEN) Test Method
EN 368, Protective clothing—Protection against liquid chemicals—Test Method: Resistance of materials to penetration by liquids

International Standards Organization (ISO) Test Method
ISO 6530, Protective clothing —Protection against liquid chemicals—Determination of resistance of materials to penetration by liquids

U.S. Federal Government (FTMS) Test Methods
FTMS 191A,5520—Water Resistance of Cloth; Drop Penetration Method
FTMS 191A,5522—Water Resistance of Cloth; Water Impact Penetration Method
FTMS 191A,5524—Water Resistance of Cloth; Rain Penetration Method

Table 3 provides a comparison of the key characteristics of each of the referenced test methods. It should be noted that some methods from different organizations are very similar to each other.

As the titles for several methods denote, the majority of these methods are intended for use with water as the liquid challenge only. Physically, many of the methods are suitable for testing with other liquids; however, the containment aspects of these test methods vary and some are clearly inappropriate for use with hazardous chemicals.

Table 3: Characteristics of Runoff-Based Penetration Tests

Test Method	Type of Delivery	Liquid Amount and Rate	Sample Orientation	Measurements
FTMS 191A,5520	Polystyrene plate with 31-0.4 mm ID capillary holes 1.73 m above sample	Determined by test end point	45-degrees, clamped onto perforated disk	Time to collect 10 mL water (from penetrating sample)
FTMS 191A,5522	Spray nozzle at end of funnel with 19-0.89 mm holes 610 mm above sample	500 mL	45-degrees, under 0.45 kg tension force	Weight of penetrating water
FTMS 191A,5524	Spray nozzle at end of funnel with 12-1 mm holes 305 mm above sample	300 seconds at selected pressure head	Horizontal	Weight of penetrating water
EN 368	Single, 0.8 mm bore hypodermic needle 100 mm above sample	10 mL at 1 mL/sec	45-degrees, over blotter in semi-circular 'gutter'	Index of repellency; index of penetration
ISO 6530	Same as EN 368			

The majority of test methods listed in Table 3 involve delivering relatively large quantities of water onto a sample and measuring the amount of water absorbed in a blotter paper placed underneath the material specimen. This approach is characteristic of AATCC 42, FTMS 191A,5522, and FTMS 191A,5524. Figure 3 shows a picture of a spray impact tester used in both AATCC 42 and FTMS 191A,5522. One method, FTMS 191A,5520, is used for materials where a significant amount of water is expected to penetrate since the time to obtain 10 mL of water is used as the test endpoint. The large

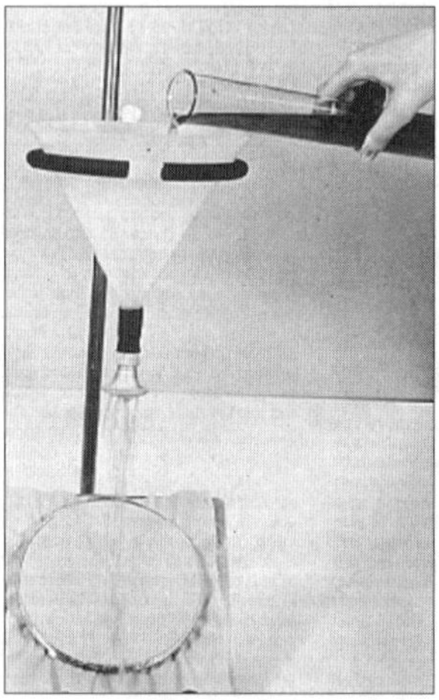

Figure 3. Spray impact testing used for testing in accordance with AATCC 42 and FTMS 191A,5222 (Courtesy of TRI/Environmental, Inc.).

quantities of water specified, and the lack of containment in the design of the apparatus make these test methods unsuitable for other liquids.

AATCC 118 was designed to measure fabric resistance to oil. However, this test is a repellency-type test, where surface appearance of the material specimen is rated after its exposure to selected hydrocarbons. This test is similar to several other tests in the literature involving water where either percent absorption of liquid by the fabric is measured or the pattern of wetness on the underside of the material specimen is rated on percent coverage.

Two of the test methods are essentially identical and are designated for use with various liquid chemicals. Both EN 368 and ISO 6530, the so-called “gutter test,” use a system where the liquid chemical is delivered by a single, small bore nozzle onto the material specimen at a distance of 100 mm (see Figure 4). The material is supported in a rigid transparent gutter that is covered with a protective film and blotter material at a set 45-degree angle with respect to the horizontal plane. A small beaker is used to collect liquid running off the sample. The two results reported in these tests are the indices of

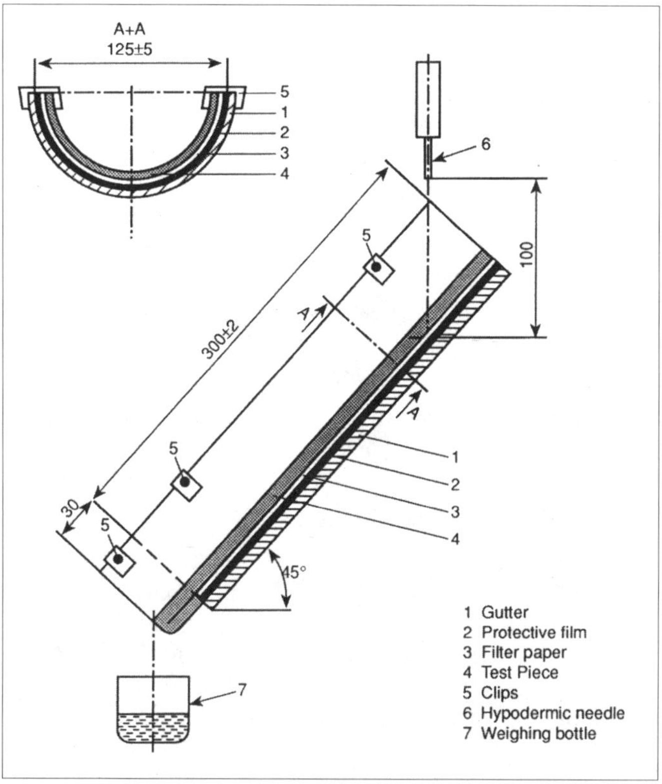

Figure 4. Specifications for CEN and ISO "Gutter Test."

penetration and repellency. The "Index of Penetration" is the proportion of liquid deposited in the blotter paper:

$$Index\ of\ penetration\ (P) = \frac{M_p\ x\ 100}{M_t} \qquad (2)$$

where:

M_p is the mass of test liquid deposited on the absorbent paper/protective film combination; and

M_t is the mass of the test liquid discharged onto the test material specimen.

The Index of Repellency is the proportion of liquid deposited in the blotter paper:

$$\textit{Index of repellency} \quad (R) = \frac{M_r \times 100}{M_t} \qquad (3)$$

where:

M_r is the mass of test liquid collected in the beaker.

A mass balance of the liquid also allows calculation for liquid retained in the material specimen.

Application of Runoff-Based Test Methods. Not all of the tests described above can be considered 'true' liquid penetration tests. Penetration with these procedures can only be characterized when some assessment or measurement of liquid passing through the material specimen is made. Typically this is done by examination of the blotter material, either visually or gravimetrically.

Runoff tests are generally used on textile materials that have surface finishes designed to prevent penetration of liquid splashes. Many of these tests easily accommodate uncoated or nonlaminate materials since the driving force for liquid penetration is relative low (when compared to hydrostatic-based test methods). As a consequence, runoff tests may be infrequently specified for chemical barrier-based clothing. The European Community, while developing a range of chemical protective clothing standards, uses EN 368 in none but its lowest clothing classification (for partial body coverings). In the United States, only a few material manufacturers use runoff-based tests for characterizing the chemical penetration resistance of their fabrics. When used in this fashion, the fabrics tested are either uncoated or have thin simple plastic films.

The relatively small amounts of liquid involved in some runoff tests are not considered a strong challenge. For this reason, ISO 6530 contains very specific limitations for its use in testing chemical protective clothing.

Clothing which has been developed from materials selected by this method of test (i.e., ISO 6530) should only be used in well-defined circumstances when an evaluation of the finished item has indicated an acceptable level of performance.

In other words, ISO recommends that the test be used only when the clothing item's overall integrity for preventing liquid penetration has been demonstrated. A discussion of liquid-based integrity methods appears later in this chapter.

Use of ISO 6530 is also subject to systemic errors. As with degradation tests described above, testing with volatile chemicals requires special handling procedures to minimize evaporation of solvent and its impact on test results. Similarly, test operators must be careful not to remove nonvolatile chemicals through the handling of the blotter material. The "open" nature of the test apparatus combined with its gravimetric basis may also be strongly influenced by environmental conditions. One testing laboratory has reported different results when tests are conducted in a hood versus those that are conducted on the laboratory bench (without ventilation).[11] For these reasons, penetration testing using ISO 6530 should be performed with uniform handling procedures and in a controlled environment that is the same from test to test.

Some in the protective clothing industry do not consider any of the runoff tests as legitimate liquid penetration tests since these methods fail to demonstrate "liquid-proof" performance for protective clothing material performance. In this context, liquid-proof performance is often defined as the ability of the material to prevent liquid penetration under conditions representative of use. These researchers contend that runoff-based or repellency penetration tests are an evaluation of the surface wettability characteristics for material finishes, and thus, not true barrier-oriented techniques.

Hydrostatic Penetration Testing

Hydrostatic-based techniques involve the pressurization of liquid behind or underneath the material specimen. It is this hydrostatic force that is the principal driver for liquid penetration. Though the term, hydrostatic, is used to describe this class of test methods, only one of the tests in this class can accommodate a wide range of liquid chemicals.

In this testing approach, liquid is contacted with the

material specimen, with at least some portion of the test period having the liquid under pressure. Different devices or test cells are available for providing this type of liquid contact with the material specimen, in essence representing the differences among representative test methods. Like runoff-based test methods, the majority of the industry tests are designed for use with water. Many of the devices described below cannot be used with other liquid or may even be damaged if anything but water is used in the respective tests.

Hydrostatic Test Methods. There are a number of liquid penetration tests that are based on runoff techniques. These include:

American Association of Textile Chemist and Colorist (AATCC) Test Method

AATCC 127, Water Resistance: Hydrostatic Pressure Test

American Society for Testing and Materials (ASTM) Test Methods

ASTM D 751, Methods for Testing Coated Fabrics, Hydrostatic Test
ASTM D 3393, Specification for Coated Fabrics—Waterproofness
ASTM F 903, Resistance of Protective Clothing Materials to Penetration by Liquids

International Standards Organization (ISO) Test Method

ISO 8096, Rubber- or plastics-coated fabrics water-resistant clothing—Specification—Part 1: PVC-coated fabrics; Part 2: Polyurethane- and silicone elastomer-coated fabrics
ISO 13994, Protective clothing—Protection against liquid chemicals—Determination of resistance of materials to penetration by liquids under pressure

U.S. Federal Government (FTMS) Test Methods

FTMS 191A,5512—Water Resistance of Coated Cloth; High Range, Hydrostatic Pressure Method
FTMS 191A,5514—Water Resistance of Coated Cloth; Low Range, Hydrostatic Pressure Method
FTMS 191A,5516—Water Resistance of Cloth; Water Permeability, Hydrostatic Pressure Method

Figure 5. Mullen Hydrostatic Tester (Courtesy of TRI/Environmental, Inc.).

Two different types of testing machines prevail for measuring hydrostatic resistance. AATCC 127, ISO 8096, FTMS 191A,5514, and FTMS 191A,5516 all use similar devices, where water is introduced above the clamped material specimen at a pressure controlled by water in a rising column. A mirror is affixed below the specimen to allow the test operator to view the underside of the specimen for the appearance of water droplets. Both the pressure and length of exposure are to be specified for the particular application. AATCC 127 and FTMS 181A,5514 define water penetration as the pressure when a drop or drops appear at three different places of the test area (on the specimen). When a specific hydrostatic head is specified, test results are reported as pass or fail. FTMS 191A,5516 also permits measurement of the amount of water penetration, collected from a funnel and drain hose underneath the material specimen (in lieu of the observation mirror).

ASTM D751, ASTM D3393, and FTMS 191A,5512 use a motor-driven hydrostatic tester (pictured in Figure 5). Water contacts the underside of the material specimen that is clamped into a circular opening. Increasing hydraulic pressure is applied to the clamped material specimen at a specified rate until leakage occurs. The pressure at which this leakage occurs is noted and reported as the test result.

Of the listed tests above, only ASTM F 903 and ISO 13994 were developed for testing liquids other than water.[12] In this test method, a 70-mm square material specimen is exposed on one side to the test chemical for a specified period of time using a special penetration test cell (see Figures 6 and 7). The test cell is positioned vertically to allow easy viewing of the material specimen. During the chemical exposure, a pressure head may be applied to the liquid for part of the test period. Penetration is detected visually and sometimes with the aid of dyes or fluorescent light. The test is generally pass/fail, that is, if penetration is detected within the test period, the material fails. Observations of material condition following chemical exposure are also usually provided. Different test specifications exist for the amount of chemical contact time and pressurization.

Application of Hydrostatic-Based Test Methods. Penetration resistance using hydrostatic-based test methods can accommodate different types of material and clothing test specimens, including:

- plastic or rubber films,
- coated fabrics,
- textiles,
- microporous films,
- clothing seam samples, and
- clothing closure samples.

For these types of material specimens there are different modes of failure. Continuous film or film coated fabrics generally only fail due to:

- imperfections in the material, such as cuts or pinholes; or
- deterioration (degradation) of the film providing an avenue for liquid penetration.

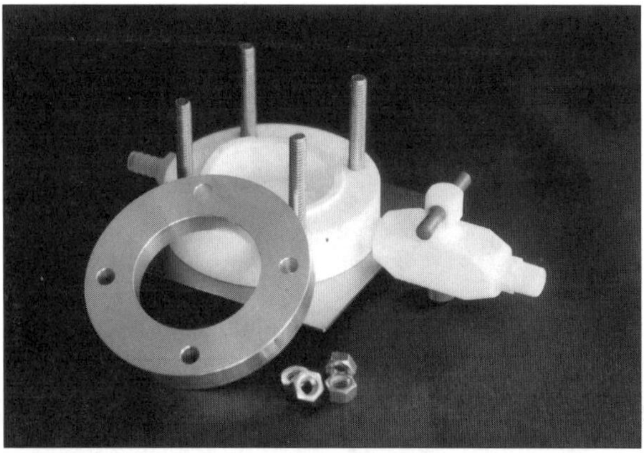

Figure 6. ASTM F 903 penetration test cell (Courtesy of TRI/ Environmental, Inc.).

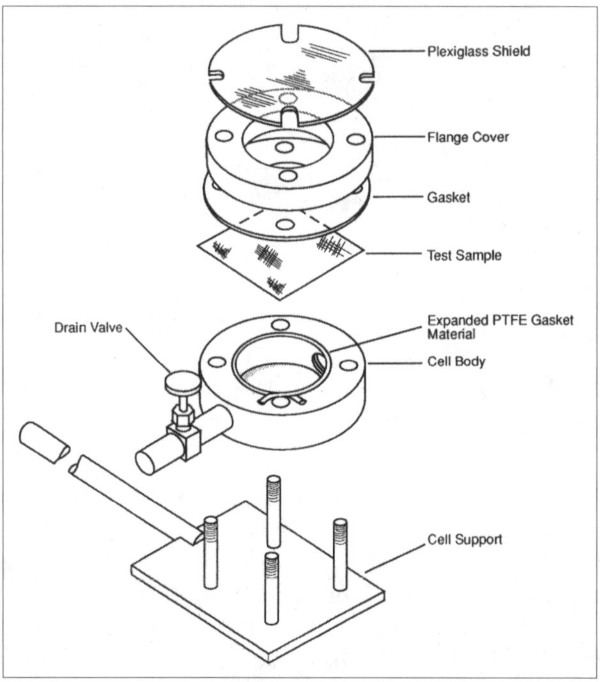

Figure 7. Exploded view of ASTM F 903 penetration test cell.

Figure 7a. Penetration testing of seam samples.

The latter type of failure often depends on the thickness of the film or coating as well as the contact time and amount of pressure applied to the specimen.

Textiles and microporous film products provide another set of possible failure mechanisms. Textiles may be considered as liquid barriers when they have been treated with water/ chemical repellent finishes. The ease of liquid penetration is therefore more a function of repellant finish quality and the surface tension of the liquid being tested. Table 4 provides the surface tension of the liquid chemicals in ASTM F 1001 (in general, lower surface tension liquids will penetrate more readily). Also, penetration may still be the result of material

Table 4:
Surface Tensions for ASTM F1001 Liquids

Chemical	Surface tension (dynes/cm)
Acetone	23.5
Acetonitrile	26.6
Carbon Disulfide	31.6
Dichloromethane	27.2
Diethylamine	26.2
Dimethylformamide	36.2
Ethyl Acetate	23.4
Hexane	17.9
Methanol	22.1
Nitrobenzene	43.3
Sodium Hydroxide	103
Sulfuric Acid	55.1
Tetrachloroethylene	31.8
Tetrahydrofuran	26.5
Toluene	27.9

degradation while in contact with the chemical.

Microporous film products represent a unique test material since by design they afford transmission of vapors but prevent liquid penetration. These materials therefore require careful observations since significant vapor penetration may occur. Like textiles, surface tension may be a factor, though most microporous films have pore sizes that preclude penetration of most common liquids at relatively low pressures (less than 12 kPa).

The integrity of seams, closures, and other clothing material interfaces are easily evaluated using penetration resistance testing.[13,14] Their uneven samples profiles must be accommodated through special gaskets or sealing techniques. For zipper closures, we have cut a groove in the test cell to provide a better seal on the protruding teeth portion of the zipper. In assessing penetration resistance for these items, failures may occur because of

- penetration of liquid through stitching holes in seams;
- solvation of seam adhesives;
- degradation of seam tapes or other seam components;
- degradation of materials joining seam, causing lifting of seam tapes or destruction of seam integrity; and
- physical leakage of closures.

Berardinelli and Cottingham[13] demonstrated the utility of this test on a number of material, seam, and closure samples. Understanding how clothing specimens may fail provides insight into identifying protection offered by the overall clothing item. The qualitative nature of the penetration test requires that test operators be familiar with failure modes so that they can correctly assess whether liquid penetration has or has not occurred.

Specific Use of ASTM F 903 and ISO 13994. Penetration testing per ASTM F 903 (and ISO 13994) provides a test for assessing the barrier performance of materials against liquid chemicals.[15] Though measuring specimen weight change is not required, this testing can also serve to measure material degradation since visual observations are required. Degradation of the material, in turn, may be a primary route for penetration by some chemicals. The difficulty in penetration testing lies in making a clear-cut determination of liquid penetration. Many high vapor pressure, low surface tension solvents spread thinly over the material and evaporate quickly. Therefore, actual liquid penetration may be difficult to observe even when enhanced by using dyes (Sudan III is recommended for most organic solvents). Still, the test serves as a good indicator of material performance against liquid contact or splashes. Since test length and pressurization periods depend on the selected procedure within the method, pass/fail determinations are clear-cut and easily define acceptable material–chemical combinations.

ASTM F 903 and ISO 13994 incorporate four types of contact time and pressure exposure formats (see Table 5). The

Table 5:
ASTM F 903 Penetration Test Variations

Procedure	Initial Contact Period (Minutes at 0 kPa)	Pressurization Period (Minutes/ Pressure)	Subsequent Contact Period (Minutes at 0 kPa)
A	5	10 min./7 kPa	none
B	5	10 min./14 kPa	none
C	5	1 min./14 kPa	54
D	60	none	none

original protocol consisted of exposing the material to the liquid for a 5-minute period at ambient pressure, followed by a 10-minute period at 13.8 kPa (2 psig). This exposure condition was selected as a test pressure to simulate the force on a protective garment of a liquid coming out a burst pipe at an approximate distance of 3 meters. A lower pressure of 6.9 kPa (1 psig) was adopted later on because many materials would "balloon" away from the test cell as pressure was applied. Some differences in material performance due to degradation effects have been noted as shown in Table 6.[12] The additional contact time/pressure formats in ASTM F903 were included to accommodate practices being used by the National Fire Protection Association in their requirements for chemical protective suit material and component penetration resistance.[16]

Many unsupported film samples cannot be tested at high pressure since they burst when the pressure is applied. Severe "ballooning" of these test specimens in testing may also give questionable test results due to the relatively large forces placing the material in tension. In these cases, the true barrier properties of the material to liquid penetration are not tested. For this reason, the optional use of screening having more than 50% open area is specified in ASTM F 903 and ISO 13994 to prevent overexpansion of the clothing material specimen.

Table 6:
Effect of Contact Time and Pressure on Penetration of Selected Material-Chemical Combinations

Material	Chemical	Penetration Test Exposure Protocol	Penetration Time (Minutes)
PVC/Nylon	Dimethyl-formamide	A	None
		B	None
		C	40
		D	50
Microporous film/nonwoven laminate	Hexane	A	None
		B	5
		C	5
		D	None

Source: Adapted from Reference 12.

Permeation Resistance

Permeation process is described in Chapte 3 as a process in which chemicals move through a material at a molecular level. Material permeation resistance is generally characterized using two test results: breakthrough time and permeation rate. Breakthrough time is the time that a chemical is first detected on the interior side of the material. As discussed below, its determination is strongly dependent on how the test is configured and the sensitivity of the detector. Permeation rate is a measure of the mass flux through a unit area of material for a unit time. Permeation rate is most commonly expressed in units of micrograms per square centimeter per minute (¼g/cm^{2}min). For a given material–chemical combination, the steady-state or maximum observed permeation rates are reported.

The measurement of chemical permeation resistance is specified in the different standard test methods offered by ASTM, CEN, and ISO:

American Society for Testing and Materials (ASTM) Test Methods

ASTM F 739, Test Method for Resistance of Protective Clothing Materials to Permeation by Liquids and Gases Under Conditions of Continuous Contact

ASTM F 1383, Test Method for Resistance of Protective Clothing Materials to Permeation by Liquids and Gases Under Conditions of Intermittent Contact

ASTM F 1407, Test Method for Resistance of Chemical Protective Materials to Liquid Permeation—Permeation Cup Method

European Committee on Standardization (CEN) Test Method

EN 369, Protective clothing—Protection against liquid chemicals—Test method: Resistance of materials to permeation by liquids

EN 374-3, Protective gloves against chemicals and microorganisms—Part 3: Determination of resistance to permeation by chemicals

International Standards Organization (ISO) Test Method

ISO 6529, Protective clothing—Protection against liquid chemicals—Determination of resistance of air-impermeable materials to permeation by liquids

These permeation tests involve either liquid or gaseous chemical contact with the material and assessment of permeation as affected by both chemical solubility and diffusion through the test material. ASTM F 739 was established in 1981 as the first standard test method for measuring material permeation resistance to liquid chemicals.[17] ASTM F 739, EN 369, EN 374-3, and ISO 6529 provide standardized procedures for measuring the resistance of protective clothing to permeation by chemicals using continuous contact of the chemical with the material's exterior surface. ASTM F 1383 is a variation of ASTM F 739 that involves testing under conditions of intermittent chemical contact. ASTM F 1407 represents a simplified form of testing where permeation is determined gravimetrically. Based on its limited sensitivity, this method is primarily used as a screening test or field method.

In each of the tests (except ASTM F1407), a similarly designed test cell is used for mounting the material specimen. The test cell consists of two hemispherical halves divided by the material specimen. One half of the test cell serves as the "challenge" side where the chemical is placed for contacting the material chamber. The other half is used as the "collection" side that is sampled for the presence of chemical permeating through the material specimen. Figure 8 shows a conceptual illustration of the permeation test cell.

The basic procedure in each test is to charge a chemical into the challenge side of the test cell and to measure the concentration of a test chemical in the test cell as a function of time. Of principal interest in permeation testing is the elapsed time from the beginning of the chemical exposure to the first detection of the chemical (i.e., the so-called breakthrough time), the permeation rate, and the cumulative amount of chemical permeated. The results reported are dependent on the test method chosen:

- ASTM F739 requires reporting of breakthrough time and maximum or steady-state permeation rate.
- ASTM F1383 specifies reporting breakthrough time and cumulative permeation.
- ASTM F1407 permits reporting either cumulative permeation or breakthrough time and permeation (maximum or steady state).
- EN 369 and ISO 6529 require reporting of breakthrough time with the total cumulative mass permeated at 30 and 60 minutes.

Other significant differences exist among the different listed test methods as described below. Table 7 provides a comparison of key characteristics for each of the different permeation test methods.

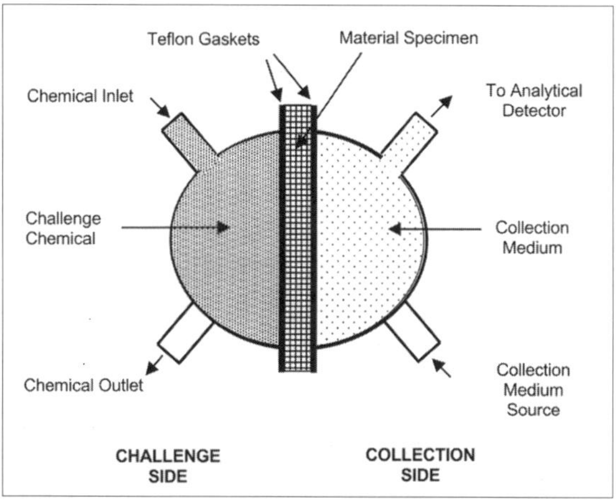

Figure 8. Conceptual diagram of ASTM F 739 permeation test cell.

Parameters Affecting Permeation Resistance Testing

Although the permeation test procedure is simple in concept and generalized procedures are specified by each of the test methods above, a number of significant variations exist in the manner in which permeation testing can be conducted. These variables include:

- the general configuration of the test apparatus,
- how the chemical contacts the material specimen in the test cell,
- the type of collection medium used and frequency of sampling,
- the type of detector and detection strategy used,
- the test temperature, and
- the effect of multicomponent solutions.

In addition, a separate class of materials—porous textile materials containing adsorbent particles—must be tested in totally different manner.

Table 7:
Differences among Permeation Test Method

Test Method	Chemicals Permitted	Type of Contact	Collection Medium Flowrate(s)	Minimum Test Sensitivity (µg/cm2min)	Test Results Reported
ASTM F 739	Liquids and gases	Continuous	50 to 150 mL/min	0.1	Breakthrough time Permeation rate
ASTM F 1383	Liquids and gases	Intermittent	50 to 150 mL/min	0.1	Breakthrough time Cumulative permeation
ASTM F 1407	Liquids only	Continuous	Not applicable	$\approx$20.0*	Breakthrough time Permeation rate Cumulative permeation
EN 374-3	Liquids and gases	Continuous	50 to 150 mL/min	1	Breakthrough time Permeation rate
EN 369	Liquids only	Continuous	520 mL/min (gas) 260 mL/min (liq.)	1	Breakthrough time Cumulative permeation at 30 and 60 minutes
ISO 6529	Liquids only	Continuous	520 mL/min (gas) 260 mL/min (liq.)	1	Breakthrough time at 30 and 60 minutes Cumulative permeation

*Depends on analytical balance, exposed specimen surface area, and time interval between measurements.

The variety of available test techniques and conditions allow several different approaches for conducting permeation testing and can provide different results for testing the same material and chemical combination.

Overall Test Apparatus. The overall test apparatus includes the:

- test cell,
- chemical delivery system, and
- collection/detector system.

This apparatus should be configured to meet testing needs and accommodate the characteristics of the chemical(s) being tested. Test cells are generally specified by the test method, but alternative designs are available and may be necessary for testing with specific chemicals or chemical mixtures. Similarly, the chemical delivery and collection/detection systems are dependent on the nature of the chemical and the requirements for running the test. The way that each part of the apparatus is operated comprises the test apparatus configuration. There are two basic modes for configuring permeation test systems: closed-loop or open-loop.

Permeation Test Cells. ASTM F 739 and ASTM F 1383 specify a 51-mm (2.0 in) diameter test cell, which is constructed of two sections of straight glass pipe to create two glass sections that are joined by flanges on both sides (Figure 9). The two sections are bolted together to provide a seal between the PTFE gaskets, glass sections, and material specimen. Collection and challenge sides of the test cell are configured to handle the types of chemicals being tested. When a liquid challenge is used, a single, relatively large inlet can be used, whereas gas challenges require both an inlet and outlet for charging gas into and out of the challenge chamber. Typically, collection sides of the test cell also contain an inlet and an outlet. When liquid is used as the collection medium, a stirring rod may be placed in one of these ports to affect uniform mixing of the collection medium with permeating chemical. Gas collection media require an inlet for charging gas into the test cell and an outlet that permits sampling near the material surface. Both of these tests permit alternative test cells, as long

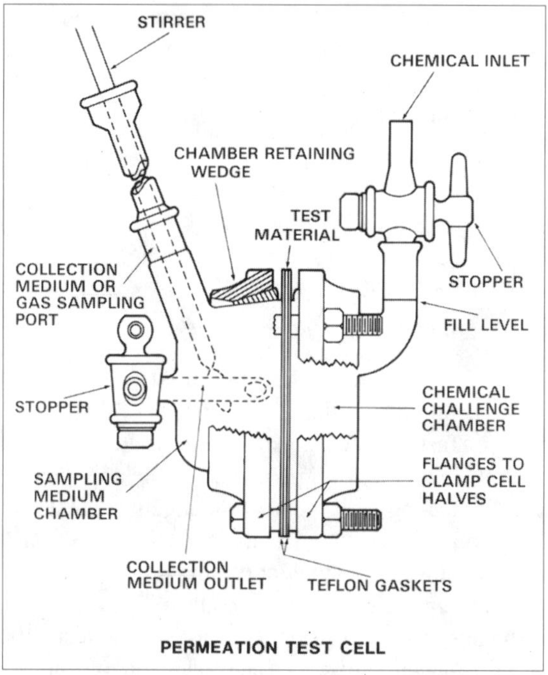

Figure 9. Specification for ASTM F 739 standard permeation test cell.

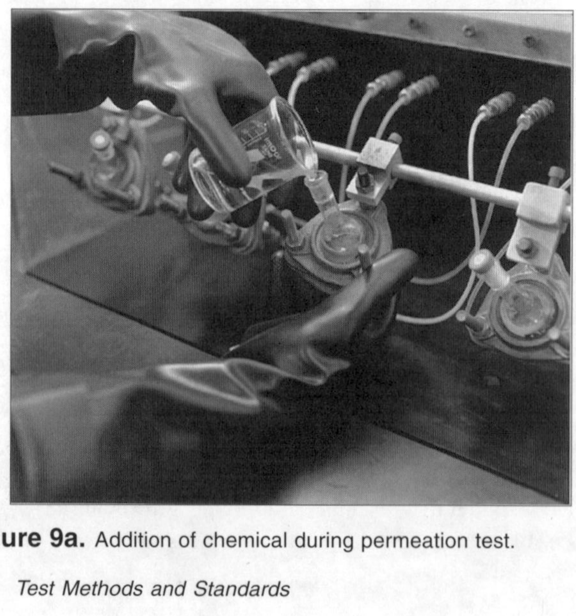

Figure 9a. Addition of chemical during permeation test.

as their equivalency can be demonstrated. While EN 374-3 specifies a similar cell as used in ASTM F 739, EN 369 and ISO 6529 specify a 25-mm diameter exposure area with a cylindrical shape, which is to be constructed of an inert material.

The test cell used for ASTM F 1407 was mentioned as part of the discussion on degradation testing. This test cell is a shallow cup, usually fashioned out of metal coated with a PTFE film. An outer ring secures the material specimen onto the cup flange by a series of thumb screws. The side opposite the challenge side is open to the atmosphere as the test cell is inverted and placed either on a stand or elevated off of the testing surface,[7,8] permitting it to be outside the test cell as depicted in Figure 2.

Test System Configuration. In closed-loop permeation systems, the volume of collection fluid is maintained throughout the test. This volume may be contained fully within the collection chamber or it may be circulated through the chamber, into a nonintrusive detector, and back into the chamber as illustrated in Figure 10. Since the total volume of collection medium remains constant, permeating chemical accumulates within the collection medium. In this system, the permeation rate must account for this accumulation of permeant as follows:

$$Rate = \frac{C_n - C_{n-1}}{t_n - t_{n-1}} \times \frac{V}{A} \qquad (4)$$

where: $(C_n - C_{n-1})$ are the changes in concentration of the challenge chemical between sampling periods,

$(t_n - t_{n-1})$ is the time between sampling periods,

V is the volume of collection medium, and

A is the exposed area of the material specimen.

In the open-loop permeation systems, a gas or liquid collection medium is passed through the collection side out of

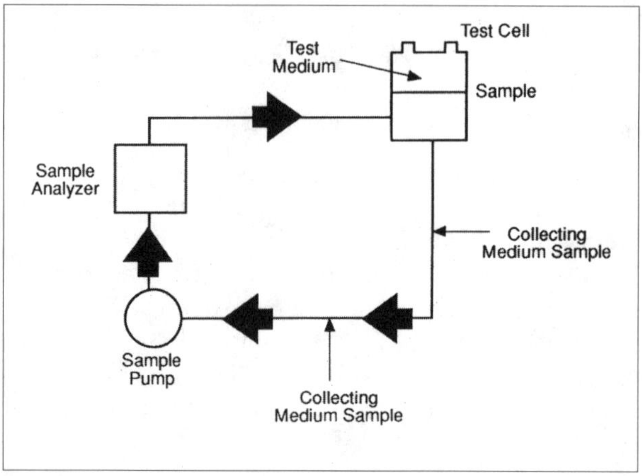

Figure 10. Closed loop permeation test system configuration. (Copyright 1983, American Industrial Hygiene Association. Reprinted with permission.)

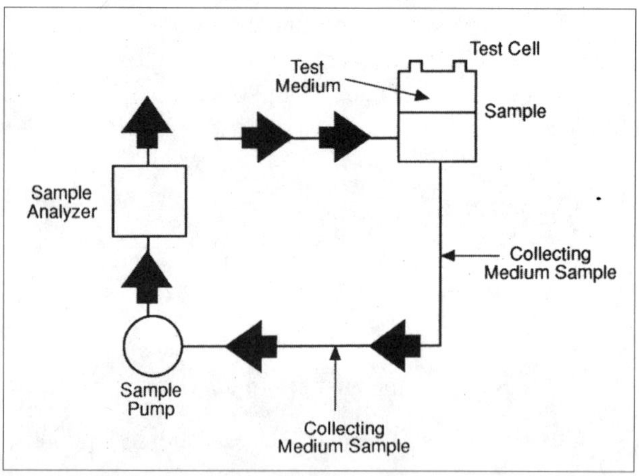

Figure 11. Open loop permeation test system configuration. (Copyright 1983, American Industrial Hygiene Association. Reprinted with permission.)

the test cell to the detection system (see Figure 11). This collection medium stream can be evaluated discretely or continuously depending on the detector selected. Therefore, collection of permeant is specific to the sample taken (over a discrete time period) and permeation rates can be directly calculated as a factor of the collection medium permeant concentration (C) and flow rate (F):

$$Rate = \frac{C \, x \, V}{A} \qquad (5)$$

The choice of a closed or open-loop system is most often determined by the properties of the chemical and the available detector. Some chemicals such as inorganic substances often require closed loop systems, particularly if ion-specific electrodes are used that have recovery time constraints. Open-loop testing is preferred for many volatile organic chemicals because these systems can be easily automated.

Both Henry[18] and Berardinelli, Mickelsen, and Roder[19] examined differences in permeation test results for closed- and open-loop systems. For the small number of material–chemical combinations investigated, both researchers could not find appreciable differences in the collection medium in either breakthrough time or permeation rate. Schwope et al.[20] looked at intrinsic differences between closed- and open-loop systems through a modeling approach based on the use of Fickian equations and assumed values for the diffusion coefficient, solubility parameter, and thickness of a material–chemical combination. In this treatment, they were able to show the affect of several system variables on permeation breakthrough time including material thickness (assuming homogeneity), specimen exposed surface area, and detector analytical sensitivity.

Continuous Contact Testing. With the exception of ASTM F 1383, each of the listed permeation test methods specify testing with neat chemicals under conditions of continuous exposure. Both EN 369 and ISO 6529 accommodate liquid chemicals only. In liquid exposures, the chemical or chemical mixture of interest is placed directly in the challenge portion of the test cell and left in contact with

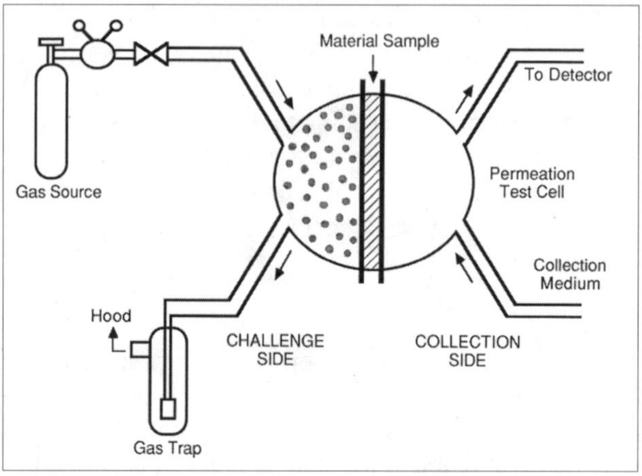

Figure 12. Configuration of permeation test system for evaluating gaseous chemical challenges.

the material specimen for the selected duration of the test.

ASTM F 739 and ASTM F 1383 also permit testing with gases, using the modifications to the test cell as described earlier and the test system configuration pictured in Figure 12. When testing gases at 100% concentrations, time zero in the test is established by passing 5 volumes of the gas through the challenge chamber within a 1-minute period. Following this initial period, the flow of gases is reduced to a rate that ensures the concentration of gas in the chamber does not change with time. Special considerations are needed for testing of gases to ensure integrity of the test cell and proper disposal of the effluent challenge gas.[21,22]

Permeation testing may also be conducted against vapors of liquid chemicals per ASTM F 739. These tests require a high level of temperature control to achieve consistency in the vapor concentration of the chemical and a different orientation of the test cell.[21] Most tests are performed with the chemical at its saturated vapor pressure at the test temperature using a test system configuration as shown in Figure 13. The test cell must be maintained in a horizontal position such that the air space above the liquid becomes saturated with vapor and the liquid does not contact

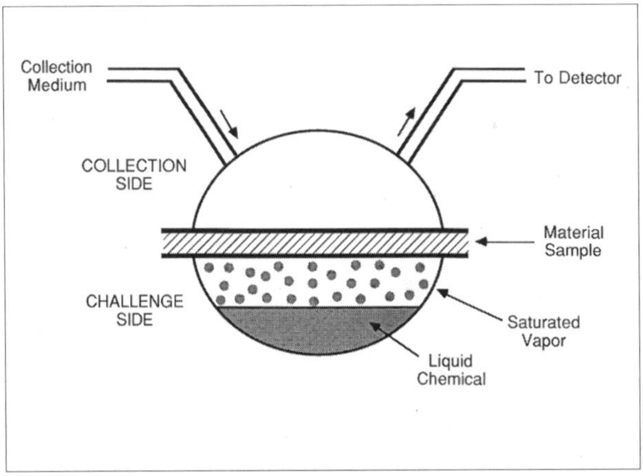

Figure 13. Configuration of permeation test system for evaluating chemical vapor challenges.

the material specimen. Unfortunately, tests conducted in this matter may incur a relatively long transition time based on volatility of the test chemical. Table 8 shows representative permeation data for two different material–chemical combinations as liquids and vapors.

Some research has also been reported for conducting permeation tests with solids. Lara and Drolet[23] describe a modified test cell where a gel-containing nitroglycerin was placed on the surface of the material's external surface for permeation testing.

Intermittent Contact Testing. Intermittent forms of chemical contact akin to splash-like exposures are prescribed in ASTM F 1383. In this test method, the time of material specimen exposure to chemical is varied in a periodic fashion. Chemical is charged into the challenge side of the test cell and then removed after a specified time. This type of exposure may be repeated in a cyclic fashion. ASTM F 1383 suggests three different exposure conditions as shown in Table 9.

The use of intermittent exposure conditions gives rise to permeation curves with a cyclic appearance (see Figure 14). As a consequence, breakthrough time with cumulative permeation is reported in lieu of permeation rate for these tests. Schwope

Table 8:
Permeation Data for Chemical Vapors for Selected Material-Chemical Combinations

Material-Chemical Combination	Chemical Challenge	Breakthrough Time (Minutes)	Permeation Rate (µg/cm2min)
Ethylene Dichloride against PVC glove	Saturated vapor @ 27°C	4	>25,000
	10 ppm in Nitrogen	4	350
Dichloromethane against Viton-Butyl suit material	Liquid	16	470
	Saturated vapor @ 27°C	28	280
	100 ppm in Air	No BT	Not applicable

Abbreviation: No BT-No breakthrough observed in 3-hour period for testing per ASTM F739.

Source: Adapted from reference 39.

Table 9:
ASTM F 1383 Recommended Exposure Conditions

Condition	Contact Time, Min	Purge Time, Min	Number of Cycles
A	1	10	10
B	5	10	7
C	15	60	3

et al.[24] illustrated this behavior for a number of material–chemical combinations and found the cumulative permeation to be proportional to the relative exposure time. Man et al.[25] compared permeation breakthrough times of protective clothing materials against specific chemicals using liquid contact, liquid splashes, and vapors. Their findings showed significant differences between the different exposure

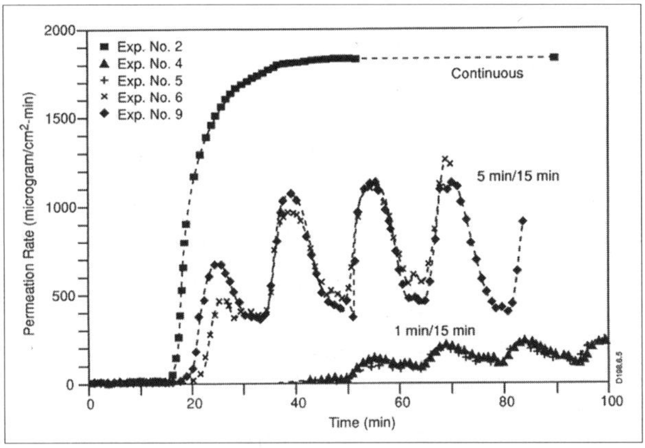

Figure 14. Cyclic permeation observed during permeation test involving intermittent contact (From Ref. 24).

conditions for some combinations of materials and chemicals, but lesser changes in breakthrough time for other material–chemical sets. They postulated that the different wetting characteristics of the test materials contributed to this phenomenon indicating that materials easily wet by a chemical may show similar permeation for liquid splash exposures as with continuous liquid exposure.

Types of Collection Media. The collection medium must have a high capacity for the permeating chemical(s), allow ready mixing, be readily analyzed for the chemical(s) of interest, and have no effect on the clothing material being tested.[26] A collection medium's capacity refers to the relative amount of chemical that can be collected. Collection media with low capacities will hinder the detection of permeation, showing lower than actual permeation rates. Schwope et al.[20] recommends that the concentration of the permeant in the collection medium, at the clothing/collection medium interface and in the bulk of the collection medium be maintained below 20% of the solubility of the permeant in the collection medium when the challenge is a neat chemical.

Air, nitrogen, helium, and water are common collection media. In general, these collection media have no effect on the clothing material and are amenable to most analytical techniques. Fricker and Hardy[27] developed a test method involving a saline collection medium to simulate sweat on skin.

In cases where the test chemical has a relatively low vapor pressure, gaseous collection media may have inadequate capacity for the permeant requiring a different choice of collection medium. Similar concerns arise for test chemicals having low water solubility when water is the collection medium. These situations may be addressed by circulating large volumes of fresh gas or water through the collection chamber; however, this practice will dilute permeant concentration in the collection medium and reduce test sensitivity.

Some test chemicals exhibit both low water solubility and vapor pressure. Chemicals with these characteristics generally include higher molecular weight chemicals such as polynuclear aromatics, polychlorinated biphenyls (PCBs), and some pesticides. One approach for conducting permeation tests with these chemicals has been to use solid collection media.[24,28] This technique involves placing a solid, highly absorbent film directly against the material specimen. Ehntholt et al.[28] designed a special test cell, successfully using a silicone rubber material for collection of pesticides. Unfortunately, this technique has also been reported to be very labor-intensive, involved with multiple replicates to determine breakthrough times, and subject to cross-contamination.[26] In addition, swelling of the test material can prevent uniform contact of the specimen with the solid collection medium. An alternative approach advocated by Pinnette and Stull[26] and Swearengen et al.[29] has been the use of a liquid splash collection. In these approaches, a solvent media is briefly contacted with the material specimen on the collection side and the extract evaluated for the chemical(s) of interest. If such an approach is followed, it first must be demonstrated that the solvent does not affect the barrier properties of the test material. For example, the absorbance and back diffusion of the solvent from the collection medium into the clothing material could swell or soften the material and thereby promote more rapid permeation of the challenge chemical. A third approach was used by Spence[30] for permeation testing with halogenated pyridines. His technique employed a method for concentrating the permeant in the collection medium by use of a trap built with the detector gas chromatograph.

Agitation of Collection Media. Agitation of the collection medium is recommended to ensure that is it is homogeneous for sampling and analytical purposes, and to prevent or minimize concentration boundary layers of permeant at the interface of the clothing material and the collection medium. It may or may not be needed depending on the medium chosen, the capacity of the medium for the test chemical(s), and the nature of the chemical challenge. ASTM F 739 specifies that a range of flow rates from 50 to 150 mL/min (ISO 6529 specifies a rate of 520 mL/min for gaseous collection media and 260 mL/min for liquid collection media). Mixing is particularly important for closed-system testing. In an open-loop system, flow of the collection medium is usually considered acceptable for test systems where the collection medium has a high capacity. Agitation of the collection medium is recommended when the concentration of the permeant is above 20% of the partition equilibrium at the clothing/collection medium interface.[20] Testing with multicomponent chemical mixtures may also require agitation to maintain a constant concentration of the challenge at the mixture/material specimen boundary.

Frequency of Sampling. Permeant sampling of the collection medium may be either continuous or discrete. Some permeation test system configurations permit continuous sampling of permeant in the collection media. In these systems, the permeant concentration or permeation rate can be constantly determined directly providing the permeation curve for the test system. For closed-loop permeation systems, nondestructive analytical methods must be used for continuous sampling. Other system configurations or test parameters dictate discrete sampling at periodic intervals dependent on the analytical detector chosen and its ability to recover prior to a new analysis. Alternatively, discrete samples may be taken for later analyses that affect the total number of separate analyses to be performed. Discrete sampling schemes may also be necessary to accommodate test systems that involve multiple test cells operated in parallel. Discrete samples influence the determination of breakthrough time, since breakthrough time must be reported at the time of the previous interval when permeant is found in a sample. For example, if discrete samples are taken at 5, 10, and 15 minutes, and a permeant is detected in the 10-minute sample, the permeation breakthrough time is 8 minutes in following the procedure established in ASTM F 739. More frequent sampling may show the permeation breakthrough time to lie sometime between 10 and 15 minutes.

Detection Strategies. The method for monitoring the collection medium for the permeating chemical is selected by the tester. In selecting an analytical detection method, the tester must consider its sensitivity and selectivity as well as its compatibility with the collection medium. Breakthrough time is totally dependent on the sensitivity of the detector. Breakthrough time has been analytically defined as a detector response twice the background level of the system. Background levels can be determined using a blank test cell, one that contains material, but not challenge chemical.

The analytical method by itself may be very sensitive, but the sensitivity of the permeation test can be orders of magnitude less. For example, consider two tests performed with the same analytical instrument and the detection limit but with different flow rates of collection media. The same permeation rate will produce a lower concentration of permeating chemical in the stream with the higher flow rate. Breakthrough will be detected at a later time for this test. In the extreme, if the flow rate were very high relative to the permeation rate, it is possible that breakthrough would go undetected. Figure 15 shows how different analytical sensitivities affect breakthrough time.

The detection of chemical breakthrough is highly dependent on the sensitivity of the analytical method. For this reason, ASTM F 739 requires that the analytical sensitivity must be reported along with breakthrough time. Reporting of analytical sensitivity alone, however, is insufficient to allow interpretation of test results. The analytical sensitivity of the detector may have little or no relevance to test method sensitivity, which is defined by the analytical sensitivity, the surface area of the

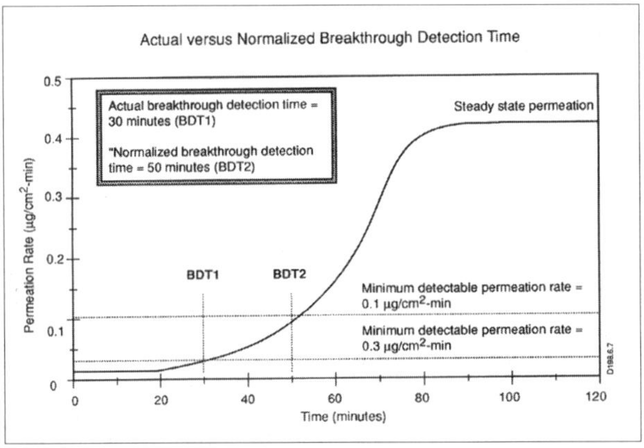

Figure 15. Impact of analytical sensitivity on interpretation of breakthrough time in permeation testing.

clothing material sample, and the collection medium flow rate (open-loop systems) or volume (closed-loop systems).

In a simplistic analysis, the sensitivity of a gravimetric system, such as one described in ASTM F 1407, shows the effect of these variables. In this case, the sensitivity of the balance, exposed surface area, and time interval between measurements directly provide a means for determining the limited sensitivity of the test method using the following equation:

$$\textit{Minimum detectable rate} = \frac{W}{(A \times t)} \qquad (6)$$

where W is the balance sensitivity in mass unit, A is the material surface exposure area, and t is the time between test measurements. Mickelsen et al.[31] showed good correlation with a gravimetric-based permeation test technique[32] as compared with gas chromatography using an infrared detector.

When analytical test measurements are made as part of the test procedure, more sophisticated procedures must be used to determine the sensitivity of the entire test system. ASTM F 739 specifies a technique for measuring minimum detection limits (or minimum detectable permeation rates) based on a modified form of the test method itself. Verschoor et al.[33] developed a technique where a specialized permeation test cell has a collection side with an additional port and aluminum foil replaces the material sample. No chemical is placed on the challenge side. The test chemical is injected near the surface of foil at a known rate using a syringe pump and the detector response is determined and compared with a known detector calibration gas response. The concentration of chemical injected is successively increased to identify the lower limit of detection. The ratio of the detector response of the test chemical with the detector response of the calibration gas can then be used in subsequent testing with the selected chemical. This approach requires determining the test sensitivity for each chemical.

Types of Detectors. A variety of detectors have been applied in tests for measuring the permeation resistance of materials. Since most inorganic chemicals involve some ionic potential, water is used as the collection media with pH meters, ion-specific electrodes, atomic absorption, or ion chromatography used as the detectors. These systems are operated in a closed-loop mode.

Permeation tests involving volatile organic chemicals usually employ gas chromatograph detectors such as thermal conductivity, flame ionization, electron capture, or photoionization detectors. In most cases when gas chromatography is used, it is not necessary to use the column unless mixture permeation studies are being performed. Because these detectors are destructive in their analysis of the sample, they can only be used in an open-loop mode or closed-loop if samples are withdrawn without replenishment of the collection medium. Perkins and Ridge[34] first described the use of infrared spectroscopy in permeation tests using a closed-system test configuration. The advantage of this system is that it allows continuous recirculation of the collection medium through the detector and test cell. Similar systems have been used by Berardinelli et al.[19]

In tests involving nonvolatile, non-water-soluble chemicals, wipe samples of material interior surfaces, solid collection media, or liquid splash collection aliquots are generally evaluated using either gas chromatography, high performance liquid chromatography, or mass spectroscopy depending on the collection/extraction solvent used and the analytes being detected. Ehntholt et al.[28] used radiochemical-labelling techniques for evaluating pesticide concentrations in isopropanol. Other techniques have been shown viable for difficult to evaluate chemicals by using small amount of collection medium[35] with UV spectroscopy.

It is important that the detector response remain linear within the range of chemical concentration to be evaluated. In some systems, rapid permeation at high rates can saturate the detector and provide meaningless data. Some laboratories have used sorbent tubes for collecting permeating chemical. This approach can also be used to determine the total or cumulative permeation when an open-loop test system is chosen, and allow for the separation and identification of specific components within challenge mixtures.

Effects of Temperature. Spence[36] first showed significant changes in the permeation resistance of protective clothing materials with increasing temperature as evidenced by shorter breakthrough times and larger permeation rates. Changes in temperature may have an influence on permeation by several mechanisms. Increased temperatures may increase the concentration of the challenge chemical adsorbed onto the material surface by increasing the solubility of the material–chemical matrix or by increasing the vapor pressure of the chemical.[37] The rate of diffusion step in the permeation process may also increase with temperature following an Arrhenius equation type of relationship.[36–40] Temperature, therefore, exhibits its effect on breakthrough time and permeation rate through the diffusion coefficient (D) and solubility (S). The expected effect manifests itself in a logarithmic-like relationship between permeation rate and temperature. Figure 16 shows this relationship for several material–chemical pairs and temperatures. Even small differences in temperature have been shown to significantly affect permeation breakthrough times as shown in Table 10. As a consequence, permeation testing must be performed under tightly controlled temperature conditions.

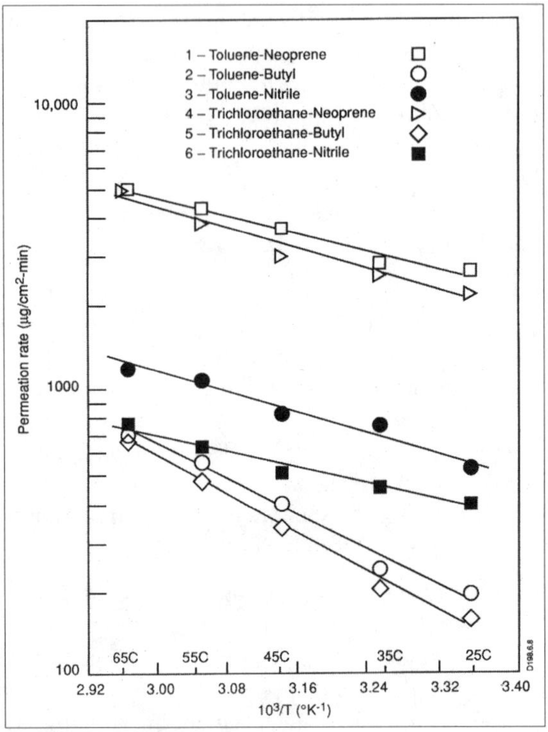

Figure 16. Plot showing effect of temperature on permeation rate for selected material-chemical combinations. (Reprinted with permission; copyright ASTM International.)

Table 10: Preliminary Indications of Temperature Effects on Breakthrough Time

Test Material	Temperature (°C)	Acetone Breakthrough Time (Minutes)
Viton/chlorobutyl laminate	20	95–98
	26.5	43–53
Chlorinated polyethylene	22	32–35
	24.5	27–31

Source: Reference 37.

Effect of Multicomponent Challenges. When permeation tests involve multicomponent chemical challenges, test configurations must employ detection techniques that permit the identification of each chemical in the permeating mixture. A number of researchers have investigated the effects of multicomponent chemical mixture permeation through barrier materials. Stampher et al.[41] investigated the permeation of PCB/paraffin oil and 1,2,4 trichlorobenzene mixtures through protective clothing. They used a small amount of isooctane in the collection medium to capture permeating PCBs. Schwope et al.[42] performed extensive testing with pesticides using different active ingredients and carrier solvents. Their tests demonstrated different breakthrough times and proportions of permeating chemicals between pesticide and carrier solvent. Bentz and Man[37] identified a case involving an acetone/hexane mixture where the mixture permeated a dual elastomer-coated material at shorter breakthrough times than either of the pure components. This testing illustrated the potential synergistic permeation of mixtures. Mickelsen, Roder, and Berardinelli[43] evaluated elastomeric glove materials against three different binary mixtures and found similar permeation behavior where mixture permeation could not be predicted on the basis the individual mixture components. Ridge and Perkins[44] attempted to model mixture permeation using solubility parameters and found the technique to be only partially successful. Goydan et al.[45] were able to predict mixture using a series of empirical rules when applied to a particular fluoropolymer laminate material.

Evaluation of Adsorptive-Based Materials. Some materials may be designed to prevent the penetration or permeation of vapors by using adsorptive components within their structure. In general, these materials are intended for use in environments where only low levels of chemical concentration are to be encountered. The evaluation of these materials may be conducted in a similar fashion as described above for gases or chemical vapors. However, these fabrics are typically evaluated by passing an air stream through the material in a technique similar to that used for assessing the service life of sorbent-based respirator cartridges.[46–48] One such technique is described by Baars, Eagles, and Emond.[49]

Their technique involves measurement of fabric adsorption rate and capacity by passing the challenge gas at a constant flow rate and concentration through the material with continuous detection of the effluent chemical concentration. When the concentration of the effluent is normalized to the inlet chemical concentration, the area above the curve for normalized concentration versus time represents the cumulative chemical adsorbed while the area under the curve is that chemical which has penetrated the fabric. Fabric saturation is reached as the normalized concentration approaches 1.

Use and Interpretation of Permeation Testing

Of the chemical resistance data used in reporting protective clothing performance, the vast majority of test results are permeation resistance data. These data accompany most product performance data sheets and are provided in a number of data compilations.[50–53] However, a review of these data generally indicates that consistent comparison cannot be made and that suppositions for discerning material performance may not be properly based unless an understanding of the test conditions is realized. Much of the data are reported generically for material classes. Yet, Michelsen and Hall[54] showed significant differences in chemical permeation through elastomers that were generically the same in composition and thickness. This illustrates that permeation data must be specific to the material–chemical combination being evaluated.

Reporting of Permeation Data. As originally indicated, breakthrough time and steady-state or maximum permeation rate are typically provided as permeation test data. ASTM F 739 as well as CEN and other test methods also require reporting of key test parameters. In general, these include a complete description of the test material, test chemical, and test system configuration. Table 11 lists test parameters that should be reported with each test. ASTM F 1194, *Guide for Documenting the Results of Chemical Permeation Testing on Protective Clothing Materials*, provides a more extensive list of testing reporting requirements.

Table 11:
Permeation Test Report Parameters

Area	Reporting Requirements
Test chemical	Components
	Concentration
	Source
Test material	Identification
	Source
	Condition at time of testing
	Thickness
	Unit area weight
Test system	Overall configuration (open or closed loop
	Type of test cell
	Type of challenge (continuous or intermittent)
	Collection medium
	Collection medium flow rate
	Detector or analytical technique
Test results	Breakthrough time
	Normalized breakthrough time
	Test system sensitivity
	Steady-state or maximum permeation rate
	Cumulative permeation

Since sensitivity significantly affects breakthrough times, ASTM, CEN, and ISO have adopted reporting requirements that are intended to normalize the effect of test system parameters on this measurement. Currently, ASTM F 739 and ASTM F 1383 specify reporting of the "normalized" breakthrough time in addition to actual breakthrough time. Normalized breakthrough time is defined as the time when the permeation rate is equal to 0.10 ìg/cm^{2}min. International and European test methods specify reporting breakthrough times at the time the rate equal 1.0 ìg/cm^{2}min, a single order of magnitude difference with U.S. test methods. Therefore, it is important that permeation breakthrough time data be only compared when the respective sensitivities of the test laboratories are the same or if data are normalized on the same basis.[55]

Interpretation of Permeation Data. Permeation resistance testing is the appropriate test when vapor protection is required. This does not mean that the test can only be applied for gas or vapor challenges, but rather that the test discriminates among chemical hazards at a molecular level owing to the sensitivity for detecting permeating chemical in its vapor form (as opposed to liquids or solids). As such, permeation testing represents the most rigorous of chemical resistance test approaches.

Within the protective clothing industry, many end users judge the acceptability of a material on the basis of how its breakthrough time relates to the expected period of exposure. Reporting of permeation rate offers a more consistent and reproducible means of representing material permeation. The inherent variability and test system dependence on breakthrough times make this data a less than satisfactory choice for characterizing material performance. Permeation rate data can be used to show subtle changes in material characteristics and determine cumulative (total) permeation when acceptable "dose" levels of the test chemical can be determined. On the other hand, some material–chemical systems take a long time to reach steady-state or exceed the capacity of the detector. In addition, the lack of widespread data on acceptable dermal exposure levels for most chemicals leads many specifiers to rely on breakthrough times exclusively.

The flexibility of most permeation tests allow testing laboratories to choose those conditions that best represent the expected performance of the material. Usually, the primary decisions in specifying permeation tests involve the following:

- the chemical and its concentration,
- the state and periodicity for contacting the chemical with the material,
- the material and its condition prior to exposure,
- the environmental conditions of the exposure,

- the length of the test, and
- sensitivity of the test system.

The majority of permeation tests in the protective clothing industry are conducted using neat chemical continuously contacting pristine material at room temperature for a period of 8 hours. Test sensitivities are at 0.10 ìg/cm^{2}min or better but may be higher for difficult to evaluate chemicals. Other barrier materials are generally evaluated against chemicals for longer period of times at slightly elevated temperatures for examining steady-state permeation rates and cumulative permeation. These test conditions are considered worst case, because constant contact of the material with the chemical is maintained, which may or may not be representative of actual use. When specific barrier product applications are identified, it is best to model the conditions of use through the selection of test parameters. If general performance is to be determined, using industry practices for test set-up are preferred so that material performance may be compared against other available data.

Overall Product Integrity

Product integrity refers to the ability of entire clothing systems to prevent inward leakage of chemicals, whether in gaseous, liquid, or particulate forms. Tests for measuring product integrity compliment material barrier tests because the clothing design has a significant effect on the overall protection to the end user. A material with good barrier characteristics against chemical vapors in clothing with a poor design (in terms of vapor or gas-tight integrity) will still result in exposure. Thus, product integrity tests evaluate the other parts of clothing such as:

- seams,
- closures,
- interfaces with other clothing (such as sleeve ends with gloves and trouser cuffs with footwear), and
- interfaces with other equipment (such as with a respirator).

Overall product integrity performance testing can also determine how well protective clothing prevents substances from leaving the clothing to contaminate a sensitive environment such as in clearness for electronics assembly or pharmaceutical manufacturing. Linting by clothing or the release of particles from the wearer can both be damaging to certain processes. In these cases, the clothing provides a shield to the environment from the person, and the clothing itself cannot contribute to particle release.

As with material chemical resistance, there are three principal types of overall product integrity testing:

- Particulate-tight integrity,
- Liquid-tight integrity, and
- Gas-tight integrity.

Particulate-Tight Integrity

Particle-tight integrity tests determine if particles enter or leave whole items of protective clothing.

In this testing, protective clothing is either placed on a manikin or worn by a test subject. The manikin or test subject is then exposed to a particulate atmosphere, usually an aerosol formed by a nontoxic, easily detectable liquid. Human subjects usually perform a series of exercises to put stress on the garment. The particulate-contaminated atmosphere and atmosphere inside the protective clothing are sampled either to:

- Determine the levels of particle intrusion inside the protective clothing, or
- Measure rates of particle release from the protective clothing when worn inside a "clean" chamber.

No current standardized tests exist for measuring the intrusion of particles (or aerosols) into protective clothing. A number of informal test methods have been established in the United States that use corn oil aerosol in a chamber to measure intrusion into the PPE by sampling air. In Europe and internationally, ISO 13982-2 (Protective clothing for use against solid particulate chemicals—Part 2: Test method of determination of inward leakage of aerosols of fine particles into suits) has been proposed as a test method. This test method uses a salt (sodium chloride) aerosol that is introduced into a chamber with a test subject wearing the protective clothing to be evaluated. Sampling of air to determine salt concentration, both inside and outside the clothing, is used to determine clothing effectiveness against particles. Garment performance is usually characterized in terms of "intrusion coefficient," which is the ratio of the outside contaminant concentration to the concentration of contaminant measured on the inside. Higher intrusion coefficients indicate garments with better resistance to penetration by particles.

Garments and gloves may also be evaluated for particle release in cleanroom applications using one of two methods:

1. Garments are tumbled in a special apparatus, which is flushed with clean air from a laminar flow hood. The air is then monitored for particle density.
2. Gloves or finger cots are immersed in water and the water is analyzed for particulate content.

Particle release methods are provided in recommended practices established by the Institute of Environmental Sciences (IES):

- IES-RP-CC003.2 (for cleanroom garments)
- IES-RP-CC005.2 (for cleanroom gloves and finger cots).

Results for particle release are given in particle counts by size on the basis of sampled air volume or PPE item surface area. Low particle counts for particulate release methods indicate "cleaner" garments for use in particle-sensitive environments.

Liquid-Tight Integrity

Liquid integrity tests determine if liquid enters to the interior side of the protective clothing or onto wearer underclothing. Methods for measuring liquid-tight integrity involve spraying protective clothing on a manikin with a liquid and observing penetration of liquid onto an inner liquid-absorptive garment or the interior of the clothing. Water often is treated with either surfactant to lower surface tension and better simulate organic liquids, and/or dyes to enhance detection of penetrating liquid.

In general, tests may be conducted at periods longer than expected exposure to assist in observing leakage. Several techniques can be used to quantify leakage such as in the use of spectophometry on dyes-based liquid challenges and electroconductivity for salt-based liquid challenges. However, the majority of testing is performed with visual detection of liquid leakage. Most results are therefore reported as detected penetration in terms of passing or failing performance. Observations of areas where leakage occurred may also be reported to help determine problems in the product that limit the clothing item's integrity. Smaller scale tests are possible on gloves and boots. Standard Test Methods include:

American Society for Testing and Materials (ASTM)

ASTM D 5151, Detection of Holes in Medical Gloves (for gloves)

ASTM F 1359, Liquid Penetration Resistance of Protective Clothing or Protective Ensembles Under a Shower Spray While on a Manikin (for garments)

European Committee on Standardization (CEN) Test Method

EN 374-2, Protective gloves against chemicals and microorganisms—Part 2: Determination of resistance to penetration (European method for gloves)

EN 463, Protective clothing—Protection against liquid chemicals—Test method: Determination of resistance to penetration by a jet of liquid (European method using liquid jet)

EN 468, Protective clothing—Protection against liquid chemicals—Test method: Determination of resistance to penetration by spray (European method using liquid spray)

International Standards Organization (ISO) Test Method

ISO 17491, Protective clothing—Protection against gaseous and liquid chemicals—Determination of integrity of protective clothing to penetration by liquids and gases (Procedures C, D, E, and F)

Table 12 provides a comparison of these methods. The four different standards for evaluating garments provide slightly different test approaches for measuring liquid-tight integrity:

- ASTM F 1359 involves placing clothing on a manikin and spraying the manikin with surfactant-treated water from five different nozzles. The volumetric flow out of each nozzle is specified. In various applications of this test method, the liquid is sprayed at the manikin from 20 minutes to 1 hour. During the exposure period, the manikin is rotated through four orientations to completely challenge the garment. An inner liquid-absorptive garment worn on the manikin underneath the clothing is used for detecting liquid penetration. The garment interior can also be examined. Test results are recorded as "pass" or "fail."
- EN 468 is similar to ASTM F 1359, in that it involves surfactant-treated water that is sprayed at the protective clothing in a specified pattern over a specified duration. However, this test involves a human test subject (who wears a liquid-absorptive garment and performs a series of exercises) and requires the use of a dye. Performance of the garment is determined by measuring the total area of dye staining on the liquid-absorptive garment and relating the area of a "calibrated" stain representing a specific volume of penetrating liquid.
- In EN 463, an aqueous jet, containing a fluorescent or visible dye tracer, is directed under controlled conditions at chemical protective clothing worn by a test mannequin or human test subject. Inspection of the inside surface of the clothing and the outside surface of absorbent clothing worn underneath allows any points of inward leakage to be identified.

ISO 17491 covers all three approaches and provides criteria for applying each test:

- Method C specifies a method for determining the

Table 12:
Characteristics of Liquid-Tight Integrity Tests for Complete Chemical Protective Clothing

Test Method	Clothing Placement	Test Liquid	Liquid Application	Method of Detection
ASTM F 1359 (also ISO 17491, Procedure E)	Clothing placed on manikin that is dressed in liquid-absorptive garment	Water treated with surfactant to 32 dynes/cm	Liquid sprayed from 5 nozzles positioned above and to sides of test clothing at 3.0 L/min from each nozzle for 60 minutes; mannequin is rotated through 4 positions during test	Interior of garment and liquid-absorptive garment is examined for signs of liquid penetration; results are pass/fail
EN 463 (also ISO 17491, Procedure C)	Human subject wears test clothing over liquid-absorptive garment; alternatively clothing may be placed on manikin	Water treated with fluorescent dye	Liquid directed at specific locations on test clothing at pressure of 180 kPa at distance of 1 m for 5 seconds; if human subject is used, subject remains stationary	Staining of liquid-absorptive garment is compared with calibrated stain (0.1 mL); approximate area of staining is reported
EN 468 (also ISO 17491, Procedure D)	Human subject wears test clothing over liquid-absorptive garment; subject stands on turn-table	Water treated with surfactant to 30 dynes/cm and fluorescent dye	Liquid sprayed from 4 nozzles positioned vertically on test stand beside test subject (at distance of 1.5 m); liquid is sprayed at 1.14 L/min from each nozzle for 1 minute; during spray, subject performs stationary exercises	Staining of liquid-absorptive garment is compared with calibrated stain (0.1 mL); approximate area of staining is reported
ISO 17491 Procedure F	Human subject wears test clothing over liquid-absorptive garment; subject stands on turn-table	Water treated with surfactant to 52.5 dynes/cm and fluorescent dye	Same as EN 468 (or Procedure D of ISO 17941) except that flow rate is at 0.47 L/min	Staining of liquid-absorptive garment is compared with calibrated stain (0.1 mL); approximate area of staining is reported

resistance of chemical protective clothing to penetration by jets of liquid chemicals (analogous to EN 463). This procedure is applicable to clothing that may comprise one or more items and that is intended to be worn where there is a risk of exposure to a forceful projection of a liquid chemical. It is also applicable to clothing that is intended to be resistant to penetration under conditions that require total body surface cover, but do not demand the wearing of gas-tight clothing. This procedure is not applicable to permeation of liquid chemicals through materials from which the clothing is made.

- Method D specifies a method for determining the resistance of chemical protective clothing to penetration by sprays of liquid chemicals (based on EN 468). This procedure applies to protective clothing that may comprise one or more items that are intended to be worn when there is a risk of exposure to slight splashes of a liquid chemical, or to spray particles that coalesce and run off the surface of the garment. It is also applicable to clothing that is intended to be resistant to penetration under conditions that require total body surface cover, but do not demand the wearing of gas-tight clothing. This procedure is not applicable to the permeation of liquid chemicals through the materials from which the clothing is made.
- Method E (based on ASTM F 1359) specifies an alternative method for determining the resistance of chemical protective clothing to penetration by sprays of liquid chemicals. This procedure applies to protective clothing that may comprise one or more items and that is intended to be worn when there is a risk of exposure to slight splashes of a liquid chemical or to spray particles that coalesce and run off the surface of the garment. It is also applicable to clothing that is intended to be resistant to penetration under conditions that require total body surface cover, but do not demand the wearing of gas-tight clothing. This procedure is not applicable to the permeation of liquid chemicals through the materials from which the clothing is made. Method E

differs from Method D in that it uses a static manikin instead of a test subject. Method E also uses a different spray configuration and duration (1 hour compared to 30 minutes for Method D). Lastly, Method E is based on a qualitative determination of observed liquid on the absorbent coverall or interior of the chemical protective clothing. In Method D, a 2-cm diameter or larger stain is used for pass/fail criteria.

- Method F is a modification of Method D where the spray has been modified to light spray or mist by use of different nozzles and spray conditions. This method is intended for partial body protective clothing where the likelihood of splash exposure is low.

Gas-Tight Integrity

Gas-tight integrity tests determine if gas can penetrate protective clothing. Gas-tight integrity testing can only be performed on items that can be sealed, including:

- Full-body garments (totally encapsulating suit)
- Gloves (non-air impermeable materials)
- Footwear (non-air impermeable materials).

The most common approach for testing gas-tight integrity of protective clothing is to inflate the item to a specified pressure and then observe for any change in pressure within the item after several minutes. Using a soapy water solution on exterior of the clothing can then identify the location of leaks. Alternative approaches involve placing PPE on a test subject in a closed environment containing a gas (such as ammonia) and measuring concentration of challenge agent inside suit. In some rare cases, full-scale testing of suits against actual hazardous materials has been performed in special projects. These tests may be modified for gloves and footwear.

American Society for Testing and Materials (ASTM) Test Method

ASTM F 1052, Pressure Testing of Gas-Tight Totally

Encapsulating Chemical Protective Suits (for totally encapsulating suits)

European Committee on Standardization (CEN) Test Method

EN 464, Protective clothing—Protection against liquid and gaseous chemicals, including liquid aerosols and solid particles—Test method: Determination of leak-tightness of gas-tight suits (European method for totally encapsulating suits)

International Standards Organization (ISO) Test Method

ISO 17491, Protective clothing—Protection against gaseous and liquid chemicals—Determination of integrity of protective clothing to penetration by liquids and gases (Procedure B2)

U.S. Government Regulations

Appendix A of OSHA 29 CFR 1910.120 (gas-tight and ammonia leakage tests for totally encapsulating suits)

Both ASTM F 1052 and EN 464 involve similar testing approaches where the protective clothing is inflated and left at a specific pressure. ASTM F 1052 uses a lower test pressure than EN 464, but the results are similar. Extensive work by Carroll[56] has demonstrated that a test pressure of 1 kPa (4 inches water gauge pressure) over a 4-minute period is sensitive enough to evaluate most encapsulating clothing for very small leaks that could contribute to poor integrity.

ISO 17491 and the procedures in OSHA 29 CFR 1910.120 involve a challenge environment where a test subject wears the clothing inside a closed chamber in contact with a test agent. The subject performs a series of exercises while wearing the clothing. The interior of the clothing is either measured during or after the exposure to determine the ratio of the concentration on the outside to the ratio of the concentration of test agent on the inside. This ratio, known as the intrusion coefficient, is used to measure clothing performance similar to the way protection factors are used to judge the performance of respirators.

Physical Properties

A number of other properties are important for judging the effectiveness of chemical protective clothing selections.

Physical properties assess or determine:

- Weight and thickness of clothing and materials,
- Strength of clothing,
- Resistance to specific physical hazards, and
- Product durability.

A number of these properties apply to different types of materials, which may include:

- Woven textiles,
- Knit textiles,

Nonwoven textiles (created by nonweaving or knitting processes),

- Rubber,
- Plastics, and
- Coated fabrics.

In general, more than one test method may be available for a performance property based on the type of material or product. In addition, many material physical property results are reported in different material directions. Results that are parallel to the direction the material is fabricated or comes off the roll are referred to as warp direction for woven textile fabrics; course direction for knit textile fabrics; and machine direction for nonwoven textile, rubber, and plastic materials. Results that are perpendicular to the direction the material is fabricated or comes off the roll are referred to as fill direction for woven textile fabrics; wales or weft direction for knit textile fabrics; and cross-machine direction for nonwoven textile, rubber, and plastic materials.

The multitude of different physical properties prevents detailed coverage of this performance area in this chapter. Typical physical properties applied to chemical protective clothing include:

- Weight—reported as the weight of a material per unit area.
- Thickness—reported as the nominal or average thickness of a material in mils (thousands of an inch) or millimeters.

- Breaking strength—measures the force required to break clothing materials or components when items are pulled along one direction; breaking strength tests can also be used to measure the strength of seams and closures.
- Burst strength—measures the force or pressure required to rupture clothing materials or components when a force is directed perpendicular to the item; this property may be related to ability of PPE materials to prevent items from protruding through garments.
- Tear or snag resistance—measures the force required to continue a tear in a clothing material once initiated or the resistance of a material in preventing a tear or snag from occurring.
- Abrasion resistance—measures the ability of clothing surfaces or materials to resist wearing away when rubbed against other surfaces.
- Cut resistance—measures the ability of clothing items or materials to resist cutting through by sharp-edged objects or machinery.
- Puncture resistance—measures the ability of clothing items or materials to resist penetration by a slow-moving, pointed object.
- Flex fatigue resistance—measures the ability of clothing items or materials to resist wear or other damage when repeatedly flexed.

Table 13 provides representative test methods for many of the above listed physical properties.

For each physical property reviewed, it is important to understand how the test method relates to the intended use of the protective clothing and how to interpret the measurement. For some properties, increasing values generally mean better performance. But physical properties that are too high may also create a tradeoff for such properties as stiffness or aspects of clothing related to use.

Table 13: List of Representative Physical Performance Test Methods

Physical property	Type of test/material application	Available test methods
Weight	Textiles	FTMS 191A,5041
	Woven textiles	ASTM D 3776 ISO 3801
	Nonwoven textiles	ASTM D 1117 ISO 9073-1
	Coated fabrics	ASTM D 751 ISO 2286
Thickness	Textiles	FTMS 191A,5030
	Woven/knit textiles	ASTM D 1777 ISO 5084
	Nonwoven textiles	ASTM D 5736 ISO 9073-2
	Coated fabrics	ASTM D 751 ISO 2286
	Plastics, rubber, foam	ASTM D 1565 ASTM D 3767
Breaking strength	Grab method – textiles	ASTM D 5034 EN ISO 13934-2 ISO 5082
	Strip method – textiles	ASTM D 5035 EN ISO 13934-1 ISO 5081
	Nonwoven textiles	ISO 9073-3
	Rubber/coated fabrics	ASTM D 412 ASTM D 751 ISO 1421
	Plastic	ASTM D 638 ASTM D 882 ISO 1421
	Seam strength	ASTM D 751 ASTM D 1683
Burst Strength	Ball method – textiles	ASTM D 3787 ASTM D 3940
	Ball method – rubber or plastic	ISO 3303
	Diaphragm method – textiles	ASTM D 3786
	Diaphragm method – rubber or plastic	ISO 2960
	Coated fabrics	ASTM D 751
Flex fatigue	Flat material	ASTM F 392 ISO 7854
	Cut growth – rubber	ASTM D 3629

Table 13 *(continued)*

Physical property	Type of test/material application	Available test methods
Tear Resistance	Elmendorf method – textiles	ASTM D 1424 ISO 13937
	Elmendorf method – plastics	ASTM D 1922
	Tongue tear – textiles	ASTM D 2261 ASTM D 2262 ISO 13937
	Tongue tear – nonwoven textiles	ASTM 5735
	Tongue tear – plastics	ASTM D 1938
	Trapezoidal method – textiles	ASTM D 5733 ISO 9073-4
	Tear resistance – plastics	ASTM D 1004 ISO 4674
	Tear resistance – coated fabrics, rubber	ASTM D 751 ISO 4676
	Puncture propagation tear resistance	ASTM D 2582 ISO 13995
Abrasion resistance	Taber method – textiles	ASTM D 3884
	Taber method – coated fabrics	ASTM D 3389
	Taber method – plastics	ASTM D 1044
	Wyzenbeek method – textiles	ASTM D 4157
	Martindale method – clothing materials	ASTM D 4966 EN 530 ISO 12974
	Martindale method – gloves	EN 388, Clause 6.1
	Flexing/abrading method – textiles	ASTM D 3885
	Flexing/abrading method – coated fabrics	ISO 5981
	Inflated diaphragm method – textiles	ASTM D 3886
	NBS method – rubber	ASTM D 1630
	Loose abradant – plastics	ASTM D 1242
	Uniform method – textiles	ASTM D 4158
Cut resistance	Dynamic method – gloves and clothing materials	ASTM F 1790 EN 388, Clause 6.2 ISO 11393
	Impact method – gloves	EN 388, Clause 6.5
Puncture resistance	General method	ASTM F 1342 EN 863 ISO 13996
	Gloves	EN 388, Clause6.4
	Footwear	ANSI Z41.1, Section3.5

Human Factors

Human factors describe how PPE affects the wearer in terms of functionality, fit, comfort, and overall well-being. Most human factor properties represent tradeoffs with protection, for example, barriers to chemicals versus thermal comfort. Common human factor properties for chemical protective clothing include:

- Material compatibility—evaluates the potential for skin irritation or adverse reactions due to contact with certain substances that may be present in or on clothing.
- Thermal insulation and breathability—evaluate the ability of protective clothing materials to allow the passage of air, moisture, and the heat associated with body evaporative cooling and environmental conditions.
- Mobility and range of motion—evaluate the effects of PPE on wearer function in performing work tasks.
- Hand function—assesses the effects of gloves on specific functions of the hand used to perform tasks or manipulations, such as dexterity (the ability for the individual glove wearers to manipulate small or large objects with their hands), tactility (the ability of the individual glove wearers to sense by touch differences in surfaces), and grip (the ability of the individual glove wearers to grasp objects under both dry and wet conditions).
- Ankle support—evaluates the ability of footwear to maintain support for the ankle under conditions of use.
- Slip resistance—measures the ability of footwear to maintain traction with a surface.
- Clarity and field of vision—evaluate the ability of an individual to see through a visor or a faceshield; field of vision testing evaluates peripheral vision for an individual wearing the visor or faceshield.
- Ease of communications—evaluates the ability of PPE to allow intelligible (understood) communications of the wearer to other persons.

- Sizing and fit—determines how well clothing fits the individual wearer.
- Donning and doffing ease—evaluate how easily or how quickly individuals can put on and take off protective clothing.

Most human factors are difficult to measure and must employ human subjects. For this reason, most human factors evaluations are conducted as wear trials where end users can rate or rank different properties related to function and comfort for different clothing items. In some cases, standardized test methods do exist for measuring human factors. Representative standard test methods for human factors described in this subsection are summarized in Table 14.

Recommended Testing Approaches

As products differ, so must the test strategies that are designed to evaluate barrier material characteristics of these products. Obviously, different tests are needed to evaluate different products based on how they are designed, what performance is intended, their required durability, and the expected application. The requirements for conducting barrier testing arise from several needs:

- demonstrating product distinctive advantages (marketing),
- meeting customer needs or demands,
- complying with appropriate standards,
- determining the viability of new products, or
- documenting product quality control.

While quality design and fabrication of product clothing items are paramount to offering a good product line, the majority of testing efforts are directed to evaluating the materials used in the construction of barrier products. This is

Table 14:
List of Representative Human Factors Test Methods

Human Factor	Type of Test/Application	Available Test Methods
Material biocompatibility	Medical device (gloves, gowns) biocompatibility tests	ISO 10993
	Natural rubber latex proteins	ASTM D 5712
	pH level in gloves	ISO 4045
Thermal insulation and breathability	Air permeability —materials	ASTM D 737
	Moisture vapor transmission—materials	ASTM E 96 CAN/CGSB-4.2 No. 49
	Thermal resistance —materials	ASTM D 1518 ISO 5085
	Thermal resistance —garments	ASTM F 1291
	Total heat loss—materials	ASTM F 1868 ISO 11092
Mobility and range of motion	Overall product evaluation	ASTM F 1154 NFPA 1991, Section 6-4 EN 943-1, Clause 7.2
Hand function	Gross dexterity	ASTM F 2010
	Fine dexterity	EN 420, Clause 6.3 NFPA 1999, Section 6-14
	Grip	NFPA 1971, Section 6-39
Clarity and field of vision	Clarity/field of vision	NFPA 1991, Section 6-4 EN 943-1, Clause 7.2
	Haze	ANSI Z87.1, Section 5.9 ASTM D 1003
	Transmittance	ANSI Z87.1, Section 5.10
Sizing and fit	Body measurements	ASTM D 5585 ASTM D 5586
	Garment sizing/dynamic fit test	ASTM F 1731 ANSI/ISEA 101
Donning and doffing ease	Full body suits	NFPA 1991, Section 6-4

ANSI—American National Standards Institute; ASTM—American Society for Testing and Materials; EN—European Norm; CAN/CGSB —Canadian General Standards Board; ISO—International Standards Organization; NFPA—National Fire Protection Association

partly because whole product performance may be difficult to assess in simulated "use" tests.

The use of standardized test methods provides advantages over in-house procedures. The latter procedures are often simple in design, are done internally, and are not always reproducible. Manufacturers tend to choose those methods that best represent their product's performance. Comparison of product performance on this basis is virtually impossible. Acceptance and use of ASTM and other recognized standards overcome these problems and help end users in their evaluation of product performance.

Selection of Test Methods to Characterize Barrier Materials

The appropriateness of a specific chemical resistance test is dependent on the product's application and expected performance. The test selected should also consider the nature of the material to be evaluated. Some materials should not be evaluated in certain barrier tests, because the methods do not allow for discrimination of their performance.[57]

- Degradation testing may show how product materials deteriorate or are otherwise affected, but will not always demonstrate retention of barrier characteristics with respect to specific chemicals. Degradation testing is most useful when retention of specific physical properties (e.g., strength) is desired or as a screening technique for other chemical resistance (barrier) tests. This type of testing may be applied to all types of materials.
- Runoff-based penetration testing should only be used if the wetting or repellency characteristics of material surfaces are to be evaluated. Like degradation testing, runoff-based penetration testing does not offer an adequate assessment of material barrier performance. This testing approach can be used with all types of

materials, but is best applied to textile materials or lightly coated fabrics for determining surface finish characteristics.

- Hydrostatic-based penetration tests are designed to evaluate water or chemical barrier performance of materials. Only ASTM F903 and ISO 13994 allow testing with chemicals other than water. As such, this test is appropriate for the evaluation of material performance against liquid chemicals and can be used to distinguish adequate chemical resistance for specific chemicals. This testing easily accommodates microporous and continuous film-based materials.

- Chemical permeation testing provides an assessment of a barrier material's total chemical resistance permitting the measurement of relatively small amounts of permeating chemical. This test is best suited when only extremely small levels of chemical are permitted to pass through a material. Permeation testing should be employed for any type of continuous barrier material.

In the protective apparel industry, the decision to apply permeation or penetration data therefore requires a careful review of the

- the hazards associated with the chemical;
- the intended duration of exposure; and
- the work environment.

There are many cases where materials that resist liquid penetration are suitable for working environments, particularly when chemical exposure is unlikely and there is relative little hazard from wearer contact with chemical vapors. Some sample situations include working with dilute acids and bases. Having both penetration and permeation data for specific clothing products and chemicals in combination with a thorough understanding of the chemicals' hazards and working environment allow the safety specialist to choose protective clothing that provides the needed level of performance. Table 15 provides sample data for three different materials including a microporous laminate. This testing

Table 15:
Comparison of Penetration and Permeation Resistance for Representative Liquid Splash-Protective Barrier Materials

Chemical	PVC/Nylon			Saranex/Tyvek Laminate			Microporous Film/Nonwoven Laminate		
	F903(C) Result	F739 B.T.	F739 P.R.	F903(C) Result	F739 B.T.	F739 P.R.	F903(C) Result	F739 B.T.	F739 P.R.
Acetone	Pass	8	>50	Pass	28	3.4	Pass	<4	>50
Acetonitrile	Pass	12	25	Pass	88	0.27	Pass	<4	>50
Carbon Disulfide	Fail (6)	4	>50	Pass	4	>50	Pass	<4	>50
Dichloromethane	Fail (6)	4	>50	Pass	4	>50	Pass	<4	>50
Diethylamine	Fail (20)	8	>50	Pass	20	20	Pass	<4	>50
Dimethylformamide	Fail (40)	28	>50	Pass	72	1.8	Pass	<4	>50
Ethyl Acetate	Pass	8	>50	Pass	20	1.5	Pass	<4	>50
Hexane	Fail (40)	20	8	Pass	None	N/A	Pass	<4	>50
Methanol	Fail (55)	16	13	Pass	None	N/A	Pass	<4	>50
Nitrobenzene	Pass	32	>50	Pass	120	6	Pass	<4	>50
Sodium Hydroxide	Pass	None	N/A	Pass	None	N/A	Pass	<4	>50
Sulfuric Acid	Pass	120	6	Pass	None	N/A	Pass	<4	>50
Tetrachloroethylene	Fail (30)	16	>50	Pass	128	1.3	Pass	<4	>50
Tetrahydrofuran	Pass	8	>50	Pass	4	>50	Pass	<4	>50
Toluene	Fail (25)	12	>50	Pass	24	40	Pass	<4	>50

Abbreviations: B.T.—Breakthrough time in minutes, P.R.—Permeation rate in ?g/cm2min, N/A—Not applicable.

Penetration results provide as pass or fail with penetration time in parentheses; Permeation tests per ASTM F739 at ambient temperature for 3 hours.

Source: Reference 25.

illustrates the how dissimilar materials can provide significantly different performance against a standard set of chemicals.

A strategy for testing a product will typically involve choosing the test method and establishing the specific conditions for the test. Many barrier tests are conducted under room temperature and humidity. Vapor transmission and permeation tests are very sensitive to these conditions, requiring a controlled environment for testing. In most cases, the duration of the test is set by the procedure, but degradation and permeation testing periods are not specified. For these tests, test duration should be set to the maximum period of expected product use and chemical exposure. No studies have established conditions where accelerated testing provides good correlation to full duration testing.

Test materials may also be subjected to various preconditions prior to testing, especially for barrier testing. Preconditions may involve certain exposures intended to simulate product use such as abrasion, flexing, cleaning, or heat aging. Use of preconditioning may help to identify product barrier performance more consistent with use expectations.

Selection of Test Chemicals

Manufacturers may be faced with a myriad of chemicals to select for testing their products. If the products are intended for broad applications among several markets, these decisions become increasingly difficult. In the past, end users in some industries were unable to compare product performance, since all manufacturers tested their materials to different sets of chemicals. The ASTM F-23 Committee sought to overcome this problem in the protective clothing industry by devising a standard battery of chemicals that represented the following[58]:

- A wide range of chemical classes;
- High volume usage in the chemical industry;
- Varying levels of toxicity; and
- Aggressive interactions with most materials.

The result was a 15 liquid-chemical battery established in ASTM Standard Guide F 1001 during 1986. This battery was expanded in 1989 to include 6 gases as well. The standard guide permits testing groups to test either the liquid battery, gas battery, or both, as applicable. These chemicals and relevant properties are listed in Table 16. Spence[59] developed procedures that permit simultaneous permeation testing with the 13 liquid organic chemicals in the battery as a means of quickly screening material performance.

In addition to barrier testing decisions, chemical protective clothing must be evaluated for the appropriate level of overall product integrity, physical properties, and human factors. The selected properties and the level of performance will depend on the specific application of chemical protective clothing. Often, specific decisions or determinations of tradeoffs in performance can be made based on a risk assessment, where the end user or the organization determine which properties are relevant and the consequences of clothing performance properties and features.

Overall Specifications

A number of overall standard specifications or classifications for chemical protective clothing have been developed. Specifications generally establish minimum performance criteria while classifications provide a means for ranking products based on standard test methods.

Protective Clothing

In the United States, the primary protective clothing standards that have been developed are related to emergency response applications. These include:

- NFPA 1991, Standard on Vapor-Protective Ensembles for Hazardous Materials Emergencies
- NFPA 1992, Standard on Liquid Splash-Protective Ensembles for Hazardous Materials Emergencies.

Table 16:
ASTM F 1001 Chemicals and Key Properties

Chemical	Class	Molecular Weight	Vapor Press. (mm Hg)	Molar Volume (cm^3/mol)	Specific Gravity	TLV® (ppm)
Acetone	Ketone	58	266	74	0.791	750
Acetonitrile	Nitrile	41	73	53	0.787	40
Ammonia	Inorganic gas	17	> 760	Ñ-	N/A	25
1,3-Butadiene	Alkene	54	910	87	N/A	10, cancer
Carbon Disulfide	Sulfur compound	76	300	62	1.26	10, skin
Chlorine	Inorganic gas	70	> 760	Ñ-	N/A	0.5
Dichloromethane	Halogen compound	85	350	63.9	1.336	50, cancer
Dimethylformamide	Amide	73	2.7	77	0.949	10, skin
Ethyl Acetate	Ester	88	76	99	0.92	400
Ethylene Oxide	Heterocyclic	44	> 760	Ñ-	N/A	1, cancer
Hexane	Aliphatic	86	124	131.6	0.659	50
Hydrogen Chloride	Inorganic gas	37	> 760	Ñ-	N/A	5
Methanol	Alcohol	32	97	41	1.329	200, skin
Methyl Chloride	Halogen compound	51	> 760	Ñ-	N/A	50, skin
Nitrobenzene	Nitro compound	123	<<1	102.7	1.203	1, skin
Sodium Hydroxide	Inorganic base	40	/ 0	N/A	2.13	2 mg/m^3
Sulfuric Acid	Inorganic acid	98	< 0.001	N/A	1.841	1 mg/m^3
Tetrachloroethylene	Halogen compound	166	14	101.1	1.631	25, cancer
Tetrahydrofuran	Ether	72	145	81.7	0.888	200
Toluene	Aromatic	92	22	106.8	0.866	50, skin

Both standards cover ensembles that are the combination of suits or garments, with gloves and footwear for complete body protection. Performance criteria are provided in terms of:

- Overall ensemble integrity
- Material chemical resistance
- Material flame resistance
- Material physical properties
- Clothing human factors
- Component function (e.g., airline pass-throughs for respiratory equipment and exhaust valves).

Two different levels are established by these standards based on overall ensemble integrity and material chemical resistance as shown below.

Type of Clothing	Material Performance	Overall Clothing Performance
Vapor Protective (NFPA 1991)	Permeation resistance	Gas-tight integrity
Liquid-Splash Protective (NFPA 1992)	Penetration resistance	Liquid-tight integrity
Particulate Protective	Particulate holdout	Particulate-tight integrity

Using this hierarchy, the NFPA attempted to define protective clothing types on the basis of needed performance as demonstrated by test methods designed to measure the type of protection provided. This scheme associated clothing gas-tight integrity and material permeation resistance with vapor protection and clothing liquid-tight integrity and material penetration resistance with liquid splash protection. To complete the hierarchy, the NFPA is considering working on a standard for particulate protective clothing.

In addition to performance criteria, these standards establish requirements for third party certification and labeling of suits and clothing for demonstrating compliance with these standards. Manufacturers must also meet certain design criteria and provide documentation to the end user such as instructions and technical data supporting claims against the standard.

In Europe, several standards have been developed related to chemical protective clothing and equipment. These include:

- EN 465 (Type 4: spray-tight chemical protective clothing)
- EN 466 (Type 3: liquid-tight chemical protective clothing)
- EN 466-2 (Type 3-ET: liquid-tight chemical protective clothing for emergency teams)
- EN 467 (Type 7: partial body chemical protective clothing)
- EN 943 (Types 1 and 2: gas-tight chemical protective clothing)
- EN 943-2 (Type 1-ET: gas-tight chemical protective clothing for emergency teams)
- EN 1511 (Type 3: limited use liquid-tight chemical protective clothing)
- EN 1512 (Type 4: limited use spray-tight chemical protective clothing)
- EN 1513 (Type 7: limited use partial body chemical protective clothing)
- EN 13034 (Type 6: chemical protective clothing with limited liquid protection)
- EN 13982-2 (Type 5: particulate-tight chemical protective clothing).

Each standard defines a specific type of chemical protective clothing with differences in the intended protection based on general clothing design, clothing integrity, and the type of chemical resistance provided. In each of the standards (with the exception of EN 466-2 and EN 943-2), clothing performance is classified in terms of material performance for chemical resistance, physical properties, and other properties. Each standard also specifies certain minimum design and integrity performance for the respective type of clothing.

Under the International Standards Organization, an effort to harmonize requirements for chemical protective clothing between Europe and North America has resulted in a proposed standard, ISO 16602, *Clothing for protection against chemicals—Classification and performance requirements*. In essence, this standard consolidates many of the clothing types and performance classifications into one standard. Table 17 provides an overview of the primary requirements for chemical protective clothing under this standard.

Gloves

In early 2000, the American National Standards Institute and the Industrial Safety Equipment Association developed a new standard on gloves, ANSI/ISEA 105-2000, *Standard for Hand Protection Selection Criteria*. This standard provides several different properties for classifying the performance of gloves, including chemical resistance gloves. The results for a specific glove style are reported according the level achieved for the respective property. Performance properties are defined in terms in 4 to 6 levels. Gloves not achieving the lowest level are reported as level 0. Properties covered in the standard include:

- Cut resistance (ASTM F 1790)
- Puncture resistance (EN 388, Clause 6.4)
- Abrasion resistance (ASTM D 3889)
- Chemical permeation resistance (ASTM F 739)
- Chemical degradation resistance (new method)

Table 17:
Type classification of chemical protective clothing in ISO 16602

General Performance Area	Specific Performance Test	Chemical Protective Clothing Type								
		1a	1b	1c	2	3	4	5	6	7
Whole chemical protective clothing item integrity	Leak tightness	X	X	X						
	Inward leakage		X[a]	X	X					
	Liquid jet test					X				
	Liquid spray test					X	X			
	Particle test							X		
	Limited liquid spray test								X	
Chemical protective clothing material chemical resistance[b]	Permeation resistance	X	X	X	X	X				
	Penetration pressure						X			
	Particulate penetration resistance							X		
	Liquid penetration resistance								X	X
	Liquid repellency								X	X

[a]Applies to Type 1b chemical protective suits when the facepiece is not permanently attached to the suit.

[b]Applies to primary material used in construction of chemical protective clothing item; may or may not apply to seams (see Section 7).

- Liquid-tight integrity (ASTM D 5151)
- Flame resistance (ASTM F 1358)
- Heat degradation resistance (ISO 17493)
- Conductive heat resistance (ASTM F 1060)
- Conductive cold resistance (ISO 5085-1).

Europe has established a number of standards for gloves. These include:

- EN 374 (protective gloves against chemical and biological hazards)
- EN 388 (protective gloves against mechanical risks)
- EN 420 (general requirements for protective gloves).

Each of these standards establishes some minimum performance requirements and then uses other properties for classifying performance.

References

1. Linch, A.L., A.L., Protective Clothing, *CRC Handbook of Laboratory Safety* (CRC Press, Boca Raton, FL, 1971, pp. 124-137).

2. Olishifski, J.B., *Fundamentals of Industry Hygiene*, 2nd Edition (National Safety Council, Washington, DC, 1971, p. 67).

3. Coletta, G.C., A.D. Schwope, I.J. Arons, J.W. King, and A. Sivak, *Development of Performance Criteria for Protective Clothing Against Carcinogenic Liquids* (U.S. Department of Health, Education, and Welfare, Washington, DC, DHHEW [NIOSH] Publication No. 79-106, 1978).

4. Stull, J.O., Performance Standards and Testing of Chemical Protective Clothing, *Proceedings for Clemson University's Sixth Annual Conference on Protective Clothing* (Greenville, SC, 1992).

5. Coletta, G.C., S.Z. Mansdorf, and S.P. Berardinelli, Chemical Protective Clothing Standard Test Method Development: Part II. Degradation Test Method. *Am. Ind. Hyg. Assoc. J.* 49, 26-33 (1988).

6. Schwope, A.D., T.R. Carroll, R. Haung, and M.D. Royer, Test Kit for Evaluation of the Chemical Resistance of Protective Clothing, *Performance of Protective Clothing: Second Symposium, ASTM STP 989*, edited by S.Z. Mansdorf, R. Sager, and A.P. Nielsen (American Society for Testing and Materials, Philadelphia, 1988, p. 314).

7. Stull, J.O., J.H. Veghte, and S.B. Storment, An Evaluation of a Permeation Field Test Kit to Aid Selection of Chemical Protective Clothing, *Performance of Protective Clothing: Fourth Volume, ASTM STP 1133*, edited by J.P. McBriarity and N.W. Henry (American Society for Testing and Materials, Philadelphia, 1992, p. 382).

8. Stull, J.O., *Early Development of a Hazardous Chemical Protective Ensemble, Final Report* (CG-D-24-86, AD A174, 885, Washington, DC, 1986).

9. Stull, J.O., The Direction of Protective Clothing Testing in the 1990's, *Proceedings of IFIA High Performance Fabrics Conference* (Industrial Fabrics International Association, St. Paul, MN, 1990).

10. van Amerongen, G.J., Diffusion in Elastomers. *Rubber Chem. Techno*l. 37, 1065 (1964).

11. Dodgen, C.R., *TRI/Environmental, Inc. Internal Test Report* (August 1994).

12. Mansdorf, S.Z., and S.P. Berardinelli, Chemical Protective Clothing Standard Test Method Development, Part 1. Penetration Method. *Am. Ind. Hyg. Assoc.* J. 49, 21-25 (1988).

13. Berardinelli,S.P., and L. Cottingham, Evaluation of Chemical Protective Garment Seams and Closures for Resistance to Liquid Penetration, Performance of Protective Clothing, ASTM STP 900, edited by R. L. Barker and G.C. Coletta (American Society for Testing and Materials, Philadelphia, 1986, pp. 263-275).

14. Stull, J.O., D.F. White, and T.C. Greimel, A Comparison of the Liquid Penetration Test with Other Chemical Resistance Tests and Its Application in Determining the Performance of Protective Clothing, Performance of Protective Clothing: Fourth Volume, ASTM STP 1133, edited by J.P. McBriarity and N.W. Henry (American Society for Testing and Materials, Philadelphia, 1992, pp. 123-138).

15. Stull, J.O., and D.F. White, Selecting Chemical Protective Clothing on the Basis of Overall Integrity and Material Performance Tests. Am. Ind. Hyg. Assoc. J. 53, 455-462 (1992).

16. Stull, J.O., Performance Standards for Improving Chemical Protective Suits, Chemical Protective Clothing Performance in Chemical Emergency Response, ASTM STP 1037, edited by J.L. Perkins and J.O. Stull (American Society for Testing and Materials, Philadelphia, 1989, pp. 245-264).

17. Henry, N.W., and C.N. Schlatter, The Development of a Standard Method for Evaluating Chemical Protective Clothing to Permeation By Liquids. Am. Ind. Hyg. Assoc. J. 42, 202-207 (1981).

18. Henry, N.W., Comparative Evaluation of a Small Version of the ASTM Permeation Test Cell Versus the Standard Cell, Performance of Protective Clothing: Second Symposium, ASTM STP 989, edited by S.Z. Mansdorf, R. Sager, and A.P. Nielsen (American Society for Testing and Materials,

Philadelphia, 1988, p. 236).

19. Berardinelli, S.P., R.L. Mickelsen, and M.M. Roder, Chemical Protective Clothing: A Comparison of Chemical Permeation Test Cells and Direct-Reading Instruments. Am. Ind. Hyg. Assoc. J. 44, 886 (1993).

20. Schwope, A.D., R. Goydan, R.C. Reid, and S. Krishnamurthy, State-of-the-Art Review of Permeation Testing and the Interpretation of its Results. Am. Ind. Hyg. Assoc. J. 45, 634 (1984).

21. Stull, J.O., and M.F.S. Pinette, Special Considerations for Conducting Permeation Testing of Protective Clothing Materials with Gases and Chemical Vapors. Am. Ind. Hyg. Assoc. J. 51, 378 (1990).

22. Stull, J.O., M.F.S. Pinette, and R.E. Green, Permeation of Ethylene Oxide through Protective Clothing Materials for Chemical Emergency Response. *App. Ind. Hyg.* **5**, 448-462 (1990).

23. Lara, J., and D. Drolet, Testing the Resistance of Protective Clothing Materials to Nitroglycerin and Ethylene Glycol Dinitrate, *Performance of Protective Clothing: Fourth Volume, ASTM STP 1133*, edited by J.P. McBriarity and N.W. Henry (American Society for Testing and Materials, Philadelphia, 1992, pp. 153-161).

24. Schwope, A.D., R. Goydan, T. Carroll, and M.D. Royer, Test Methods Development for Assessing the Barrier Effectiveness of Protective Clothing Materials, *Proceedings of the Third Scandinavian Symposium on Protective Clothing Against Chemicals and Other Heath Risks* (Norwegian Defence Research Establishment, Gausdal, Norway, September 1989).

25. Man, V.L., V. Bastecki, G. Vandal, and A.P. Bentz, Permeation of Protective Clothing Materials: Comparison of Liquid Contact, Liquid Splashes, and Vapors on Breakthrough Times. *Am. Ind. Hyg. Assoc. J.* 48, 551 (1987).

26. Pinette, M.F.S., J.O. Stull, C.R. Dodgen, and M.G. Morley, A New Permeation Testing Collection Method for Non-Volatile, Non-Water Soluble Chemical Challenges of Protective Clothing, *Performance of Protective Clothing: Fourth Volume, ASTM STP 1133*, edited by J.P. McBriarity

and N.W. Henry (American Society for Testing and Materials, Philadelphia, 1992, pp. 339-349).

27. Fricker, C., and J.K. Hardy, The Effect of an Alternative Environment as a Collection Medium on the Permeation Characteristics of Solid Organics through Protective Glove Materials. Am. Ind. Hyg. Assoc. J. 55, 738-742 (1994).

28. Ehntholt, D.J., D.L. Cerundolo, I. Dodek, A.D. Schwope, M.D. Royer, and A.P. Nielsen, A Test Method for the Evaluation of Protective Glove Materials Used in Agricultural Pesticide Operations. Am. Ind. Hyg. Assoc. J. 51, 462-468 (1990).

29. Swearengen, P.M., J.S. Johnson, and S.J. Priante, A Modified Method for Fabric Permeation Testing (Paper presented at 1991 American Industrial Hygiene Conference and Exposition, Salt Lake City, Utah, 1991).

30. Spence, M.W., An Analytical Technique for Permeation Testing of Compounds with Low Volatility and Water Solubility, Performance of Protective Clothing: Second Symposium, ASTM STP 989, edited by S.Z. Mansdorf, R. Sager, and A.P. Nielsen (American Society for Testing and Materials, Philadelphia, 1988, pp. 277-285).

31. Mickelsen, R.L., R.C. Hall, R.T. Chern, and J.R. Myers, Evaluation of a Simple Weight-Loss Method for Determining the Permeation of Organic Liquids through Rubber Films. Am. Ind. Hyg. Assoc. J. 52, 445-447 (1991).

32. Al-Hussaini, H.S.A., W.J. Koros, M. Howard, and H.B. Hopfenberg, A Simple Apparatus for Measurement of Liquid Permeabilities through Polymeric Films, *Ind. Eng. Chem. Proc. Res.* Dev. 23, 317-320 (1984).

33. Verschoor, K.L., L.N. Britton, and E.D. Golla, Innovative Method for Determining Minimum Detectable Limits in Permeation Testing, Performance of Protective Clothing: Second Symposium, ASTM STP 989, edited by S.Z. Mansdorf, R. Sager, and A.P. Nielsen (ASTM, Philadelphia, 1988, pp. 252-256).

34. Perkins, J.L., and M.C. Ridge, Use of Infrared Spectroscopy in Permeation Tests, Performance of Protective Clothing, ASTM STP 900, edited by R.L. Barker and G.C. Coletta (American Society for Testing and Materials, Philadelphia,

1986, pp. 22-31).

35. Moody, R.P., Analysis of 2,4-D Glove Permeation Under Controlled Environmental Conditions, Performance of Protective Clothing: Fourth Volume, ASTM STP 1133, edited by J.P. McBriarity and N.W. Henry (American Society for Testing and Materials, Philadelphia, 1992, pp. 189-197).

36. Spence, M.W., Chemical Permeation through Protective Clothing Materials: An Evaluation of Several Critical Variables (Presented at American Industrial Hygiene Conference and Exposition, Portland, Oregon, 1981).

37. Bentz, A.P., and V.L. Man, Critical Variables Regarding Permeability of Materials for Totally Encapsulating Suits (Proceedings of the 1st Scandinavian Symposium on Protective Clothing Against Chemicals, Copenhagen, Denmark, 1984).

38. Nelson, G.O., B.Y. Lum, and G.J. Carlson, Glove Permeation by Organic Solvents, Am. Ind. Hyg. Assoc. J. 42, 217-224 (1981).

39. Huang, R.Y., and V.J.C. Lin, J. App. Polymer Sci. 12, 2615 (1968).

40. Vahdat, N., and M. Bush, Influence of Temperature on the Permeation Properties of Protective Clothing Materials, Chemical Protective Clothing Performance in Chemical Emergency Response, ASTM STP 1037, edited by J.L. Perkins and J.O. Stull (American Society for Testing and Materials, Philadelphia, 1989, pp. 132-145).

41. Stampher, J.F., M.J. McLeod, M.R. Bettis, A.M. Martinez, and S.P. Berardinelli, Permeation of Polychlorinated Biphenyls and Solutions of These Substances through Selected Protective Clothing Materials. Am. Ind. Hyg. Assoc. J. 45, 634-641 (1984).

42. Schwope, A.D., R. Goydan, D.J. Ehntholt, U. Frank, and A.P. Nielsen, Permeation Resistance of Glove Materials to Agricultural Pesticides, *Performance of Protective Clothing: Fourth Volume, ASTM STP 1133*, edited by J.P. McBriarity and N.W. Henry (American Society for Testing and Materials, Philadelphia, 1992, pp. 198-209).

43. Mickelsen, R.L., M.M. Roder, and S.P. Berardinelli, Permeation of Chemical Protective Clothing by Three Binary Solvent Mixtures. *Am. Ind. Hyg. Assoc.* J. 47, 236-240 (1986).

44. Ridge, M.C., and J.L. Perkins, Permeation of Solvent Mixtures through Protective Clothing Elastomers, Chemical Protective Clothing Performance in Chemical Emergency Response, ASTM STP 1037, edited by J.L. Perkins and J.O. Stull (American Society for Testing and Materials, Philadelphia, 1989, pp. 113-131).

45. Goydan, R., J.R. Powell, A.P. Bentz, and C.B. Billing, Jr., A Computer System to Predict Chemical Permeation through Fluoropolymer-Based Protective Clothing, Performance of Protective Clothing: Fourth Volume, ASTM STP 1133, edited by J.P. McBriarity and N.W. Henry (American Society for Testing and Materials, Philadelphia, 1992, pp. 956-971).

46. Nelson, G.O., and A.N. Correia, Respirator Cartridge Efficiency Studies: VIII. Summary and Conclusions. Am. Ind. Hyg. Assoc. J. 37, 514-525 (1976).

47. Yoon, Y.H., and J.H. Nelson, Application of Gas Adsorption Kinetics I. A Theoretical Model for Respirator Cartridge Service Life. Am. Ind. Hyg. Assoc. J. 45, 509-516 (1984).

48. Wood, G.O., Estimating Service Lives of Organic Vapor Cartridges. Am. Ind. Hyg. Assoc. J. 55, 11-15 (1994).

49. Baars, D.M., D.B. Eagles, and J.A. Emond, Test Method for Evaluating Adsorptive Fabrics, *Performance of Protective Clothing, ASTM STP 900*, edited by R.L. Barker and G.C. Coletta (American Society for Testing and Materials, Philadelphia, 1986, pp. 39-50).

50. Schwope, A.D., P.P. Costas, J.O. Jackson, J.O. Stull, and D.J. Weitzman, *Guidelines for the Selection of Chemical Protective Clothing*, 3rd ed., Volumes I and II (American Conference of Government Industrial Hygienists, Akron, OH, 1988).

51. Forsberg, K., and L.H. Keith, *Chemical Protective Clothing Performance Index* (Wiley-Interscience, New York, 1989).

52. Forsberg, K., and S.Z. Mansdorf, *Quick Selection to Chemical Protective Clothing*, 2nd ed. (Van Nostrand Reinhold, New York, 1993).

53. *CPC Base* (Arthur D. Little, Cambridge, Massachusetts, 1990).

54. Mickelsen, R.L., and R.C. Hall, A Breakthrough Time Comparison of Nitrile and Neoprene Glove Materials Produced by Different Glove Manufacturers. Am. Ind. Hyg. Assoc. J. 48, 941-947 (1987).

55. Jamke, R.A., Understanding and Using Chemical Permeation Data in the Selection of Chemical Protective Clothing, *Chemical Protective Clothing Performance in Chemical Emergency Response, ASTM STP 1037*, edited by J.L. Perkins and J.O. Stull (American Society for Testing and Materials, Philadelphia, 1989, pp. 11-22).

56. Carroll, T.R., C.J. Resha, C.T. Vencill, and J.D. Langley, Determining the Sensitivity of International Test Methods Design to Assess the Gas-Tight Integrity of Fully Encapsulating Garments, *Six Volume, Performance of Protective Clothing, ASTM STP 1273*, edited by J.O. Stull and A.D. Schwope (American Society for Testing and Materials, 1997, pp. 3-15).

57. Stull, J.O., Developing Test Programs to Demonstrate Protective Clothing Material Performance, *Proceedings of INDA IMPACT '90 Conference* (Association of the Nonwoven Fabrics Industry, Cary, NC, February 1990).

58. Spence, M.W., A Proposed Basis for Characterizing and Comparing the Permeation Resistance of Chemical Protective Clothing Materials, *Performance of Protective Clothing, ASTM STP 900*, edited by R.L. Barker and G.C. Coletta (American Society for Testing and Materials, Philadelphia, 1986, pp. 32-38).

59. Spence, M.W., *Streamlined Methodology for Permeation Testing Using the ASTM F-1001 Standard Test Chemical Battery* (Paper presented at 1987 American Industrial Hygiene Conference and Exposition, Montreal, Quebec, May 1987).

6

Full-Body Protection and Standards

By Jeffrey O. Stull

Introduction

The purpose of this booklet is to provide information that will help with the understanding and specification of full-body protective clothing, including totally encapsulating suits and nonencapsulating suits, sometimes referred to as "splash suits." The combination of suits and other clothing components forms ensembles for true full-body protective clothing.

The booklet explains differences among types of protective suits, particularly their features and characteristics. It provides a detailed description for this type of clothing, specifically pointing out variations in suit design and properties of the materials that contribute to different types of performance. These performance characteristics are discussed relative to requirements in the National Fire Protection Association standards on chemical protective suits and garments. Supporting guidelines are provided to allow end users to compare different commercially available full-body protective clothing, some of which are certified to the respective NFPA standards.

Classification of Full-Body Chemical Protective Clothing

EPA Levels of Protection

The type of chemical protective clothing required for any given situation depends on:

- the volume, composition, and toxicity of the chemical(s);
- the duration of exposure; and

- the physical environment.

The U.S. Environmental Protection Agency (EPA) has established levels of protection that create ensembles of chemical protective clothing and other equipment commensurate with the types of hazards present. These include Levels A through D as described in Table 1. These levels of protection also appear in Appendix B of the Occupational Safety and Health Administration (OSHA) regulations of Title 29, Code of Federal Regulations Part 1910.120, *Hazardous Waste Operations and Emergency Response*.

In extreme conditions, or in cases where the above factors are poorly defined, the highest degree of skin and respiratory protection is required. The EPA has categorized such protection as Level A protection. As defined, Level A protection includes a totally encapsulating, single-piece, chemically resistant ensemble with "gas-tight" closures and interfaces between pants/boots and sleeves/gloves. It also includes a pressure-demand, self-contained breathing apparatus (SCBA), or pressure-demand, airline respirator (with escape bottle). It is important to recognize that suits that simply encapsulate the wearer may still not provide integrity against gases and vapors.

In less severe situations, Level B or Level C protection is appropriate. Specifically, Level B is used when chemical skin absorption hazards are less, while Level C is used when there are diminished hazards for both skin and respiratory chemical contact. Level B protection requires the same respiratory protection as Level A, but specifies single or multipiece "splash" suits with chemical-resistant gloves and boots. Level C protection requires the same protective clothing as Level B, but an air-purifying respirator is worn instead of the self-contained breathing apparatus.

It is important to realize that the EPA levels were established principally for hazardous material response and cleanup operations. While these levels specify by design the items needed to form an ensemble for full-body protection, they do not specify minimum performance. Definitions are not provided for "totally encapsulating protective suit" or even for "chemically resistant." Thus, it is left up to the user to determine what is appropriate protective clothing under the various EPA levels of protection. As a consequence, the EPA levels do not in of themselves provide adequate guidance for protective clothing selection.

National Fire Protection Association Standards

In 1986, a subcommittee was formed by the National Fire Protection Association to develop standards on hazardous chemical protective clothing. The subcommittee chose to develop three standards based primarily on the needs of emergency responders:

- NFPA 1991, Standard on Vapor-Protective Suits for Hazardous Chemical Emergencies;
- NFPA 1992, Standard on Liquid Splash-Protective Suits for Hazardous Chemical Emergencies; and
- NFPA 1993, Standard on Support Function Protective Garments for Hazardous Chemical Operations.

Since their introduction in 1991, the standards have now been updated and revised twice, once in 1994 and more recently in early 2000. As the result of these changes, there are now just two standards with slightly different titles:

- NFPA 1991, Standard on Vapor-Protective Ensembles for Hazardous Materials Emergencies
- NFPA 1992, Standard on Liquid Splash-Protective Ensembles and Clothing for Hazardous Materials Emergencies.

Most recently, the NFPA released a new standard, NFPA 1994, Standard on Protective Ensembles for Chemical/Biological Terrorism Incidents, which became available August 2001. Due to its emphasis on chemical and biological agents involved in terrorism incidents, this standard is described later in this booklet.

At the time this booklet was prepared, these documents represented the only comprehensive standards on chemical protective suits. Separate efforts for full-body chemical protective clothing covering a broader range of applications were being developed, but their adoption remains uncertain. Other standards have been established or proposed in Europe and internationally (described later in this section).

Table 1:
EPA Levels of Protection

Level	Equipment	Protection Provided
A	**RECOMMENDED:** • Pressure-demand, full-facepiece SCBA or pressure-demand supplied-air respirator with escape SCBA. • Fully-encapsulating, chemical resistant suit. • Inner chemical-resistant gloves. • Chemical-resistant safety boot/shoes. • Two-way radio communications. **OPTIONAL**: • Cooling unit. • Coveralls. • Long cotton underwear. • Hard hat. • Disposable gloves and boot covers.	• The highest available level of respiratory, skin, and eye protection.
B	**RECOMMENDED:** • Pressure-demand, full-facepiece SCBA or pressure-demand supplied-air respirator with escape SCBA. • Chemical resistant clothing (overalls and long-sleeved jacket; hooded, one- or two-piece chemical splash suit; disposable chemical-resistant one-piece suit. • Inner and outer chemical-resistant gloves. • Chemical-resistant safety boot/shoes. • Hard hat. Two-way radio communications. **OPTIONAL:** • Coveralls. • Disposable boot covers. • Face shield. • Long cotton underwear.	• The same level of respiratory protection, but less skin protection than Level A. • It is the minimum level recommended for initial site entries until the hazards have been further identified.
C	**RECOMMENDED:** • Full-facepiece, air-purifying canister-equipped respirator. • Chemical resistant clothing (overalls and long-sleeved jacket; hooded, one- or two-piece chemical splash suit; disposable chemical-resistant one-piece suit. • Inner and outer chemical-resistant gloves. • Chemical-resistant safety boot/shoes. • Hard hat. • Two-way radio communications. **OPTIONAL:** • Coveralls. • Disposable boot covers. • Face shield. • Escape mask. • Long cotton underwear.	• The same level of skin protection as Level B, but a lower level of respiratory protection.
D	**RECOMMENDED:** • Coveralls. • Safety boots/shoes. • Safety glasses or chemical splash goggles. • Hard hat. **OPTIONAL:** • Gloves. • Escape mask. • Face shield.	• No respiratory protection. Minimal skin protection.

Should Be Used When	Limiting Criteria
• The chemical substance has been identified and requires the highest level of protection for skin, eyes, and the respiratory system based on either: – measured (or potential for) high concentration of atmospheric vapors, gases, or particulates; or – site operations and work functions involving a high potential for splash, immersion, or exposure to unexpected vapors, gases, or particulates of materials that are harmful to skin or capable of being absorbed through intact skin. • Substances with a high degree of hazard to the skin are known or suspected to be present, and skin contact is possible. • Operations must be conducted in confined, poorly ventilated areas until the absence of conditions requiring Level A protection is determined.	• Fully-encapsulating suit material must be compatible with the substances involved.
• The type and atmospheric concentration of substances have been identified and require a high level of respiratory protection, but less skin protection. This involves atmospheres: – with IDLH concentrations of specific substances that do not represent a severe skin hazard; or – that do not meet the criteria for use of air-purifying respirators. • Atmosphere contains less than 19.5 percent oxygen. • Presence of incompletely identified vapors or gases is indicated by direct-reading organic vapor detection instrument, but vapors and gases are not suspected of containing high levels of chemicals harmful to skin or capable of being absorbed through intact skin.	• Use only when the vapor or gases present are not suspected of containing high concentrations of chemicals that are harmful to skin or capable of being absorbed through intact skin. • Use only when it is highly unlikely that the work being done will generate either high concentrations of vapors, gases, or particulates, or splashes material that will affect exposed skin.
• The atmospheric contaminants, liquid splashes, or other direct contact will not adversely affect any exposed skin. • The types of air contaminants have been identified, concentrations have been measured, and a canister is available that can remove the contaminant. • All criteria for the use of air-purifying respirators are met.	• Atmospheric concentration of chemicals must not exceed IDLH levels. • The atmosphere must contain at least 19.5 percent oxygen.
• The atmosphere contains no known hazard. • Work functions preclude splashes, immersion, or the potential for unexpected inhalation of or contact with hazardous levels of any chemicals.	• This level should not be worn in the Exclusion Zone. • The atmosphere must contain at least 19.5 percent oxygen.

Application

Each NFPA standard provides performance-oriented requirements for a specific protective suit ensemble (including gloves for hand protection and boots/footwear for foot protection) within an intended application area. NFPA 1991 covers protective suits that offer protection from all forms of chemicals, including vapors, gases, liquids, and solids. NFPA 1992 pertains to clothing that prevents wearer exposure to liquid splashes or solid chemicals. NFPA 1991 and 1992 have been prepared specifically for emergency response applications where:

- the hazards range in severity,
- the physical environment varies, and
- many aspects of the response remain uncharacterized.

As a result, only NFPA 1991 and 1992 compliant chemical protective suits are intended for hazardous chemical emergency response. NFPA 1992 was also developed to cover nonemergency situations and applications outside the "hot zone," such as waste site cleanup, industrial situations, and decontamination where the type and extent of hazards present are more commonly known. For both standards, the requirements were written to reflect a single use. This does not mean that suits cannot be reused, but simply the exposure of suits to chemicals may require disposal as there are no provisions that address decontamination for reuse.

Organization

The NFPA standards are divided into seven sections. These include chapters on:

- Administration, which contains the scope, purpose, and definitions used in the standard;
- Certification, which details manufacturer and third party organization responsibilities for product certification;
- Labeling and information requirements, setting out the information that the manufacturer must report to the end user;
- Design requirements that indicate the minimum features required for suits and suit components;
- Performance requirements, where specific criteria are provided

for the overall suit, suit materials, and suit components;

- Test methods, the section defining test procedures, parameters, and results to be reported; and
- Referenced publications.

There is also an appendix that provides supplementary information and recommendations to end users of the standard.

Certification and Documentation

A key attribute of the NFPA standards is the requirement for outside, third-party certification. Each product must be tested by an independent laboratory to demonstrate compliance with each of the requirements. To be certified, a product must pass all requirements. No partial certifications are permitted. In addition, certification entails independent, periodic audits of manufacturer quality assurance practices. The latest editions of the NFPA standards require the manufacturer to have a quality assurance program that is registered to ISO 9001 requirements. Both follow-on testing and quality assurance audits are conducted annually for "listing" the products by the certification organization. Products meeting these requirements are then labeled showing compliance with the respective NFPA standard. Each NFPA standard also specifies the minimum information that must appear on the label, such as the compliance language and product descriptions.

In addition, each NFPA standard requires that the manufacturer provide a "Technical Data Package." This package provides a complete description of the product as well as all data showing compliance of the product with the standard. The Technical Data Packages, therefore, offer a basis for end users or purchasers to easily compare products, since the NFPA standards specify exactly which tests should be conducted and what information should be reported.

Hierarchy of Protection

The NFPA standards define a hierarchy of hazard protection as shown in Table 2. Using this hierarchy, the NFPA has attempted to define protective clothing types on the basis of needed performance, as demonstrated by test methods designed to measure the type of protection provided. This scheme associates clothing gas/vapor integrity and material permeation resistance with vapor protection, and clothing liquid integrity and material penetration resis-

Table 2:
Definitions of Protective Clothing Types by Performance Tests in NFPA Standards

Performance-Based Clothing Classification	Required Material Performance	Required Overall Clothing Performance	EPA Level Replaced
Vapor-Protective (NFPA 1991)	Permeation resistance	Gas/vapor integrity	Level A
Liquid Splash-Protective (NFPA 1992)	Penetration resistance	Liquid integrity	Levels B and C

tance with liquid splash protection. These performance-based definitions are intended to replace the design-oriented EPA levels of protection, which define what clothing should look like but not how it should perform.

Overview of Requirements

Each NFPA standard sets performance requirements that include:

- overall suit or garment integrity,
- primary material, seam, and closure chemical resistance,
- primary material flame resistance,
- material physical properties, and
- component function.

Table 3 lists the requirements, test methods, and basis of performance criteria in the current edition of each standard. A summary of these requirements is provided below:

- ***Overall suit and garment integrity***—NFPA 1991 products are tested for gas-tight and liquid-tight integrity, whereas NFPA 1992 suits and garments are evaluated for liquid-tight integrity only. Products under both standards are also evaluated for functionality and then assessed for their maintenance of integrity following a simulated use. Vapor-protective suits are also evaluated for airflow capacity in the event that a self-contained breathing apparatus worn inside the suit vents at a very high rate.

Table 3:

NFPA 1991 and 1992 Requirements, Test Methods, and Performance Criteria

Requirement	NFPA Standard 1991	NFPA Standard 1992	Test Method and Description	Application and Criteria
Overall suit shall provide water-tight integrity	5-1.1	5-1.1	ASTM F 1359: A sample suit is placed on mannequin dressed with an inner water-absorptive garment. The suited mannequin is placed in a special shower stall that has 5 different water nozzles. Surfactant-treated water is sprayed at the suited mannequin in several orientations for 1 hour. The suit is then removed from the mannequin and the inner garment and suit interior examined for signs for water penetration.	In NFPA 1991, this test is used to evaluate the effectiveness of exhaust valve covers in preventing water penetration through the valves. In NFPA 1992, the test is used to assess water-tight integrity of the suit or garment, particularly seams, closures, and interfaces with other clothing or equipment (e.g., suit sleeve to glove attachment).
Gloves and footwear shall provide water-tight integrity	Does not apply	5-2.1 5-3.1	ASTM D 5151: Flexed sample glove and footwear are filled to within 25 mm (1 in.) of the opening with surfactant treated water and then observed after 2 minutes to determine if leakage has occurred.	Gloves and footwear cannot leak when separately tested. The testing is intended to determine if the items can prevent inward leakage of hazardous liquids.
Overall suit shall provide gas-tight integrity and permit functionality following simulated use	5-1.2	5-1.2	ASTM F1154/F 1052: Sample suits are worn during exercises simulating work tasks; suits are tested for integrity by being inflated to 127 mm (5 in.) $H2O$ water gauge pressure that is then reduced to 102 mm (4 in.). Pressure in the suit observed after 5 minutes at the test pressure.	The internal suit pressure must not drop more than 20% (it must remain above 81 mm or 3.2 in. $H2O$ water gauge pressure). Lower ending test pressures indicate air leakage out of the suit and lack of overall suit gas-tight integrity. The suit must allow the test subjects to complete all tasks, clearly see through the visor, permit the removal and reinsertion of their hands into the glove systems, and accommodate head protection.

Table 3 *(continued)*

Requirement	NFPA Standard 1991	NFPA Standard 1992	Test Method and Description	Application and Criteria
Overall suit shall provide adequate air flow in the event of SCBA failure	5-1.3	Not required	NFPA Test Protocol: Air is pumped into a sample suit at 500 L/min. for 5 minutes; the pressure is monitored during the airflow and the suit is tested for gas-tight integrity to ensure that the suit is not ruptured.	The test only applies to NFPA 1991. During the simulated high airflow, the pressure cannot exceed 38 mm (1.5 in.) water gauge pressure. The suit must meet the same pressure test criteria in 5.1.2 above. If higher pressures are observed during testing, more exhaust valves must be added to the suit design.
Overall suit must provide gas-tight integrity fitting installed	5-1.4	Does not apply	ASTM F 1052: Suits are tested for integrity by being inflated to 127 mm (5 in.). Water gauge pressure is then reduced to 102 mm (4 in.). Pressure in the suit observed after 5 minutes at the test pressure.	The internal suit pressure must not drop more than 20% (it must remain above 81 mm or 3.2 inches H2O water gauge pressure).
Exhaust valves shall be mounted on suits such that they do not easily pull out	5-1.5	Does not apply	NFPA Test Protocol: A tensile testing machine is used to measure the force required to pull the exhaust valve out of the suit.	Suit exhaust valves must withstand a force of 135 N (30 lb.) without being pulled out. The test was designed so that exhaust valves would not become unseated and create a large hole in the suit.
External fittings shall be installed in suits such that they do not easily pull out	5-1.6	Not required	NFPA Test Protocol: A tensile testing machine is used to measure the force required to pull the external fitting out of the suit.	Suit external fittings must withstand a force of 1000 N (225 lb.) without being pulled out. The criteria applies to fittings used to connect external airlines for supplied air systems.

Table 3 *(continued)*

Requirement	NFPA Standard 1991	NFPA Standard 1992	Test Method and Description	Application and Criteria
Exhaust valves shall not leak inward when a vacuum is created inside the suit	5-1.7	Does not apply	NFPA Test Protocol: An exhaust valve is placed onto a rigid plate that divides a special test cell into two chambers. In the chamber representing the suit interior a 25 mm (1-in.) water gauge vacuum is pulled, and the airflow into the valve from the chamber representing the suit exterior is measured.	Each exhaust valve can only show inward leakage of 30 milliliter of air per minute or less. This requirement is the same as used for qualifying valves used in respirators under National Institute for Occupational Safety and Health certification regulations.
Garment, glove, and footwear materials shall resist chemical permeation after repeated flexing and abrasion; Garment seams, visors, and visor seams shall resist chemical permeation	5-2.1 5-4.1 5.5.1 5-2.6 5-3.1 5-3.6	Does not apply	ASTM F 739: A material specimen is placed in a test cell dividing the cell into two separate chambers. The chemical is introduced into one chamber while air or water in the other chamber is sampled for chemical that permeates the material specimen. The time that chemical is first detected and the rate at which chemical permeates is reported. The test is conducted for 21 liquid chemicals listed in ASTM F 1001. (See below for procedures related to repeated flexing and abrading.)	The breakthrough time must be one hour or greater for all test chemicals. The one-hour criterion was selected to represent a worst-case exposure for the emergency responder. While permeation rate must be reported, no criterion is established for this measurement.

Table 3 *(continued)*

Requirement	NFPA Standard		Test Method and Description	Application and Criteria
	1991	1992		
Garment, visor, glove, and footwear materials shall resist chemical penetration	Does not apply	5-1.3 5.1.7.1 5-2.2	ASTM F 903: A material specimen is placed in a test cell forming a closed chamber on the external side of the material. Chemical is introduced into the closed chamber contacting the material specimen. After 5 minutes exposure, the pressure is raised to 76 kPa (2 psi) for one minute and then reduced to ambient pressure for the remainder of one hour. The opposite side of the material is then viewed for evidence of visible liquid penetration. The test is conducted for the 7 liquid chemicals in ASTM F 1001 that are not suspected/actual carcinogens or skin toxic chemicals.	Materials must show no penetration within the one-hour test period for all battery chemical. The contact of the chemical with the material for one hour was chosen to represent a worst case exposure. Pressure is applied at one minute to simulate liquid under pressure from a burst pipe. The test is intended to demonstrate how well the material prevents liquid transfer. **Materials showing liquid penetration may still allow chemical permeation.**
Garment and glove materials resist permeation (or penetration) after flexing	5-2.1	5-1.3	ASTM F 392/F 739 or F 903: Garment and glove materials are flexed using a Gelbo flex test device. A rectangular sample of material is formed into a cylinder along its long axis and placed on a Gelbo flex machine. The machine then reproducibly twists and compresses the sample for 100 cycles at a rate of 45 cycles per minute. Circular specimens are then cut out of the flexed material and tested for chemical permeation (or penetration) resistance, as described above.	Flexed primary materials must show breakthrough times (NFPA 1991) greater than one-hour or no penetration (NFPA 1992) for all test chemicals. Flexing of materials is intended to simulate stress on the suit during a single wearing. The Gelbo test device offers a method for consistently flexing materials.

Table 3 *(continued)*

Requirement	NFPA Standard 1991	NFPA Standard 1992	Test Method and Description	Application and Criteria
Footwear materials shall resist permeation after flexing	5-5.1	5-5.1	FIA 1209/ASTM F739 or F 903: A footwear specimen is placed on a footwear flexing machine and is flexed for 10,000 cycles in a manner simulating walking. Samples are cut from the flexed footwear upper and then are tested for chemical permeation (or penetration) resistance, as described above.	Flexed footwear upper material must show breakthrough times (NFPA 1991) greater than one-hour or no penetration (NFPA 1992) for all test chemicals. Flexing of footwear materials is intended to simulate stress on the footwear during a single wearing. The footwear flexing device offers a method for flexing in a manner simulating walking.
Garment, glove, and footwear materials shall resist permeation (or penetration) after abrasion	5-2.1 5-4.1 5-5.1	5-1.3 5-2.2 5-3.2	ASTM D 4157F 739 or F 903: A material specimen is placed on an oscillating drum-like abrading machine. A special coarse sandpaper is used to abrade the material for 25 back and forth cycles at a specified tension and pressure. Circular specimens are then cut out of the abraded material and tested for chemical permeation resistance as described above.	Abraded primary materials must show breakthrough times one hour or greater for all test chemicals. The type of abrasion provided by the test simulates abrasion that may occur from crawling on an asphalt surface. The amount of abrasion was selected to be consistent with a single suit wearing.
Closures shall resist chemical penetration	5-2.8	See below	ASTM F 903: Samples of the suit closure plus the seam attaching the closure to the garment material are tested for chemical penetration resistance as described earlier. NFPA 1991 seams are tested against the 15 ASTM F 1001 liquids while NFPA 1992 seams are tested against the 7 liquid chemicals in ASTM F 1001 which are not suspected/actual carcinogens or skin toxic chemicals.	Closures must show no penetration within the one-hour test period for all battery chemicals. Closures represent one of the "weak" links in chemical protective suits or garments. Closures are tested for chemical penetration resistance for both NFPA 1991 suits since permeation testing per ASTM F 739 does not accommodate closure type samples. The requirement in NFPA 1992 ensures consistency of primary material, seam, and closure chemical resistance.

Table 3 *(continued)*

Requirement	NFPA Standard 1991	NFPA Standard 1992	Test Method and Description	Application and Criteria
Garment seams and closures, and visor seams, shall resist penetration		5-1.7.1 5-1.8.1	ASTM F 903: Test is performed as above, except it is limited to isopropanol and sulfuric acid.	Seam and closures must show no penetration within the 1-hour test period for both chemicals. The contact of the chemical with the material for one hour was chosen to represent a worst case exposure. Pressure is applied at one minute to simulate liquid under pressure from a burst pipe. The test is intended to demonstrate how well seams and closures prevent inward liquid leakage. **Seams and closures showing liquid penetration may still allow chemical permeation.**
Garment, visor, glove, and footwear materials shall resist ignition and burning resulting from flame impingement	5-2.2 5-3.2 5-4.2 5-5.2	Not required	ASTM F 1358: A folded material specimen is placed in a vertical flame chamber. A flame of specified characteristics is applied to the material's folded edge for 3 seconds. If the sample ignites, the test is discontinued. If the sample does not ignite, the flame is reapplied to the same specimen for 12 seconds. If the sample ignites during this subsequent exposure, the time the material continues to burn, and the distance burned are measured. Observations of burning behavior are also recorded.	Materials fail if they ignite during the first 3-second flame exposure. Samples that ignite during the subsequent 12-second exposure can only burn for 10 seconds or less and have burn distances less than 102 mm (4 in.). This requirement is intended to eliminate the use of materials that contribute to the suit wearer's hazards. **This requirement does not imply protection from open flame or other severe heat exposures.**

Table 3 *(continued)*

Requirement	NFPA Standard 1991	NFPA Standard 1992	Test Method and Description	Application and Criteria
Garment and visor materials shall not burst when a force is applied perpendicular to the material surface	5-2.3 5-3.3	5-1.4 5-1.7.2	ASTM D 751 (Ring and Clamp): This test measures the force required to burst a material. A circular material specimen is clamped in a fixture on a tensile testing machine. The tensile testing machine measures the force to push a 25-mm (1-in. ball) through the material.	Burst strength is a measure of how easily a material can be penetrated by a hard round object. NFPA 1991 requires a garment material burst strength of 200 N (45 lbs.), while NFPA 1992 requires a 134 N (30 lbs.) burst strength. These criteria were based on field trials demonstrating acceptable and unacceptable performance for selected garments.
Garment and visor materials shall resist snagging and tearing by sharp protruding objects	5-2.4 5-3.4	5-1.5 5-1.7.3	ASTM D 2582: A material specimen is draped over the material holder platform on a puncture propagation tear tester. A carriage on which a puncture probe is mounted slides down a vertical track, striking the material sample, causing it to tear. The length of the tear is related to the force required to propagate that tear.	This test simulates how material may be snagged and torn by a protruding nail or other similar sharp objects. NFPA 1991 requires a puncture/tear resistance of 49 N (11 lbs.) while NFPA 1992 requires a puncture/tear resistance of 135 N (5.5 lbs.). These criteria were based on field trials demonstrating acceptable and unacceptable performance for selected garments.
Garment and glove materials shall not become objectionally stiff at cold temperatures	5-2.5 5-4.5	5-1.6 5-2.5	ASTM D 747 (modified): A small material specimen is placed in a stiffness testing machine. The machine is placed in an environmental chamber where the temperature is set at -30°C (-13°F). The machine measures the force (moment) required to bend the material at this temperature.	The test measures how materials become stiff at cold temperatures, which can lead to reduced wearer mobility. Materials must have a bending moment of less than 0.57 N-m (0.50 in.-lb.). Higher bending moments represent stiffer materials. This requirement is based on materials with known stiffness problems.

Table 3 *(continued)*

Requirement	NFPA Standard		Test Method and Description	Application and Criteria
	1991	1992		
Visor materials shall not crack at cold temperatures	5-3.5	5-1.7.4	ASTM D 2136: A material specimen is placed in a bending device that in turn is placed in an environmental chamber where the temperature is set at -30°C (-13°F). The device rapidly bends the sample and it is then examined for cracking.	The test determines if the visor material will crack at the selected temperature. Some visor materials that are very rigid may not bend at all during this test.
Garment and visor seams shall resist breaking apart when pulled in tension	5-2.7 5-3.7	5-1.8.2	ASTM D 751 (Seam Strength): The strength of a seam is measured in the same way as material tensile strength. In this test, a material seam sample is placed between two grips in a tensile testing machine and pulled until it breaks.	A minimum seam strength requirement prevents separation of suit seams while under stress. NFPA 1991 requires that garment seams have a breaking strength of 135 N/50-mm (30 lbs. per 2-inch width of seam). NFPA 1992 requires that garment seams have a breaking strength of 67 N/50-mm (15 lbs. per 2-in.).
Closures shall resist breaking apart when pulled in tension	5-2.9	5-1.9.2	ASTM D 751 (Seam Strength): The strength of a closure is measured in the same way as seam breaking. In this test, a closure assembly sample is placed between two grips in a tensile testing machine and pulled until it breaks.	A minimum closure strength requirement prevents separation of the suit closure while under stress. NFPA 1991 requires that garment closures have a breaking strength of 135 N/50-mm (30 lbs. per 2-inch width of seam). NFPA 1992 requires that garment closures have a breaking strength of 67 N/50-mm (15 lbs. per 2-in.).

Table 3 *(continued)*

Requirement	NFPA Standard 1991	NFPA Standard 1992	Test Method and Description	Application and Criteria
Glove and footwear upper materials shall protect against sharp edges	5-4.3 5-5.3	5-2.3 5-3.3	ASTM F 1790: A material specimen is placed on a metal mandrel of a cut testing device. A blade on a weighted pivot arm is pulled across the specimen until it makes contact with the mandrel. The device measures how far the blade travels before cutting through the specimen.	Cut resistance is based on whether cut-through is achieved within 25 mm (1-in.) or not. If the length of blade travel is further than 25 mm, then the material passes the cut resistance test. If less than 25 mm, it fails. The weight on the pivot arm is changed based on the product being tested. Gloves are tested at 400 grams in NFPA 1991 and 200 grams in NFPA 1992; footwear is tested at 800 grams in NFPA 1991 and 400 grams in NFPA 1992.
Glove materials and upper shall protect against punctures	5-4.4 5-5.4	5-2.4 5-3.4	ASTM F 1342: A material specimen is paced on a fixture that is attached to a tensile testing machine. The tensile testing machine, which is outfitted with a compression cell, has a probe provided on the opposite end similar to a 6-penny nail. The probe is pushed into the material at a fixed rate of speed, and the force required to puncture the material is recorded.	This test is intended to simulate damage from a protruding nail. Glove materials must not yield to puncture under a force of 2.3 kg (5 lbs) while footwear material must not allow puncture at 3.6 kg (8 lbs.). These requirements were based on a review of field experience for glove and footwear physical hazard resistance.
Gloves shall provide adequate dexterity	5-4.6	5-2.5	NFPA Test Protocol: Test subjects are timed in the length of time required to place 25 pins in a pegboard with and without gloves. The percentage is different in test times while wearing gloves is compared to test times for barehanded tests.	NFPA 1991 permits a 600% increase in the test times while NFPA 1992 allows a 300% increase. All gloves effect hand function. These criteria were based on trials using different gloves determined to offer acceptable protection and performance.

Table 3 *(continued)*

Requirement	NFPA Standard 1991	NFPA Standard 1992	Test Method and Description	Application and Criteria
Footwear toes shall protect against falling objects	5-5.5	5-3.5	ANSI Z41 (1.4): A footwear toe specimen is placed in a tensile test machine outfitted with a special device. An impact assembly is then dropped on the toe specimen. Following impact, the toe specimen is observed for signs of compression by measuring the clearance inside the specimen.	This test is used to measure how well the footwear will protect the wearer's toes from impact. Requirements for impact and compression resistance prevent injury to the toe when heavy objects are dropped on the wearer's foot. ANSI Z41 requires that protective footwear must resist compression forces of 11,121 N (2500 lbs.) and impact forces of 101.7 J (75 ft.-lbs.) or it fails the test. NFPA 1991 and NFPA 1992 footwear have the same requirement.
Footwear soles shall protect against bottom punctures	5-5.6	5-3.6	CSA Z195-M: Footwear sole and heel specimens are placed in a fixture. The fixture is attached to a tensile testing machine that has been outfitted with a compression cell. A probe with dimensions to a 6-penny nail is attached to the other end of the tensile testing machine. The probe is pushed into the material at a fixed rate of speed, and the force required to puncture the material is recorded.	This test is designed to measure the resistance of the sole/heel material to sharp objects. Footwear must protect the responder from sharp objects that might puncture the sole/heel material during the fire ground activity. NFPA-compliant footwear must not exhibit signs of puncture during this test. NFPA 1991 and NFPA 1992 footwear have the same requirement.

Table 3 *(continued)*

Requirement	NFPA Standard 1991	NFPA Standard 1992	Test Method and Description	Application and Criteria
Footwear soles and heels shall resist excessive wear	5-5.7	5-3.7	ASTM D 1630: Footwear sole/heel specimens are subjected to an NBS abrader. A special abradant wheel is used to abrade the specimen until 0.1 inch of material is removed. The number of revolutions required to perform the specified 2.5 mm (0.1 in.) abrasion is determined and compared to a standard reference sample. The number of revolutions required in the test sample is compared to the number required in the control sample. This ratio yields a sole/heel abrasion resistance rating.	This test is a measure of the durability of the sole/heel material in protective footwear. Higher ratings indicate higher sole wear (abrasion) resistance. A resistance rating of at least 65 is recommended. NFPA 1991 and NFPA 1992 footwear have the same requirement.
Footwear soles shall resist bending	5-5.8	5-3.8	NFPA Test Protocol: Footwear sole specimens are placed on a stand supporting both front and back ends. A 182 kg (400 lb.) force is applied to the middle of the sole and the amount of deflection is measured.	The footwear sole is not allowed to deflect more than 6 mm (¼ inch). This requirement was created to ensure that footwear provide support to the wearer while transversing uneven ground or climbing a ladder. NFPA 1991 and NFPA 1992 footwear have the same requirement.
Footwear soles shall provide good traction	5-5.9	5-3.9	ASTM F 489: Footwear sole specimens with tread are placed on a friction testing machine. This device moves the sole bottom against a specified surface and measures the coefficient of friction for both surfaces. The test is performed under dry conditions.	This test is intended to measure traction offered by chemical protective footwear. The test simulates worst cases by being performed wet. Footwear sole must have a coefficient of 0.75 or greater in a dry condition. NFPA 1991 and NFPA 1992 footwear have the same requirement.

Table 3 *(continued)*

Requirement	NFPA Standard		Test Method and Description	Application and Criteria
	1991	1992		
Overall suits shall provide sufficient protection for escape during chemical flash fires	5-6.2 Option	5-6.1 Option	NFPA Test Protocol: A suit is placed on a manikin inside a closed chamber that is filled with propane at a concentration established to produce a 6–8 second chemical flash fire. Any flaming on the test is observed. The suit is also tested for visor clarity and gas-tight integrity following the simulated chemical flash fire. NFPA 1992 suits are tested for liquid-tight integrity.	This test allows the entire ensemble to be evaluated for the effects of a long-duration chemical flash fire. Flaming on the suit cannot continue for more than 2 seconds following the flash fire. The suit must allow visual clarity of at least 20/100 through the visor. NFPA 1991 suits must have minimal gas-tight integrity showing and ending pressure of 13 mm (.5 in.) while NFPA 1992 garments must pass a liquid-tight integrity test. **The test is intended to demonstrate the survivability of the responder. Responders should not intentionally enter situations where chemical flash fire is expected.**
Garment, glove, and footwear materials shall provide sufficient insulation during a chemical flash fire	5-6.3 Option	5-6.2 Option	NFPA Test Protocol: A material specimen is placed on a thermal protective performance (TPP) testing device. The specimen is horizontally exposed to a combination of quartz lamps and burners that simulate both radiant and convective heat. A weighted block containing a sensor is placed on top of the specimen and measures the rate of heat transfer through the specimen. The rate of heat transfer is compared with experimental measurements of skin response to high heat and a prediction for the time of a second-degree burn is made. This time, multiplied by the heat exposure energy, is the TPP rating.	TPP is a measure of material insulation to heat. This means that the test measures how quickly heat is transferred from the outside of the material to the inner material surface, which may contact the wearer's skin. The test is intended to simulate the environment of a chemical flash fire. Ensemble materials must have a TPP rating of at least 12, which is equivalent to a second-degree burn time of 6 seconds. NFPA 1991 and NFPA 1992 footwear have the same requirement.

Table 3 *(continued)*

Requirement	NFPA Standard 1991	NFPA Standard 1992	Test Method and Description	Application and Criteria
Garment, glove, and footwear material shall resist ignition during a chemical flash fire	5-6.4 Option	5-6.3 Option	ASTM F 1358: Material specimens are tested as described above. A folded material sample is placed in a vertical flame chamber. A flame of specified characteristics is applied to the material's folded edge for 3 seconds. If the sample ignites, the test is discontinued. If the sample does not ignite, the flame is reapplied to the same specimen for 12 seconds. If the sample ignites during this subsequent exposure, the time the material continues to burn (afterflame) and the distance burned are measured. Observations of burning behavior are also recorded.	More stringent criteria are applied to materials used for optional chemical flash fire escape. Materials must not ignite during the initial 3-second exposure and are only permitted to have an afterflame time of no more than 2 seconds and burn distance no more than 100 mm (4-in.) during the subsequent 12-second exposure period. **While more stringent than the base requirement, this requirement also does not imply protection from open flame or other severe heat exposures.**
Garment and glove materials shall not produce or hold static electricity after being rubbed	5-6.5 Option	5-6.4 Option	NFPA Test Protocol: A material specimen is placed in a specimen holder and is contacted with a rotated wheel that is covered with Teflon felt. After 10 seconds, the sample holder falls in front of an electrostatic detector, and the potential on the material is measured. The test is applied to both surfaces of the material at a temperature of 75°F and relative humidity of 45%.	This test evaluates the charge produced on materials by rubbing action. Materials that show a high voltage after several seconds are likely to produce and hold static charge. NASA requires that materials must not show potentials greater than 350 volts 5 seconds after being released in front of the detector for their fuel handlers' ensembles.

Table 3 *(continued)*

Requirement	NFPA Standard		Test Method and Description	Application and Criteria
	1991	1992		
Garment, visor, glove, and footwear materials shall resist permeation of liquefied gases	5-7.2 Option	Does not apply	ASTM F 739 (modified): Permeation testing is conducted as described above, except the liquefied gas is maintained on the challenge side of the cell as liquid (at its liquefaction temperature). The testing is conducted using a minimum of the 6 ASTM F1001 gases as liquids.	Ensemble material must exhibit a normalized breakthrough time of 15 minutes or more. This time was chosen by end users as the maximum exposure time likely for a response involving liquefied gases.
Garment, visor, glove, and footwear materials shall resist permeation of gases following embrittlement conditions	5-7.3 Option	Does not apply	ASTM D 2136 (modified)/ASTM F 739: Material specimens are placed in contact with dry ice for 15 minutes, placed is a bending test fixture, and subjected to a rapid bending force. Specimens that show no damage are then subjected to permeation testing against the 6 ASTM F 1001 gases (as gases).	This test is intended to simulate the cold contact that might occur when liquefied gases come in contact with chemical protective suit materials. Dry ice represents the low-end temperature for most liquefied gases. A 15-minute contact period is consistent with the requirement above.
Garment, visor, glove, and footwear materials, and seams, shall resist permeation by chemical warfare agents	5-9.1 5-9.2 5-9.3 Option	Does not apply	ASTM F 739/CRDC-SP-84010: Similar procedures as described above are used for permeation testing of chemical warfare agents. Cyanogen chloride is tested as a gas and breakthrough time is measured. Cumulative permeation of Sarin (GB), V-agent (VX), Sulfur Mustard (HD) and Lewisite (L) is measured when contacted with the material at surface concentration density of 100 g/m2.	This optional testing is intended to apply to vapor-protective suits also used for protection against weapons of mass destruction. The breakthrough time for cyanogen chloride must be 1-hour or more. The cumulative permeation cannot except 1.25 micrograms for Sarin and V-agent and 4.0 micrograms for Sulfur Mustard and Lewisite. These exposure levels were established based on military material acceptance criteria.

Table 3 *(continued)*

Requirement	NFPA Standard		Test Method and Description	Application and Criteria
	1991	1992		
Overall suits shall not leak under user conditions	5-9.4 Option	Does not apply	NFPA Test Protocol: Sample suits are worn by a test subject inside a chamber filled with sulfur hexafluoride at a concentration of 1000 ppm. The test subject performs 3 sets of stationary exercises in the chamber. Samples of air inside the suit are taken during the testing and analyzed for penetrating sulfur hexafluoride.	The test evaluated the effectiveness of the suit in preventing the inward leakage of outside contaminants while the suit is in use. The percent of sulfur hexafluoride leaking into the suit cannot exceed 0.02 percent.

- ***Primary material, seam, and closure chemical resistance***—Primary materials include as a minimum, the materials of the garment, visor, gloves, and boots, and any other major materials used in the construction of the chemical protective suit. Permeation resistance requirements are applied to all materials against a 21-chemical battery for NFPA 1991; penetration resistance requirements are applied against a 7-chemical battery for NFPA 1992. Seams and closure samples are also evaluated for chemical resistance to the same chemical batteries using the same testing approach.
- ***Primary material flame resistance***—Suit or garment primary materials are tested for resistance to flame impingement. NFPA 1991 requires specific minimum performance of primary materials; whereas, NFPA 1992 includes an option for flash fire escape protection that entails a flame resistance test. The requirement in NFPA 1991 is intended to prevent the use of material in a chemical protective suit that could contribute to the further injury of the responder. Products that pass are not proven to provide flame resistance in fire fighting or other high heat or flame protection applications.
- ***Material physical properties***—A number of physical properties are measured for the garment, visor, glove, and footwear materials. The specific physical property measurements required are listed in the later section on physical properties and in the series booklet, "Test Methods." Both garment and glove materials are evaluated for durability by measuring chemical resistance following both abrasion and repeated flexing in NFPA 1991 and 1992.
- ***Component function***—Visor materials are evaluated for clarity to assess whether the visor provides clear visibility for the wearer. Exhaust valves are tested for inward leakage resistance as a measure of preventing the penetration of outside contaminants into the suit.

Chemical Batteries

Different chemical batteries are used in each standard for evaluating the chemical resistance of primary materials (garment, visor, gloves, and footwear materials), seams, and closures. The list used for NFPA 1991 is based on ASTM F 1001, *Standard Guide for Chemicals to Evaluate Protective Clothing Materials*. This standard guide

Table 4:
Standard Chemical Batteries for Evaluating Chemical Protective Clothing

Chemical	ASTM F 1001 Liquids	ASTM F 1001 Gases	NFPA 1991	NFPA 1992
Acetone	X		X	X
Acetonitrile	X		X	X
Ammonia (Anhydrous)		X	X	
1,3 - Butadiene		X		
Carbon Disulfide	X		X	
Chlorine		X	X	
Dichloromethane	X		X	
Diethylamine	X		X	
Dimethylformamide	X		X	
Ethyl Acetate	X		X	X
Ethylene Oxide		X		
Hexane	X		X	X
Hydrogen Chloride		X		
Methanol	X		X	
Methyl Chloride		X		
Nitrobenzene	X		X	
Sodium Hydroxide (50% w/w)	X		X	X
Sulfuric Acid (conc.)	X		X	X
Tetrachloroethylene	X		X	
Tetrahydrofuran	X		X	X
Toluene	X		X	

includes 15 liquid and 6 gaseous chemicals representing a range of chemical classes, high volumes of transport, known permeation aggressiveness, and other factors. A shorter list is used for NFPA 1992, taking into account that these suits and garments are intended to provide protection against liquids (and not gases) that do not have vapors that are skin toxic or carcinogenic. Eight of the 15 ASTM F 1001 liquid chemicals do not meet these criteria, leaving 7 chemicals that form the chemical battery for NFPA 1992. Table 4 provides the lists of chemicals used in each standard.

Options

Both standards include optional criteria for additional product certification. NFPA 1991 provides three different options:

1. Liquefied gas protection,
2. Chemical flash fire escape protection,
3. Chemical and biological terrorism protection.

NFPA 1992 provides a single option for chemical flash fire escape protection.

Liquefied gas protection entails further testing of primary suit materials against 6 chemicals under liquefaction conditions with an assessment of material embrittlement. Because the application involves potential exposure to gases, this option only applies to NFPA 1991 ensembles. The chemical flash fire escape protection requirements, as the subtitle implies, are intended to permit responders to exit a chemical flash fire safely (the requirements are not intended for entry into known fire situations). The additional criteria in this certification option include a more stringent flame resistance test, a test of material insulation against convective and radiant heat, and an overall evaluation of the suit survivability in a simulated chemical flash fire. The additional, optional criteria offered for chemical and biological terrorism protection include permeation testing requirements against chemical warfare agents as a means of determining the appropriateness of the suit for domestic preparedness applications. This option was limited to NFPA 1991 given the known vapor hazards of these substances. Nevertheless, with the existence of NFPA 1994, less reliance is placed on those criteria. Instead, manufacturers use NFPA 1994 for certification of clothing ensembles for chemical/biological terrorism incidents.

Comparison of Full-Body Clothing

A number of factors can be used to distinguish full-body chemical protective clothing (ensembles). These include ensemble materials, features, and physical performance data for the following parts of the ensemble:

- product,
- suit (garment),
- seams,
- gloves,
- footwear,
- closure,
- fittings, and
- other items.

This list has been created to address complete ensembles representing the highest level of protection (totally encapsulating chemical protective suits). Therefore, not all areas will apply to different commercial full-body chemical protective clothing (such as splash suits).

In addition, chemical resistance, flame resistance, and physical property data can distinguish primary materials used in the construction of chemical protective clothing.

Product

Each manufacturer assigns a specific product name or model number to their chemical protective suits. For some products, several model names may exist based on variants in the clothing design, e.g., front-positioned closure versus rear-positioned closure. Each product can be further classified by its:

- overall design type,
- EPA level of protection,
- type of integrity offered, and
- compliance with industry NFPA standards.

Design

Design classifications include different forms of full body clothing described below:

- *Totally encapsulating protective suit*—These suits fully encapsulate the wearer and his or her respiratory protective equipment, and include temporarily or permanently attached gloves, and boots. Many suits, however, may have soft booties instead of attached boots.
- *Partially encapsulating protective suit*—These suits are similar to totally encapsulating suits, but do not encapsulate the wearer's breathing apparatus, which is worn outside the suit. However the facepiece of the respirator provides coverage of the wearer's face. These suits also provide wearer encapsulation with gloves and boots. These items, though not necessarily attached, must be worn together with the suit as part of the overall suit ensemble.
- *Hooded splash protective suit*—One or multipiece garments covering the wearer's torso, head, arms, and legs. The face area is left open with a hood that interfaces with the facepiece of the wearer's breathing apparatus. Gloves and boots similarly interface with suit sleeves and trouser cuffs, respectively. However, these items are not normally provided by the same manufacturer and may be obtained separately. Two-piece clothing is much like a rain suit or foul weather gear, while one piece garments are typically constructed as coveralls.
- *Splash protective suit*—Same as above, however, the suit does not have a hood. Since these chemical protective suits do not provide full-body protection, they are not further considered in this booklet. Other forms of partial body clothing such as smocks, aprons, and sleeve protectors fall into the same category (see series booklet, "Types of Chemical Protective Clothing").

Figure 1 shows examples for each type of chemical protective suit.

EPA Level

Each chemical protective suit can be further classified into the respective EPA level. These levels are similarly established in Appendix B of Occupational Safety and Health Administration regulations that appear in Title 29 U.S. Code of Regulations Part 1910.120, "Hazardous Waste Operations and Emergency Response." The EPA levels, while design oriented, do not always correlate with the design classifications provided above. For example, EPA Level A describes an encapsulating suit as part of its ensemble. However, the EPA guidelines do not define what encapsulation means (though Appen-

Figure 1.
Examples of different full-body chemical protective clothing: (a) Totally Encapsulating Chemical Protective Suit (courtesy of DuPont™ Tychem®); (b) Partially Encapsulated Chemical Protective Suit (courtesy of Trelleborg); (c) Hooded Chemical 2-Piece Splash Suit (courtesy of Tingley); (d) Chemical Protective Coverall (courtesy of DuPont™ Tychem®).

dix B of OSHA 29 CFR Part 1910.120 does provide two tests that measure the ability of the suit to prevent inward leakage of gases or vapors). Moreover, different users and clothing manufacturers have varying concepts on what constitutes an encapsulating suit. Some believe that a hooded coverall may be considered an encapsulating suit. However, industry practice has moved toward defining only totally encapsulating protective suits as EPA Level A. Similarly, partially encapsulating and hooded splash suits are deemed consistent with EPA Levels B and C. For these levels, there is less definition as to the configuration of the clothing. However, differences in splash suit design can have a significant effect on clothing performance. For example, seams and closures that are not taped or otherwise covered may provide pathways for liquid penetration.

Integrity

The integrity of the chemical protective suit offers a performance-oriented classification of products. Two levels of integrity are defined in the United States:

- *Gas/vapor integrity.* This type of integrity implies that the suit is capable of forming a gas- or vapor-leak-free envelope around the wearer. Such suits resist *penetration* of chemical contaminants in a gaseous state. Gas/vapor integrity is demonstrated for suits when they are tested and pass the criteria established in ASTM F 1052, *Standard Test Method for Pressure Testing Vapor-Protective Ensembles*. The test involves the inflation of the garment to a specified pressure, and either measuring a pressure drop over time or applying a soap solution to the outside of the garment to detect leaks (by the appearance of bubbles). The suit exhaust valves must be closed to perform the test and a provision must be made for attaching a pressure gauge. This procedure is shown in Figure 2. Some manufacturers specify the pressure to which the suit should be inflated. The ASTM method specifies a minimum inflation pressure (5 inches water gauge pressure), a minimum test pressure (4 inches water), and an allowable pressure drop (20%) over a 4-minute period. It also requires using the soap solution to locate leaks if the suit does not meet the pass/fail criteria. Suits that show less than a 20% pressure drop are considered to be "gas-tight ."

- *Liquid integrity.* Chemical protective suits that have liquid integrity prevent the penetration of liquids into the chemical protective suit or suit ensemble, which may occur as the result of chemical splashes or similar forms of liquid contact. As with gas-tight integrity, suit liquid-tight integrity may be demonstrated by a standardized procedure, ASTM F 1359, *Standard Test Method for Liquid Penetration Resistance of Protective Clothing or Protective Ensembles Under a Shower Spray While on a Mannequin.* In this test, a suit is placed on a human form mannequin that is already dressed in a liquid-absorptive coverall. Water treated with a surfactant to a surface tension of approximately 34 dynes/cm is sprayed at the suited

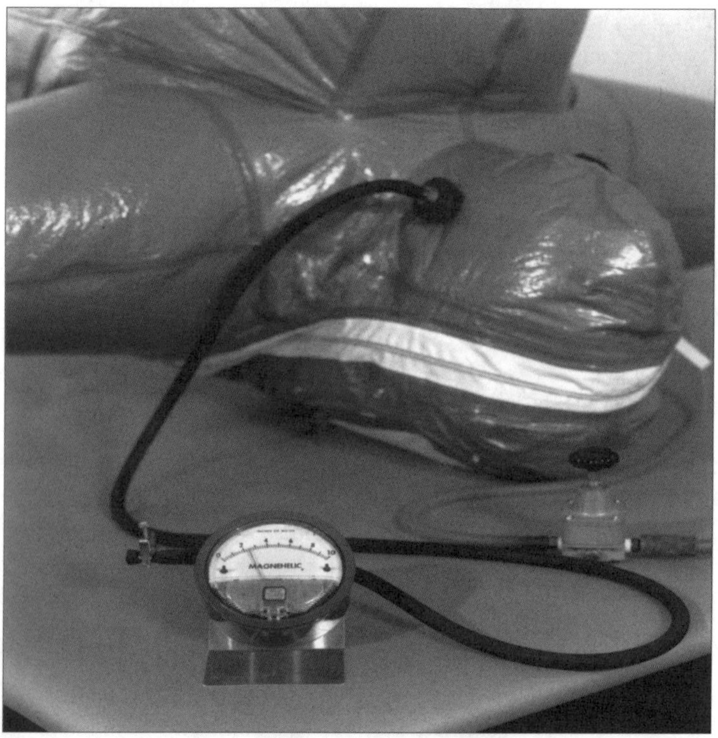

Figure 2.
Pressure (inflation) testing of totally-encapsulating chemical protective suit.

mannequin from several different directions for a period of 1 hour (see Figure 3). The suit is then removed, and the inner liquid-absorptive garment and interior of the suit are examined for signs of water penetration. Suits that show no evidence of water penetration are considered to offer "liquid" integrity. The length of this test is not intended to simulate exposure but rather to permit sufficient liquid to leak through the suit for easy detection of holes or gaps in the garment's protection.

Many types of chemical protective suit may neither offer gas/vapor nor liquid integrity. This may be because the suits have low barrier seam construction, lack sufficient closures, or fail to provide adequate interfaces with other ensemble components such as respirators, gloves, or boots. Only totally encapsulating chemical protective suits can have complete gas/vapor integrity. This does not mean that all totally encapsulating chemical protective suits have this integrity, but simply that a full envelope around the wearer, including attached gloves and boots, is necessary to achieve gas/vapor integrity. Similarly, a garment that

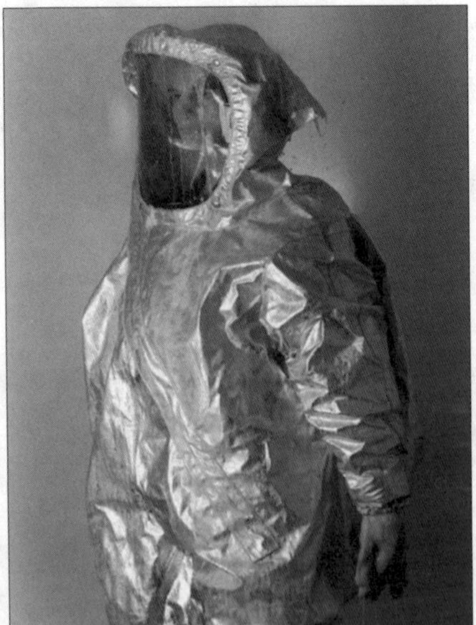

Figure 3.
Overall suit liquid-tight integrity test (courtesy of Intertek Testing Services).

completely covers the wearer does not mean that it will prevent inward penetration of liquid chemicals. The clothing must be designed to shed liquids, prevent liquid accumulation, and provide seams, closures, and interfaces with liquid integrity.

There are no agreed upon standards or methods of classification in the United States for clothing intended to block penetration by solid particles (clothing with particulate integrity).

Compliance

Full-body chemical protective clothing can only comply with three different standards:

1. NFPA 1991, Standard on Vapor-Protective Ensembles for Hazardous Materials Emergencies,
2. NFPA 1992, Standard on Liquid Splash-Protective Ensembles and Clothing for Hazardous Materials Emergencies,
3. NFPA 1994, Standard on Protective Ensemble for Chemical/Biological Terrorism Incidents.

Some products out of Europe may have additional certifications appearing as a "CE" mark on the clothing, but few North American users have an understanding of the various product categories and property classifications used in European standards. The European system is based on seven types of chemical protective categories, which similarly differentiate clothing types by their integrity and material performance. In Europe, suits are type classified as follows:

- Type 1—gas-tight encapsulating suits—suits must meet a gas-tight integrity pressure test and materials are required to be permeation resistant (further distinctions are made between a Type 1a suit that completely encloses the wearer and their breathing apparatus, a Type 1b suit that uses the facepiece of a respirator to complete the seal around the wearer in the face area, and a Type 1c suit that provides air to the wearer only from a supplied air source; there is also a Type 1-ET suit for emergency responders).
- Type 2—non–gas-tight encapsulating suits—suits must meet an inward leakage test, but are not gas tight; materials must be permeation resistant.

- Type 3—liquid "jet" protective clothing—clothing that prevents inward leakage of liquid under pressure as demonstrated by a jet test with materials that are permeation resistant.
- Type 4—liquid "splash" protective clothing—clothing that prevents inward leakage of liquid as demonstrated by a spray test with materials that are permeation resistant (there is also a Type 4-ET suit for liquid splash protection during emergency operations).
- Type 5—particulate protective clothing—clothing that prevents inward leakage of particles during an overall aerosol test.
- Type 6—light liquid "splash" protective clothing—clothing that prevents inward leakage of a liquid mist as demonstrated by a light spray test with materials that are repellent and penetration resistant.
- Type 7—partial body protective clothing—clothing that offers no integrity with materials that are repellent and penetration resistant.

The correspondence of these European-type classifications of chemical protective clothing with U.S. representation of these products is difficult because the European standards use different test methods and criteria than those found in the NFPA standards. The eventual development of either a U.S. or international standard may remedy this situation to provide classification of all chemical protective clothing.

Available Sizes

There are currently no North American standards that apply to the uniform sizing of chemical protective suits, except for disposable/limited use coveralls (ANSI/ISEA 101). In Europe, product sizing must be based on at least two key body dimensions (e.g., height and chest size). Many chemical protective suits are available in at least three sizes, usually small, medium, and large, although NFPA 1991 requires that compliant suits be provided in at least 4 sizes. Since suit designs vary significantly, there is no consistency from one product to another for the same generic "size." That is, one manufacturer's medium-sized suit may fit well, while a medium suit from a different manufacturer may not fit at all.

In general, sizing becomes more uniform for clothing that is closer fitting, as with normal clothing (e.g., trousers, pants, coveralls) and is more variable with complex, unusual clothing designs, such as encapsulating suits.

Suit

The suit comprises the major (and sometimes only) portion of the full-body protective clothing, generally providing protection to the wearer's torso, limbs, and sometimes head area. Suits can be described in terms of:

- material of construction,
- thickness,
- weight,
- tensile strength,
- burst strength,
- tear resistance,
- puncture/tear resistance, and
- stiffness.

Material

The suit material can either be listed by generic or trade name. An example of generic name is "chlorinated polyethylene," which indicates the key polymer in the material's construction. Trade names are often used for complex materials that are proprietary and include several different polymer components.

In general, materials are composed of a film or coating in combination with a substrate fabric. The film or coating may be on one side or both sides. Usually, the film or coating provides the barrier properties of the material, while the substrate fabric mainly provides material strength and support. Films and coatings comprise a number of materials, either elastomeric or plastic in polymer composition. Examples of elastomeric films include Neoprene (chloroprene), butyl rubber, chlorobutyl rubber, and Viton®. Plastics and thermoplastics are increasingly being used for chemical protective clothing. Traditionally, polyvinyl chloride and polyethylene have been used. The diversity of these plastics is expanding as manufacturers produce a variety of laminate materials that include different polymer layers.

A different class of chemical protective clothing materials involves microporous films. Microporous films are engineered with pores ranging in size from 0.01 to 10 microns. These materials generally offer "breathability" in terms of water vapor transmission and sometimes, air permeability. These materials are designed to offer liquid penetration resistance, but because of their structure, they cannot provide an effective barrier to permeation by most chemicals. They may also be designed with the film sandwiched between fabric layers for additional protection from wear.

For low levels of protection, mainly against solid particles and aerosols, garment materials may use only substrate fabrics. In these applications, the material must provide an effective filter to particles, but being porous, will provide only limited protection against liquids and no protection against gases or vapors.

Substrate fabrics may be of woven or nonwoven type. Nylon, polyester, Nomex®, and fiberglass fabrics are common examples of woven supporting fabrics used in chemical protective suits. Nonwoven materials include polyester, polypropylene, and spun-bonded polyethylene. The substrate fabrics are either laminated to the plastic or rubber film/sheet under heat and pressure or coated with a solution of the plastic/rubber material. In addition to supplying strength, the fabric layer may serve as a wear indicator. When it can be seen through the outer layer, it warns of excessive abrasion and loss of thickness of the chemical barrier layer.

In some cases, a wear indicator layer is incorporated into the outer plastic/rubber material. Upon abrasion, the wear layer, which is a different color, becomes visible, alerting the user to take appropriate action (i.e., repair or discard the suit).

Some NFPA 1991- and NFPA 1992-compliant materials are actually a combination of two materials. Typically, an outer material is used to furnish flame resistance and physical protection for the garment, while the inner material provides chemical barrier properties. The outer material is usually provided as a separate cover without the same level of integrity as the suit constructed from the inner material. Both clothing items—cover and inner suit—must be worn together in order to meet NFPA requirements, if the combination of materials is tested together for demonstrating compliance.

Thickness

Material thickness is usually provided in units of mils (one mil is equivalent to 0.001 inches or 0.025 mm). When combinations of garment materials are used in suit construction, the overall thickness should be provided. As discussed in an earlier series booklet, the breakthrough time and permeation rate of a chemical through a primary material is strongly influenced by the material thickness. Material thickness can range from approximately 5 mils for the thinnest materials to 75 mils for very thick, single-layer materials. Combination materials, such as those used in NFPA 1991 compliant suits may be over 100 mils in thickness.

Weight

The strength and durability of a primary material is partly determined by its weight (i.e., weight per unit area). For combination materials, the overall weights per unit area are reported for all layers. Materials used in chemical protective clothing may range in weight from 2.5 to over 40 ounces/square yard. The supporting materials (substrate fabrics) range in weight from about l to 10 ounces/square yard.

Physical Properties

The strength of chemical protective suit materials can be characterized with measurement of several different properties. The most common of these, and those required for measurement by NFPA standards, include:

- tensile strength,
- burst strength,
- tear resistance, and
- puncture/tear propagation resistance.

Each of these properties is further described below:

- ***Tensile strength*** is a measurement of the force required to break a material, when the material is pulled linearly between two grips of a tensile testing machine. Tensile strength results are usually reported parallel and perpendicular to the direction in which the material was machined. These material sample orientations are also known as machine (warp) direction, sometimes designated as "M," and cross machine (fill) direction, indicated by "XM." Most commonly, these results are obtained by testing with ASTM D 751, *Standard Methods for Testing Coated Fabrics*, using the Grab Technique or ASTM D 5034, *Standard Test Method for Breaking Strength and Elongation of Textile Fabrics (Grab Strength).* Larger forces indicate stronger materials.

- Testing for ***burst strength*** measures the force to rupture a material when the force is applied perpendicular (normal) to the material's outside surface. The force is applied by an expanding oil-filled diaphragm (Mullen burst) or 25-mm (1-inch) steel ball. Results in this testing are typically generated per ASTM D 751, *Standard Methods for Testing Coated Fabrics*, using the 25-mm (1-inch) steel ball technique and reported in Newtons or pounds force. However, the Mullen burst test is also used with results reported in kilo Pascals or pounds per square inch. A larger force or pressure indicates a stronger, more rupture-resistant material.
- ***Tear resistance*** is a measurement of the force required to continue a tear once a notch or initial tear has occurred in the material. Several different techniques exist for measuring material tear resistance. The most common test method employs a trapezoidal test specimen, which has a 13 mm (½- in.) notch on the short parallel side of the trapezoid. A tensile testing machine is used to pull one end of the notched specimen in a parallel direction to the initiated notch. Other tear tests include the Elmendorf and tongue tear methods, which both measure the force to continue the tear in a notched material specimen, but use different specimen configurations and test equipment for this testing. Tear resistance results are reported in Newtons or pounds force for both machine and cross machine material orientations. Larger forces indicate a stronger, more tear-resistant material.
- The measurement of material ***puncture/tear propagation resistance*** entails determining the force required to initially

puncture and then tear the material. The test process is akin to walking by a protruding nail that first "snags" and then "rips" the clothing material. Testing is conducted in accordance with ASTM D 2582, *Standard Test Method for Puncture/Tear Propagation Resistance of Plastic and Thin Sheeting*. The test uses a puncture probe that is mounted on a wheeled carriage. The puncture probe carriage rolls down a vertical track striking a material sample suspended over a holder. The length of the total tear is used to measure the puncture/tear propagation force, reported in Newtons or pounds force. Test results are reported in both machine and cross machine directions. Larger forces indicate a stronger, more snag-resistant material.

A number of physical properties are associated with garment materials, including hydrostatic resistance, abrasion resistance, and flex durability. Hydrostatic resistance measures the pressure required for water to leak through a material film. Abrasion resistance testing involves an evaluation of the damage caused to a material as the result of repeated abrasion. The results for this testing can be reported as either number of abrasion cycles to cause a particular type of damage (e.g., rub through the coating on a supported material) or as the change in a different material characteristic for an established set of abrasion cycles. In the latter case, the barrier effectiveness of a material can be evaluated for a material specimen that has been abrasion to determine if liquid penetration occurs or if there is a reduction in the breakthrough time for the material. Similarly, flex durability for a material is measured by subjecting the material sample to a flexing device that repeatedly folds and bends the sample and then testing some other property afterwards.

Seams

Seams are formed when two pieces of the suit material are joined. Other seams, such as those around visors, closures, and fittings, are discussed later. A seam should, at a minimum, equal the physical strength and chemical resistance of the garment material. However, many products use seam construction methods that do not provide the same barrier qualities as the material since lower levels of performance are expected. If the garment material is very strong, a seam with lesser strength may be acceptable. Seams may be described by the type (method of construction) and by their breaking strength.

Type

Basic seaming techniques and components include:

- ***Stitching***—Stitching, typically with nylon or other types of thread, holds sections of the primary material together. In some cases, the stitching is essential to the strength of the seam. Double stitching is usually used in these cases. For additional strength, the garment material is sometimes lapped (folded over) before stitching. In other cases, the stitching is not intended to provide physical strength, but rather to hold the sections in place while they are cemented or welded together. Needle holes produced by stitching are potential routes for chemical entry. These holes must be covered or sealed. For garments that are only sewn, the most common seam type is a serge seam. In this type of seam, the two edges of fabric are aligned and sewn with a stitch that covers the raw edges of the fabric to a depth of about an eighth of an inch. Then the garment is inverted to place the rough fabric edges inside the garment. A variant of this technique involves sewing placing a third piece of fabric over the rough edge of the seam and not inverting the garment. This practice is known as a bound seam.
- ***Strapping***—Strapping involves a narrow strip of plastic/rubber material that is cemented or welded over the seam. Strapping covers the needle holes of the stitching and provides additional strength to the seam.
- ***Cement***—Some seams use cements or adhesives, which are solutions of plastics or rubbers in a solvent. Cements are used to attach strapping or to form the seam directly. When the solvent evaporates, the previously dissolved plastic/rubber remains. This residue should be equal in chemical resistance to the primary material.
- ***Welding***—Welding involves joining materials on a molecular level and contrasts with the mechanical approaches of stitching and adhesives. There are two types of welding: thermal and solvent. Thermal welding is achieved either by applying heat directly to the plastic or by passing an electric current through

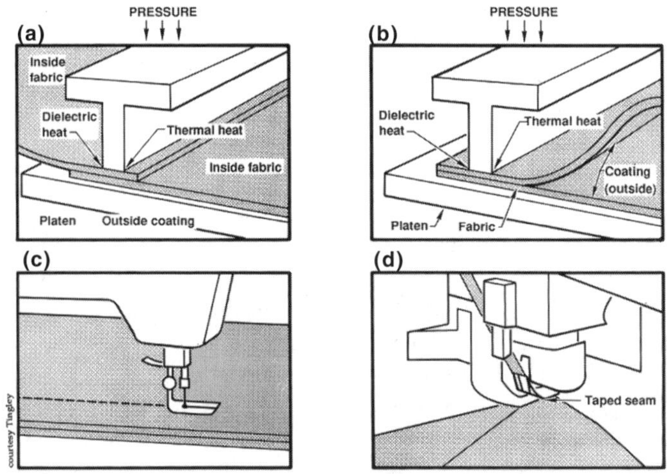

Figure 4.
Sample seam constructions: (a) Welded lap seam; (b) Welded "butt" seam; (c) Sewn seam; and (d) Sewn and taped seam.

the material to generate heat (dielectric welding). Solvent welding is achieved by partially dissolving the surfaces of the plastic material and joining the surfaces under pressure. As the solvent dries, the molecules of the two surfaces become entangled and bonding occurs. In many cases, a welded seam has the inherent chemical resistance of the suit material since no other material composes the seam. However, in some instances, the configuration of the seam and use of multilayer laminate materials may preclude a seam that forms the same barrier as the individual suit material. Not all primary materials are susceptible to welding. For example, many elastomeric materials cannot be welded.

- ***Sealant***—Sealants (top coatings) are plastics/rubbers that are dissolved in a solvent. Sealants are used to fill the holes created by stitching. The chemical resistance of the sealant should equal that of the garment material.

Most seams are constructed using a variety of the above components. Some representative seam constructions are shown in Figure 4. To achieve chemical resistance equal to the garment material, sealing or taping both sides of the seam is required. The majority of rubber-like materials use stitching in combination with strapping, whereas plastic materials using stitching and welding or simply welding.

Seam Strength

The breaking strength of seams is measured by placing a seam sample in a tensile testing machine and pulling it until it breaks. Seam strength is usually measured in accordance with ASTM D 751, *Methods for Testing Coated Fabrics, using Seam Strength Technique*. The seam in the sample runs perpendicular to the force applied by the tensile testing machine, and the force required to break the sample is measured and reported in Newtons or pounds force (sometimes on the basis of the seam sample width in millimeters or inches). The mode of seam failure may also be recorded indicating whether the breakage occurs at the seam or in the material. Breakage in the seam indicates the seam is weaker than the garment material. Figure 5 shows a seam strength test being performed.

Visors

Visors or faceshields are the part of the chemical protective suit used to protect the wearer's face and eyes, while allowing clear vision for the wearer. Visors must allow clear vision with a wide field of view. The materials should resist scratches and be chemically resistant. Visors can be described in terms of their:

- field of vision,
- material of construction,
- thickness,
- interface, and
- clarity.

Field of Visor

The visor provides a transparent window in the suit through which the wearer can see. The size and positioning of the visor greatly affects the field of vision provided by the visor. Visors

Figure 5.
Measurement of seam strength.

often limit the ability of the users to see their feet, particularly if a short person is wearing an encapsulating suit, and to the sides. Moreover, the field of view should not be reduced significantly when the wearer turns his or her head. In recent years, many encapsulating suit manufacturers have endeavored to enlarge the size of the visors to provide a wide field of vision.

Materials

Visors are often made from polycarbonate, urethanes, acrylics, fluorinated ethylene propylene (FEP), and polyvinyl chloride. Depending on the thickness and nature of the visor material, it may be flexible or rigid. Many visors today are considered "semi-rigid" having some flexibility to accommodate movement, but are stiff enough to maintain their shape and remain

Figure 6.
Example of semi-rigid suit visor in totally encapsulating suit. (Courtesy of Trelleborg)

upright in use (see Figure 6). All these materials should have good clarity and strength. Clarity is described below. Visor material strength can be determined by measuring its burst resistance and puncture propagation tear resistance (as measured for garments – see above) or impact resistance in the case for rigid visors. Impact resistance of rigid visor materials can be measured using test methods that are specified in ANSI Z87.1, *American National Standard Practice for Occupational and Educational Eye and Face Protection*.

Some visor materials such as acrylic and polycarbonate are susceptible to crazing (i.e., surface cracking) upon exposure to certain chemicals. Crazing severely degrades vision clarity and can cause catastrophic failure of the visor. For these reasons, splash visors are sometimes worn over the primary visors. Splash visors are clear, flexible, disposable films of highly chemically resistant materials, such as FEP. They are intended to prevent liquid chemicals from coming into direct contact with the visor, and thereby increase the useful life of the visor. Nevertheless, splash visors do not prevent visor exposure to vapors, which in certain situations can also cause crazing.

Thickness

The overall thickness of the visor may be reported in mils, or thousandths of an inch. The thickness of the visor is a consideration in the selection of chemical protective suits. Thin flexible visors may be subject to damage from creasing, whereas thick rigid visors are generally heavier, but can provide better physical integrity.

Interface

Visors are either permanently attached to the suit or are removable. Replacement of a permanently attached visor usually requires returning the suit to the vendor. Removable visors employ a mechanical clamping device and a rubber gasket to attach the visor in a liquid-tight or gas-tight manner.

Clarity

Visor clarity is represented in the ability of the wearer to see clearly through the visor portion of the suit without distortion. From a practical standpoint, visor clarity can be measured by having the end user attempt to read an eye chart of some other printing while wearing the suit (in many cases, the end user will also be wearing a respirator full facepiece. Thus the wearer has to be able to see clearly through at least two transparent windows. Visor clarity can also be assessed by measuring the percent light transmission and haze of the visor material. Testing is performed per ASTM D 1003, *Standard Test Method for Haze and Luminous Transmittance of Transparent Plastic*. In this test, a sample of the visor material is placed in front of a specified light beam and the amount of light transmitted through the sample is measured at a range of wavelengths relative to the light incident on the visor sample. Light transmission decreases as visor polymers or other substances absorb light. The higher the light transmission, the better the clarity of the visor. Haze is a measure of how a visor scatters light. Visors that are hazy make it difficult for the wearer to see clearly. Haze is measured using the same test method as used for light transmission; however, the amount of light scattered is reported. Visors with lower haze have better clarity.

Gloves

Gloves provide chemical and physical protection to the wearer's hands and wrists for the ensemble. A variety of different gloves can be used with chemical protective suits to form a complete clothing ensemble around the wearer. As with the garment material, gloves can be described in terms of material and thickness. In addition, the method for attaching the glove to the suit can vary, as well as available sizes of the gloves for use with chemical protective suits. In general, the gloves should have equivalent or better chemical resistance than the garment material. Gloves should also be as durable as the garment material; however, this is not essential since the gloves can be replaced more readily than other parts of the ensemble.

Material

Gloves may be composed of any of the materials previously described in Chapter 4. Overgloves are sometimes used with totally encapsulating suits to provide an additional layer of chemical permeation protection and to help minimize physical abuse to attached gloves. Because NFPA 1991 imposes rigorous chemical permeation requirements and requires that gloves offer some flame resistance, glove combinations are common with many NFPA 1991-compliant chemical protective suits. Usually, the outer glove is a flame resistant textile or elastomeric glove with an inner flat film glove (constructed of a plastic laminate material). In some cases, NFPA 1991 vapor-protective suits use up to four different gloves, with a loose inner cotton glove worn to aid in donning and absorption of hand sweat. For splash suits, which may or may not have attached gloves, single gloves are typically used.

Thickness

The average overall thickness of the glove or glove combination should be available information. Thicker gloves can make hand function difficult resulting in lower dexterity and tactility. Very thick gloves may prevent the wearer from picking up fine objects or operating small switches.

Note that in this type of application the overall thickness cannot be used as an indication of relative chemical resistance.

Thickness is commonly given in glove manufacturers' catalogs for unsupported styles, that is, gloves that consist entirely of hand-shaped pieces of rubber or plastic. A thicker rubber film will be more protective than a thin one, all else being equal, so the this can be useful information when unsupported gloves are the only protective items being worn. But the correlation between thickness and chemical resistance breaks down for supported gloves, which have a layer of woven or knitted fabric on the inside supporting the rubber or plastic film. The boundary between film and fabric is typically very rough and uncertain, so it is hard to say exactly how thick the chemical-protective film is. And when two to four gloves are worn together as parts of a full-body ensemble, the overall thickness correlates only with dexterity.

Interface

Gloves may be attached or loosely worn with chemical protective suits. Chemical protective suits with liquid-tight or gas-tight integrity require an interface of equal integrity. Liquid-tight interfaces may be achieved by simply including an elastic band at the end of the suit sleeve. Gloves used with vapor-protective suits or totally encapsulating ensembles must be attached to the suit in a gas-tight manner. Several physical configurations of the glove/sleeve interface are discussed below.

Gas-tight glove/sleeve assemblies are of two types: integral and detachable. An ***integral*** glove is directly and permanently attached to the sleeve by seaming at the interface. Some manufacturers use a rigid ring at the interface to aid this attachment. Most often the seaming technique of the primary material seams (including stitching, strapping, etc.) is used. The glove can only be removed by cutting or debonding at the interface. The glove replacement procedure is not simple, and it is sometimes required that the suit be returned to the vendor. As with all seams, the glove/sleeve interface should be as strong and as chemically resistant as the garment material.

Detachable gloves are secured to the sleeve by mechanical means and can readily be replaced by the user. Alternative gloves can be used depending on the situation. There are four basic types of detachable glove/sleeve interfaces:

1. ***Concentric Rings***—This interface is based on two tapered rings about 1 to 2 inches wide and 4 inches in diameter. One ring is slightly smaller in diameter than the other and fits snugly into the larger one. The larger ring is positioned at the end of, and inside, a tapered sleeve end. The smaller ring is placed inside the glove at the glove cuff. The glove is then pushed down the garment sleeve from the shoulder. As the glove passes through the sleeve opening, the smaller insert is pressed inside the larger one, and a compression seal is formed.
2. ***Ring/Clamp***—This interface uses a support ring in the sleeve and a clamp around the glove. The ring, about 4 inches in diameter and 1 inch wide, is placed in the garment sleeve near the cuff. The glove is then pulled over the ring on the outside of the garment. A glove clamp, similar to a hose clamp, is placed around the glove over the ring. The clamp is tightened to form the seal. Some ensembles include an additional piece of primary material at the end of the sleeve that folds over the clamp.
3. ***Connect Rings***—In this type of interface, rings installed by the manufacturer at the glove and the sleeve cuffs are mechanically joined.
4. ***Elastic Bands***—Like some of the other interfaces described above, a hard ring is installed into the end of the suit sleeve. The gauntlet end of the glove is pulled over the ring and secured in place with a wide elastic band.

Three of these glove interface designs are pictured in Figure 7.

Overgloves, which are large and bulky, may not be attached to the suit in a liquid-tight or gas-tight manner. These gloves may be simply worn over attached inner gloves or attached with an elastic band. The purpose of overgloves are to provide physical protection to a weaker inner glove.

Some chemical protective suits provide splash guards, as part of the interface, to prevent entrapment of liquid chemicals between the inner and outer gloves. A splash guard is a flap of suit material attached to the sleeve that can be pulled over the end of an overglove.

Many organizations use duct tape to seal off interface areas between gloves and suit or garment sleeves. This practice should not be used as the primary interface. Duct tape may be useful in

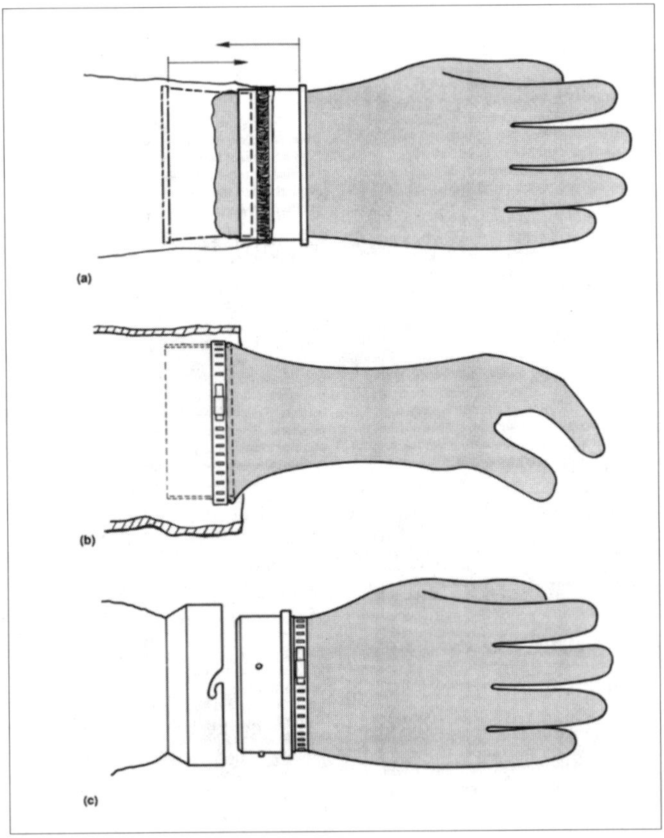

Figure 7.
Examples of glove different interface designs: (a) concentric rings; (b) ring clamps; and (c) connect rings.

helping to secure loose portions of material, but should not be relied upon for providing liquid-tight integrity. Ordinary duct tape also has poor chemical resistance and may lose adhesion once contacted with chemicals.

Available Sizes

The size or range of size for gloves provided with the chemical protective suit will probably be limited if provided with the chemical protective suit. While some suits are sized, the gloves attached to the suit may be of only one size, unless specific sizes are requested at the time of purchase. Since the fit of the glove has a significant impact on wearer hand function, it is important that gloves fit correctly.

Footwear

Footwear provides chemical and physical protection to the wearer's feet and ankles. Footwear is described by the following features:

- type,
- material,
- thickness,
- interface,
- toe type,
- sole type, and
- available sizes.

Type

There are three types of boots used with chemical protective suits: booties, standard boots (i.e., a boot having sole, heel, and closure), and overboots. ***Booties*** are integral, sock-like extensions of garment legs. Booties are the most common form of chemical protection provided for wearer feet on vapor-protective or totally encapsulating suits. ***Standard boots*** are conventional one-piece footwear usually extending above the ankle, and including an attached sole. ***Overboots*** are designed to be worn over standard boots, booties, or regular industrial footwear. Overboots for standard boots are similar to booties (in that they have no firm sole or heel), but are much larger. Overboots worn over booties will generally provide all of the needed physical protection to the foot. If the overboot is worn over another boot or standard footwear, then the overboot may provide some of the chemical resistance and physical protection. Figure 8 shows examples of different footwear worn with full-body chemical protective clothing.

Most NFPA 1991-compliant vapor-protective suits use booties made of garment material because the bootie material chemical resistance data are the same as for garment material. However, use of

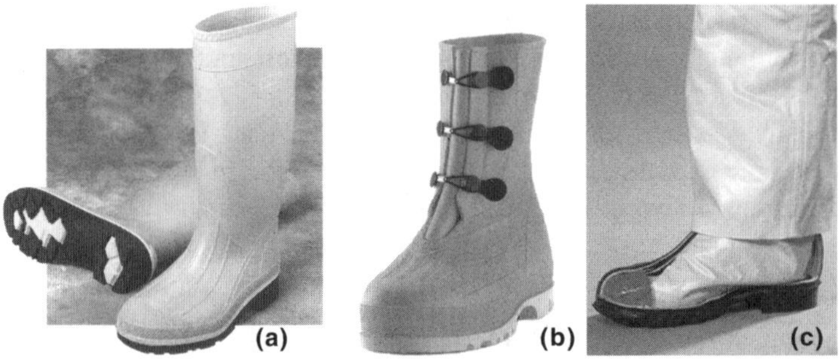

Figure 8.
Examples of different footwear used with full-body chemical protective clothing: (a) Standard rubber boot (courtesy of Norcross Safety Products LLC); (b) Overboot (courtesy of Tingley); (c) Suit bootie shown inside overboot (courtesy of Trelleborg).

booties requires an overboot for adequate physical protection and to meet the various foot protection requirements of the standard.

Material

Booties are typically fabricated of the garment material. Both standard or overboots are constructed of elastomeric or plastic materials. Common boot materials include Neoprene, polyvinyl chloride, butyl rubber, or special rubber or plastic blends. The forthcoming series booklet, "Types of Chemical Protective Clothing," provides additional detail on the construction of chemical protective boots.

Thickness

The overall thickness of the footwear or footwear combination is typically provided in mils. Since thickness varies for some types of boots, the average thickness of the upper boot is usually reported. Thicker boots will generally mean heavier boots and create more wearing stress on the wearer. As with gloves, the thickness of the footwear material is not a predictor of its chemical resistance.

Interface

Booties are attached to the garment with a seam (usually of the same construction as the primary material seam). Because these materials are usually susceptible to abrasions, booties are designed for wearing inside standard boots. Since the booties are attached to the garment with a gas-tight seam, the standard boots worn over them do not need a gas-tight interface. Typically, encapsulating suits with bootie leg terminations have splash guards. A splash guard is a flap of primary material attached to the garment leg that can be pulled over the top of the overboot in order to minimize the entry and accumulation of liquids in the overboot.

Standard boots may be integral or detachable. Gas-tight boot/pant interfaces are achieved by means similar to those described above for the glove/sleeve interfaces. Overboots can be made integral to the garment by seaming, in which case they are used in conjunction with standard boots that do not have a gas-tight interface. Alternatively, overboots can be of the pullover type with an elastic cuff, in which case they are used with standard boots that have gas-tight interfaces.

Toe Type

Footwear should have steel toes or other toe reinforcements to protect the foot from falling objects. Most industrial footwear is equipped with steel or composite toes that resist impact and compression in accordance with ANSI Z41, *Standard for Safety-Toe Footwear*. When booties are used, this protection must be provided in the overboot. Alternatively, this requirement can be provided by standard industrial footwear when a soft overboot or overshoe is used.

Sole Type

Outer soles should be provided a slip resistant tread. NFPA 1991 and 1992 set specific requirements for a coefficient of friction test for the footwear outer sole. In addition, the sole must provide resistance to puncture as defined in the ANSI Z41 standard. This is generally accomplished by a metal plate inside the sole of either the overboot or primary footwear.

Available Sizes

The range of boot sizes is provided for boots or other footwear provided with the chemical protective suit. As with gloves, some manufacturers provide only one size boot only. Boot size is not applicable to booties, since booties are not sized in conformance with conventional footwear sizing standards. Nevertheless, wearing comfort of the suit will be heavily dependent on the footwear size. With bootie and overboot configurations, end users must select a larger size than normally worn to accommodate the extra material in the sock-like booties.

Closures

Closures are used to seal the opening through which the ensemble is donned and doffed. Factors affecting closures include:

- type,
- length,
- orientation,
- interface,
- cover, and
- crosswise strength.

Type

There are seven types of closures currently being used in chemical protective garments and suits:

- conventional plastic or metal chain (teeth) zippers,
- snaps,
- hook and loop tapes,
- adhesive tapes,
- pressure-sealing zippers,
- extruded track closures, and
- thermally welded cofferdam closures.

Of these, the first four (conventional zippers, snaps, hook and loop tapes, and adhesive tapes) offer no barrier protection when used alone in chemical protective suits. They may offer some resistance to penetration by liquid splashes, when the suit is designed to channel liquid from the closure area. However, they are most often used as a secondary closure in conjunction with the primary closure. The last three closures (pressure-sealing zippers, extruded track closures, and thermally welded cofferdam closures, illustrated in Figure 9) provide at least liquid-tight integrity and sometimes gas-tight integrity. These closures are further described below:

- ***Conventional plastic or metal teeth zippers*** are the common, everyday zippers used in most clothing and apparel. These zippers consist of a chain (the metal or plastic teeth), a base fabric or tape to which the chain is attached, a metal or plastic slider (used to open and close the zipper), and stops at either end of the zipper. These zippers offer no barrier performance due to the open nature of the chain design, but are used instead for physical strength of the closure in combination with another closure or series of protective flaps.
- ***Snaps*** are used to provide a positive means for securing material flaps, and may be the primary closure on non-liquid-tight protective garments. Snaps are most often encountered on outer covers worn in conjunction with chemical protective suits.
- ***Hook and loop tape*** consists of two separate plastic tapes. One of the tapes has numerous hooks, while the opposing tape has loops. The two tapes stick together when the hooks "catch" the loops of the opposing tape. The most common form of hook and loop tape is Velcro®, although several other commercial forms of hook and loop tape are available. Hook and loop tapes offer neither liquid-tight nor gas-tight integrity for the garment, and therefore must be used with other closures to provide barrier protection in chemical protective suits. Hook and loop tapes are most often used to secure protective flaps over other closures or for outer garments.
- ***Adhesive tapes*** are most often used on garments intended to be disposable or for limited use. The tapes are placed on the inside edge of a cover flap with a paper tape that is removed, with the flap edge stuck to the garment. Adhesive tapes are generally used in conjunction with conventional zippers. Because the tapes require a flat surface for effective sealing, this type of closure can be subject to gaps from wrinkles in the fabric produced by the contour of the wearer's body. As a consequence, adhesive tapes often fail to produce a liquid-tight closure.
- The ***pressure-sealing zipper*** is the most common closure with liquid- and gas-tight integrity. These zippers are designed such that the teeth of the zipper protrude only slightly through a

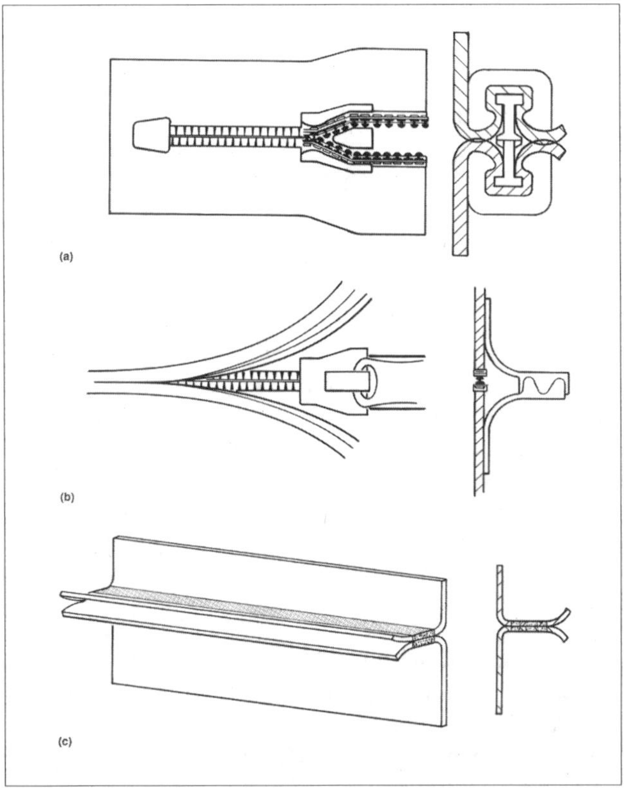

Figure 9.
Examples of different totally-encapsulating chemical protective suit closure systems: (a) pressure sealing zipper; (b) two-track closure; and (c) thermally sealed cofferdam.

plastic/rubber coated fabric. When the zipper is closed, the teeth lock, tightly pressing the plastic/rubber on the two sides of the zipper together, forming a gas-tight seal. Current pressure-sealing zippers are manufactured using either Neoprene or polyvinyl-chloride-coated nylon or polyester fabric, although a new zipper is available that uses a Viton-coated neoprene for the zipper tape. Neoprene is more commonly used because of the poor aging characteristics of polyvinyl chloride. The zippers are typically equipped with nickel-silver teeth and a brass slider. Zippers can be ordered with stainless steel components for additional chemical resistance.

- Another type of closure found on totally encapsulating chemical protective suits is based on ***extruded tracks***. This closure is similar in principal to the Ziploc™ closure of plastic bags. The sealing tracks, when pressed together, form a liquid-tight seal for the closure. These closures are sometimes used in conjunction with a conventional zipper. These closures can be difficult to seal along the extended length required in chemical protective suits. Some materials used in their construction become stiff under cold working conditions.
- A last type of closure utilizes flaps of material (along both sides of the opening), which are ***thermally welded*** together with a portable heat sealer after the suit is donned. Unsealing the suit is accomplished by cutting off the welded portion of the flaps. The flaps are initially sufficiently large to permit the closure to be sealed, cut, and resealed several times. This type of closure is only used in specialized, noncommercial chemical protective suits.

Length

The length of closure varies with the suit and type of garment design. Closure length is particularly important for totally encapsulating chemical protective suits. The greater the closure length, the easier the donning and doffing; however, cost increases with length.

Orientation

Closures for most splash suit and coverall designs are in the front of the garment. For totally encapsulating suits, closures may be located along one side or diagonally across the front of the garment, or along the center of the back. Frontal and some side closures allow the user to seal or unseal his or her own suit. Rear closures provide easier donning and doffing, but require an assistant to seal and open the suit. However, rear closures are also less susceptible to liquid splashes.

Interface

The closure is typically attached to the garment material by one of the methods described above for seaming. As with seams, this interface should be of equal strength and chemical resistance to that of the primary material. However, the expectation is that closures often provide less chemical resistance than the suit or garment's other materials. In NFPA 1991, suit closures are only tested for penetration resistance not permeation resistance, recognizing the less chemical-resistant nature of this component.

Cover

Some garments also have a protective flap that covers the closure. The flap provides protection to the closure from physical abuse and chemical splashes. This protective flap is generally constructed of garment material, and may include snaps, hook and loop tape, or adhesive tape to secure the end of the flap to the garment along the end of the flap.

Crosswise Strength

The breaking strength of closures can be measured much the same way as seam strength. Results for closure crosswise strength are produced using a modified form of ASTM D 751, *Methods for Testing Coated Fabrics, using Seam Strength Technique*. Closures should have the equal or better strength than suit seams. Closures in totally encapsulating suits that bend around respirator-compressed air cylinders (in the back of the suit) may be subject to higher stresses capable of causing their failure as compared to closures located elsewhere on the suit.

Exhaust Valves

Exhaust valves are designed to vent exhaled air from the wearer's breathing apparatus from the inside of gas-tight or encapsulating suits to the outside environment. Their use may also serve two other purposes: (1) creating a positive pressure in the suit, and (2) lifting the suit off the wearer's shoulders. The philosophy behind the use of exhaust valves is that a positive pressure in the suit counters inward leakage of gases and liquids through any openings or pinholes in the suit. However, the build-up of excessive pressures in the suit can restrict movement and stress the suit seams. Encapsulating suits are therefore equipped with exhaust valves that vent at sufficiently low pressures while still maintaining a positive pressure during suit wearing. The valves are also designed to prevent the inward flow of air as the internal suit pressure varies.

The use of exhaust valves in totally encapsulating suits are affected by:

- the number of valves,
- type of valves,
- location,
- installation,
- cover, and
- inward leakage.

Since only encapsulating suits usually have exhaust valves, this section does not apply to many types of liquid-tight or nonencapsulating chemical protective suits.

Number

The number of exhaust valves in encapsulating suits range from 1 to 4 valves. Most manufacturers prefer to use 2 valves, because in the pressure testing of suits, valves must be sealed and too many valves result in chance that a valve will be incorrectly installed after pressure testing. Increasing the number of valves also increases the potential points of leakage into the suit. For NFPA 1991-compliant suits, the number of valves are determined by the performance of the suit in an airflow test that simulates the high rate of release of air inside the suit during a self-contained breathing apparatus failure.

Type

There are two basic types of exhaust valves used in encapsulating ensembles. The most common type, a ***flapper valve***, consists of a rubber diaphragm or disk supported under tension in a housing such that the valve is closed when there is equal pressure on both sides or less pressure on the inside of the ensemble. When an overpressure exists inside the suit, the diaphragm is pushed outward, opening the valve. The second type of exhaust valve employs a spring to hold closed the diaphragm pressed against the housing and the valve. These valves are sometimes referred to as ***spring-operated***. When the pressure in

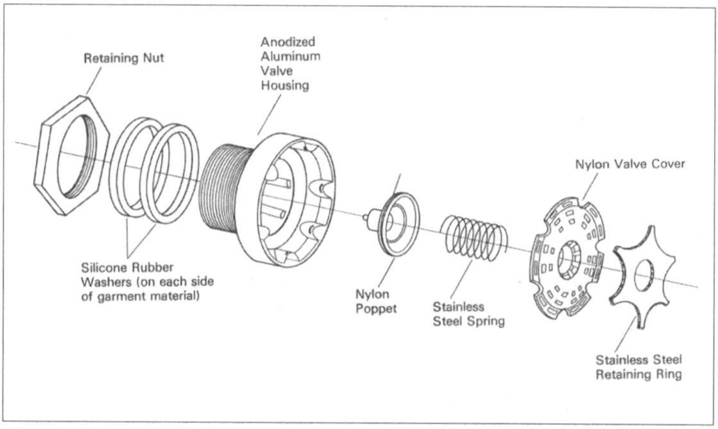

Figure 10.
Diagram showing the design of spring-loaded chemical protective suit exhaust valves.

the suit is greater than the force of the spring, the valve opens. One feature of this type of exhaust valve is that the cracking pressure can be closely controlled. However, spring-loaded valves tend to hold greater positive pressure inside the suit. A representative spring-loaded exhaust valve is pictured in Figure 10.

Location

Exhaust valves may be located in different locations on the suit. Typically, valves are located on the top of the suit, either on the hood or upper torso, although valves are sometimes placed on the suit legs. Valves are almost always placed on the rear of the suit to prevent possible exposure to splashes with liquid chemicals.

Installation

The design of the valve must provide a tight seal to prevent backflow or back diffusion of outside air that may be contaminated. Both spring-controlled valves and flapper valves typically have diaphragms of natural or silicone rubber. The housings are either metal or plastic. These materials, as well as the metallic spring, are susceptible to chemical attack. Valves may be removable or permanently installed. Moreover, the valves should routinely be inspected for material degradation and performance.

However, permanently installed valves cannot be easily examined. The exhaust valve must also be mounted securely so that it cannot be pushed into the suit by bumping up against a wall. Most values use a combination of gases and rings on both sides of the suit to accomplish a secure and gas-tight seal. Under NFPA 1991, the mounting strength of the exhaust valve is also evaluated.

Cover

Both flapper and spring-controlled valves are typically installed with covers designed to prevent direct exposure of the valve to liquid chemicals. These covers may be constructed of garment material or may be of molded or formed plastic or rubber materials that fit over the top of the valve. Some valve covers are designed to seal themselves unless air is flowing out.

Inward Leakage

Inward leakage resistance is a measure of how well a valve prevents airflow in the direction opposite its normal operation, that is, the ability of the valve to prevent contaminated outside air from getting inside the suit. Inward leakage is measured using an apparatus that applies a vacuum to the interior side of the valve while measuring flow into the valve. For a 25 mm (1-inch) water gauge vacuum, NFPA 1991 requires that the inward flow be less than 30 milliliters per minute. This is the same requirement applied to respirator exhalation valves as applied to their certification by the National Institute for Occupational Safety and Health (NIOSH).

In certain situations, such as when squatting, pressure can rapidly build inside the ensemble because of the sudden loss of free volume. This pressure must be relieved to prevent restricting the movement of the wearer and to prevent overstressing the seams. The valve must have sufficient airflow capacity to allow rapid pressure relief.

Fittings

Certain ensemble configurations include umbilical or other external air supply, cooling, and communication systems. Fittings are required where the lines for these systems enter the ensemble, usually through the garment material in the torso area. Fittings are

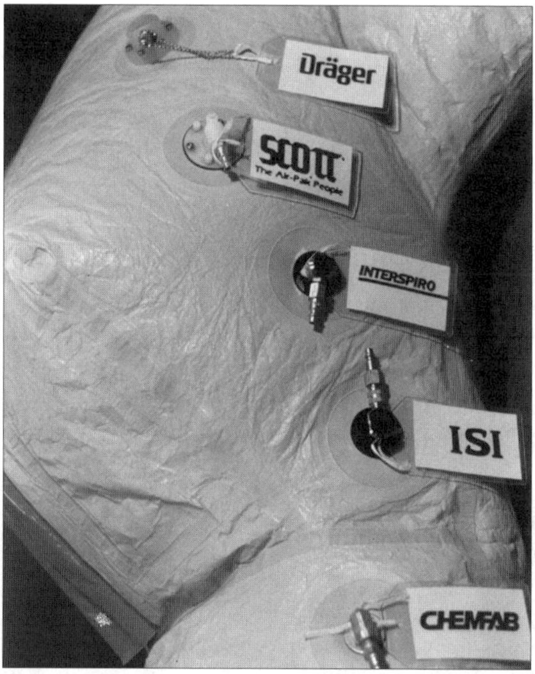

Figure 11.
Different external fittings installed on an NFPA 1991-compliant vapor-protective suit.

used for:

- connecting respirator airlines;
- providing a passthrough for cooling liquids as part of an overall cooling system;
- allowing external wiring of communications sets; or
- for inflating the suit during pressure testing.

Suit wall fittings are most common on vapor-protective or totally encapsulating protective suits (examples are shown in Figure 11). They may also be found on liquid splash-protective or non-encapsulating suits. The fittings should have standard threads (e.g., Schrader- or Hansen-style connections) or quick disconnects for attaching the lines. These fittings and the means with which they are attached to the ensemble must be selected with full consideration to chemical resistance. Furthermore, their attachment to the ensemble must be sufficiently strong to prevent their being pulled out. NFPA 1991 imposes a test for the pull-out strength of fittings installed in totally encapsulating chemical protective suits.

Air Supply Systems

OSHA regulations in 29 CFR 19101.20 and EPA guidelines on Levels A and B suits specify the use of a pressure demand, supplied-air respirator approved by the Mine Safety and Health Administration (MSHA) and National Institute for Occupational Safety and Health (NIOSH). Supplied-air respirators are of two general types:

- pressure-demand, self-contained breathing apparatus (SCBA), or
- pressure-demand, airline respirator (with escape bottle).

As discussed earlier in this chapter, not all encapsulating ensembles are either EPA Level A or B. Several encapsulating ensembles are gas-tight but do not have a supplied-air respirator. The air supply for these suits is umbilically fed to an air distribution system in the ensemble, but not to a respirator (what is considered a Type 1c chemical protective suit in Europe). In other ensembles, the respirator or the self-contained breathing apparatus may not be fully isolated from the external environment.

Egress bottles (or escape bottles) supply air to the wearer in the event that the airline is severed. Most vendors supply egress bottles as an option. In order to meet the criteria of Level A or B protection, egress bottles are required for an ensemble that is supplied air by umbilical hose.

A major consideration for selecting an air supply system is how well the garment accommodates the space requirements of the air supply system. The garment should provide sufficient space in its interior to allow a comfortable fit to the user for wearing a breathing apparatus. Components of the air supply system should not chafe or cause internal wear of the garment.

For Level C protection, air-purifying respirators are specified.

Selection of the air-purifying respirator and chemical protective clothing must account for how the two items provide protection to the face. Protective clothing should be selected that provides a liquid-tight interface of the garment hood around the respirator facepiece. Many hoods will not offer this protection, and thus, in practice, end users use duct tape to seal the interface area. While this practice is not recommended, it may be one of the few ways to prevent direct liquid contact from splashes to areas of the face and head. Duct tape has poor chemical resistance and may loose its adhesion when contacted by many chemicals.

Communications Sets

Voice communication, when using full-body encapsulating ensembles, can be achieved by three principal means:

- voice amplification,
- hard-wire systems, and
- wireless devices.

Voice amplification systems, as the name implies, consist of a microphone attached through an amplifier to a belt-mounted speaker. This system is useful when all persons are working within talking distance of one another. ***Hard-wired systems*** consist of a microphone and speaker headset that are connected by wire directly to other workers or to a central, base unit and then to other workers. This system is most conveniently used when the air supply to the ensemble is by umbilical hose. ***Wireless systems*** are based on radio transmitters/receivers and allow the greatest range of operations.

Whichever system is used, it must be located and operate in a manner that minimizes its impact on worker performance and safety. The total weight of the unit and the weight distribution of the unit, especially if it is worn on the head, must be considered. Also, the system must not abrade the inside of the ensemble.

Microphones may be held in place against the neck (i.e., throat microphones), mounted on or in the respirator facepiece, suspended in front of the mouth on a boom or, in the case of bone conduction microphones, mounted somewhere on the head. Neck microphones are well accepted and function well in high noise areas; however, some users report problems due to slippage of the microphone. Readjustment is difficult when wearing an encapsulating ensemble.

Three modes of operation are common:

- continuous,
- push-to-transmit, and
- voice-activated.

Continuous systems are convenient to use and require virtually no training. However, they are continuously consuming power, requiring frequent replacement of batteries. ***Push-to-transmit*** systems reduce battery consumption, but require a free hand to operate. ***Voice-activated*** systems combine the advantages of both, are low-energy consumption, and allow hands-free operation. A slight drawback of voice activated systems is that some training is required in their use since the first syllable of the first word is lost in activating the system.

Wireless systems operate on the basic principles of radio communication. So-called ***simplex*** systems transmit and receive on the same frequency. Thus, only one person can talk at a time and a fixed interval of time must elapse before the next person can begin talking. In a ***duplex*** system, two frequencies are utilized so that there can be rapid give-and-take conversation, e.g., the telephone is a duplex system. Through the use of a base station, duplex capability can be extended to large groups of people. Half-duplex systems allow one person (e.g., the foreman) to have full duplex capability, but all others to have only simplex capability. Thus, the foreman can talk to and hear all members of his group, but the group members can only talk to and hear the foreman. Obviously, each system has advantages and disadvantages in any given situation.

Cooling Systems

In isolating the worker from the environment, encapsulating ensembles can produce a significant thermal burden on the wearer. Cooling due to natural convection and sweat evaporation is prevented. Thermal burden can be reduced by a variety of methods as follows:

- ***Passive systems*** such as vests containing ice, dry ice, frozen gels, or other heat sinks. These systems require equipment for prefreezing the heat sink contents prior to use. Typically, such systems have a service life of 1 to 4 hours, depending on the workload and the external temperatures. One new technology that has been recently introduced involves materials that have a finite water absorption capacity. The garments are wetted out and then provide a heat sink for heat transfer. However, there is no control over the cooling rate or cooling distribution.
- ***Powered fluid circulating systems***, in which a chilled fluid is pumped in tubes in contact with the head, neck, chest, and other body regions. The fluid may be air, water, or some other heat transfer medium. Similar to the passive systems, ice, dry ice, and frozen gels are the principal heat sinks. These systems are battery powered and typically have operating periods of 1 to 4 hours. Some systems can be controlled by the wearer.
- ***Umbilical air cooling systems*** of which there are two versions. The first distributes air to the head, neck, and other body regions through a tubular manifold system. The air is exhausted through the pressure relief valves. Airflow rate can be controlled by the wearer. The second uses pressurized air that enters the ensemble through an expansion valve (i.e., vortex tube). The ensemble may contain one or more vortex tubes so that the cooling air is evenly distributed. However, vortex tubes should not be placed near the head because of the noise from the expanding air. With this system the wearer can control the rate of cooling.
- ***Powered cooling compressor systems***, based on Sterling cycle and conventional compressors. Fuel- and battery-powered systems are available. To date these systems have not proven practical because of their weight. However, new developments in high-efficiency, high-storage capacity batteries may change this situation.

In addition to the above systems, cooling can also be achieved from the evaporation of water that is sprayed over the surface of the ensemble.

As with air supply systems, the selected cooling system (if any) should not chafe the inside of the garment and consider the impact of overall weight burden to the wearer.

Ensemble Evaluation and Selection Factors

The selection of full-body chemical protective clothing depends on:

- the nature and severity of the hazards,
- the type and duration of the tasks to be performed,
- the performance features and limitations of the available clothing, and
- cost (not discussed in this booklet).

Protective clothing and equipment must be selected based on a risk assessment. The risk assessment must account for all hazards, their severity, the likelihood for exposure, and the consequences of exposure (see series booklet, "PPE Selection Criteria," for an approach for conducting the risk assessment). For chemical protective clothing applications, the hazards present will be mainly chemical in nature, but there are also concerns related to the work environment in terms of physical hazards, heat stress, and other factors.

Once the risks have been determined, the selection of an ensemble for a specific application typically involves tradeoffs among physical and chemical properties, design features, and human factors considerations. The series booklet on selection of chemical protective clothing provides several simplified flowcharts for chemical protective suit selection based primarily on the identification of chemical hazards. Once the type of suit has been determined, specific performance information on the clothing must be reviewed to determine its appropriateness for the pertinent application. The only standards available for setting minimum performance criteria are the NFPA standards for chemical emergency response.

Selection considerations can be matched with performance information on respective chemical protective suits. The performance of a protective ensemble depends on several properties including:

- chemical resistance of primary materials,
- overall integrity,
- physical properties,
- flame resistance,
- component function,
- design features, and
- comfort, fit, and function.

Considerations for each area are provided below. Criteria specific to these areas are also discussed relative to those required in the NFPA standards.

Chemical Resistance

Ideally, specific chemical resistance should be available on the selected chemical protective clothing. However, data may not be available for many products or the data that are available provide little guidance to end user such as degradation ratings provided with many splash suits. In some cases, generic data may be present from collective sources of information. Yet, these data may only approximate the materials being used in the chemical protective garments. For example, all butyl rubber materials regardless of thickness, source, and formulation are usually classified together in order to form a single recommendation for each chemical. Generic classifications provide general guidance as to chemical resistance but, as noted elsewhere, chemical resistance can vary significantly within a generic class.

While chemical resistance has been extensively discussed previously, it is important to point out that chemical resistance data should be thoroughly reviewed in evaluating chemical protective suits. Specifically:

- The chemical resistance of all primary materials (garment, visor, glove, and footwear), seams, closures, and interfaces should be considered.
- The appropriate type of chemical resistance data should be

used based on the expected chemical hazards. Permeation data should be used for suits intended to provide vapor protection or protection from highly hazardous chemicals. Penetration data may be acceptable for liquid splash exposures, where chemicals are known, and are neither toxic by skin absorption nor carcinogenic. Penetration data are also appropriate only in those cases where the vapors from the chemical are not hazardous for skin contact at the expected concentration.

- The breakthrough time, or length of the penetration test with "passing" performance should be as long as the expected duration of the suit wearing.

Penetration test results are shown as "pass" or "fail." As required in NFPA 1992, testing is performed using ASTM F 903, *Standard Test Method for Resistance of Protective Clothing Materials to Penetration by Liquids*. The specific exposure protocol used involves a 1-hour exposure at the following pressures:

- 5 minutes at ambient pressure,
- 1 minute at 2 psig, and
- 54 minutes at ambient pressure.

The determination of chemical penetration is made at the end of the 1-hour test period.

The NFPA standards set specific requirements for chemical resistance for primary materials, seams, and closure of chemical protective suits. NFPA 1991 requires that primary suit materials and seam exhibit permeation breakthrough times of 1 hour or longer for all 21 chemicals in ASTM F 1001, *Standard Guide for Chemicals to Evaluate Protective Clothing Materials*. NFPA 1992 requires that primary suit materials and resist penetration for 1 hour by those liquid chemicals in ASTM F 1001 that are neither carcinogenic (actual or suspected) or toxic by skin absorption. NFPA 1992 also requires that seams resist penetration for the same length of time and list of chemicals. NFPA 1991 and 1992 pose the same chemical resistance requirements for garment, glove, and footwear materials following abrasion and repeated flexing. If the suit or clothing is not intended for emergency applications, then test data specific to the anticipated exposure chemical(s) should be used.

Complete Ensemble Integrity

As described earlier in this booklet, two methods exist for evaluating the performance (integrity) of the complete ensemble. These methods involve measurement of suit gas/vapor integrity and liquid integrity.

An alternative method for measuring the gas/vapor integrity is specified in 29 CFR 1910.120. This test involves leak-testing chemical protective suits with a gas challenge agent in a manner simulating actual use. This testing involves exposing a test subject wearing the ensemble in a closed chamber to a challenge agent and measuring the agent concentration both inside and outside the suit. The proposed ASTM method employs ammonia gas at a concentration of 1000 to 2000 ppm and length of stain detection tubes. Other methods may use different gases or aerosols (at nontoxic concentrations) and appropriate detection methods. Dependent on the means used to measure the presence of gas, the test can be qualitative, semi-quantitative or quantitative. When concentration measurements are precise, the results can be used in a semi-quantitative manner by calculating the ratio of the external to internal gas challenge agent concentrations to determine a protection factor. Large protection factors are indicative of high ensemble integrity. More sophisticated forms of this testing are possible such as similar test using sulfur hexafluoride that is specified in NFPA 1994 (chemical/biological terrorism incidents).

Careful consideration should be given to the integrity of a chemical protective suit or garment. If chemicals are going to be encountered that are skin-toxic, carcinogenic, or otherwise hazardous by skin contact, the selected protective clothing should have gas-tight integrity. Simple encapsulation of the wearer is not sufficient. Gas-tight integrity must be demonstrated by pressure testing the suit in accordance with ASTM F 1052 or 29 CFR 1910.120 (these methods are explained in the section on product design in this booklet).

If chemicals are encountered where vapor hazards do not exist and the principal form of exposure expected is by liquid splashes, then clothing should be chosen that offers liquid-tight integrity. In general, the most significant problems for "splash suits" are the integrity of closures and interfaces with respirators, gloves, and boots.

Flame Resistance

Except in emergency situations, flame resistance is likely not required for many chemical protective clothing applications. Nevertheless, end users must realize that when incidental flame contact or high heat exposures are possible, the materials used in chemical protective clothing should not contribute to their hazards.

An appropriate test method is ASTM F 1358, *Standard Test Method for Effects of Flame Impingement on Materials Used in Protective Clothing Not Designed Primarily for Flame Resistance*. This test uses the same vertical flame test apparatus required in Federal Test Method Standard 191A, 5903.1 for "Vertical Flame Resistance of Textiles." A folded material specimen is placed in a specimen holder, and suspended over a 38 mm (1½ inch) high methane flame. The folded area of the material specimen is placed 19 mm (¾ inch) into the flame. Two flame exposures are conducted. The sequence of the test is as follows:

1. The sample is first impinged for 3 seconds.
2. After the flame is removed, a determination is made to determine if the material has ignited. If the material ignites, the test is terminated.
3. If the material does not ignite, the same material specimen is re-exposed to the flame for a 12-second period.
4. After the flame is removed, a determination is made to determine if the material has ignited.
5. If the material has ignited after the second flame exposure, the time (afterflame) and distance the material continues to burn is recorded along with any observation of the material's burning behavior (e.g., melting, dripping, consumption).

The test was included in the series of NFPA tests to eliminate use of suit materials that could increase wearer hazard in the event of flame contact. Passing materials or compliant suits do not offer protection against flame contact and these criteria by themselves do not address performance of the suit when exposed during chemical flash fires. The test is intended to prevent users from becoming "walking torches" under the worst of conditions.

The test is designed to answer two basic questions:

1. How easily does the material ignite? and
2. Does the material self-extinguish once ignited?

Some NFPA 1991-compliant suits use what are commonly called flash covers (mainly by users), which appear to use materials similar to proximity clothing. Many end users assume that these suits do indeed provide flame and chemical flashover protection. Unfortunately, in making this assumption, they decide that they can wear the inner suit alone despite the warning of the manufacturer. When they do this, they are wearing a noncompliant suit. These suit flash covers are often provided so that the complete product (both inner and outer suits together) can meet *all* the requirements of NFPA 1991.

There are few current ensembles that provide both flame and chemical protection. Nearly all chemical protective clothing materials will ignite and burn when contacted by flame. These ensembles should not be used for situations where the probability of contact with flame or flashover conditions is high. Combining thermal and chemical protective suit systems should be considered carefully. Thermal protection may be worn either inside or outside the chemical protective suits. When externally worn, thermal protective clothing is subject to contamination and must be disposed of after use. When worn inside (e.g., aramid fiber jumpsuit), the integrity of the chemical protective clothing will be lost in a flame situation, resulting in a potentially life-threatening situation. Alternatively, thermal protection can be worn both inside and outside the suit. In any of these cases, the additional burden (weight/heat stress) to the wearer must be considered.

Physical Properties

Several standard tests exist for measuring many of the physical properties of ensemble materials and in many cases these tests are specific to an ensemble component. More than one test method exists for measuring some physical properties. Table 5 lists a number of test methods for measuring these properties. In general, ASTM standard test methods are recommended because they are readily available, regularly updated to reflect changes in the state of the art, and actively supported by an identifiable group of persons who are available for consultation.

Table 5:
Physical Property Test Methods as Applied to Different Parts of the Full-Body Chemical Protective Ensembles

Physical Property	Test Method(s)	Applicable Materials					
		Garment	Visor	Glove	Boot	Seam	Closure
• Abrasion Resistance	ASTM D 3884 ASTM D 4157	X		X	X		
• Antifog	MIL-A-1542B		X				
• Blocking	ASTM D 1893	X		X	X		
• Burst Strength	ASTM D 751 ASTM D 3786	X	X				
• Coating Adhesion	ASTM D 751	X			X		
• Cold Temperature Bending	ASTM D 2136	X	X	X	X		
• Cold Temperature • Brittleness	ASTM D 2137	X	X	X	X		
• Crosswise Strength	ASTM D 2061						X
• Cut Resistance	ASTM F 1790	X		X	X		
• Distortion	ASTM D 881		X				
• Durometer	ASTM D 2240	X		X	X		
• Flammability	ASTM F 1358	X	X	X	X		
• Flex Fatigue	ASTM F 392	X		X			
• Hydrostatic Resistance	ASTM D 751	X					
• Impact Strength	ASTM D 881		X				
• Light Transmission and Haze	ASTM D 1003		X				
• Ozone Resistance	ASTM D 3041	X	X	X	X	X	X
• Puncture Propagation Tear	ASTM D 2582	X	X				
• Puncture Resistance	ASTM F 1342	X		X	X		
• Scratch Resistance	ASTM D 1044		X				
• Seam Strength	ASTM D 751					X	
• Static Charge Resistance	NFPA 1991	X	X	X	X		
• Stiffness	ASTM D 737	X		X	X		
• Tear Resistance	ASTM D 751 ASTM D 5733	X	X	X			
• Tensile Strength (supported materials)	ASTM D 751 ASTM D 5034	X					
• Tensile Strength • (Unsupported materials)	ASTM D 412			X	X		
• Thickness	ASTM D 1777	X	X	X	X		
• UV Light Resistance	ASTM G 26	X	X	X	X	X	X
• Weight	ASTM D 751	X	X	X	X		

The importance of material physical properties is often underestimated. Physical properties provide a means for estimating and comparing the strength and durability of various ensemble components. Information on certain of these properties is considered essential to the evaluation of ensembles. Although no vendor provides data on all of the properties, it is useful for the user to be familiar with the terms:

Abrasion Resistance—The amount of material removed from a surface during relative motion against another surface. Test is sometimes conducted as a precondition to a chemical resistance test to determine barrier properties following simulated wear.

Antifog—Measurement of the effect of water vapor condensation on the read-through characteristics of clear materials.

Blocking—The adherence of materials to one another or themselves.

Bursting Strength—The multidirectional force required to break a material.

Coating Adhesion—The force required to peel a coating from a substrate fabric.

Cold Temperature Bending—The degree of cracking that occurs at a specified temperature in a flexible material when the material is bent around a cylindrical rod.

Cold Temperature Brittleness—Evaluation of the lowest temperature at which flexible materials will not exhibit fractures or cracks when subjected to specified impact conditions.

Crosswise Strength—The force required to pull a zipper apart in a direction perpendicular to the zipper direction. This can be applied to other types of garment closures.

Cut Resistance—The force required to cause a sharp-edged blade to cut through a material.

Distortion—The deviation of the line of sight through flat or curved sections of transparent materials.

Durometer—The degree to which materials resist indentation by a specified probe under a specified load. Thus, durometer is a measure of material hardness.

Flammability—The ignition, propagation, and self-extinguishing characteristics of a material.

Flex Fatigue—The number of repeated bends required to cause cracking or other damage in a flexible material. Test is sometimes conducted as a precondition to a chemical resistance test to determine barrier properties following simulated wear.

Hydrostatic Resistance—The pressure required to cause the penetration of water through a coated fabric.

Impact Resistance—The force required to cause materials to exhibit fractures or cracks when subjected to impact.

Light Transmission and Haze—Properties of transparent materials; light that passes into one of these materials is either transmitted through in a straight or refracted line, scattered by haze, or absorbed.

Ozone Resistance—The degradation effects on a material as the result of ozone exposure.

Puncture Propagation Tear Resistance—The force required for a sharp object to snag and subsequently tear a material.

Puncture Resistance—The force required to penetrate a material with a cylindrical probe in a direction normal to the surface of the material.

Scratch Resistance—The degree to which clear materials are scratched or marred as a result of motion against another surface.

Seam Strength—The force required in tension to break a seam with the force applied perpendicular to the direction of the seam. This type of testing is similar to tensile testing of the primary material.

Static Charge Properties (Electrostatic Dispersion, or ESD)—The tendency to accumulate or disperse a static electrical charge. In some cases, manufacturers report surface resistivity or electrostatic decay as measures of static charge resistance.

Stiffness—The force required to twist or bend a material a specified amount or the amount of bending occurring as a result of a specific applied force.

Tear Strength—The force required to initiate or propagate a nick or cut in a direction normal to the direction of the applied force. Tear strength results are usually reported for both the warp and fill directions of coated fabrics. Warp refers to threads in a fabric running in the lengthwise direction (machine direction), and fill to the fibers in the crosswise direction (cross-machine direction).

Tensile Strength—The force in tension required to break a material. Tensile strength results are usually reported for both the warp and fill directions of coated fabrics.

UV Light Resistance—The degradation effects on a material as the result of exposure to UV light.

Weight—There are two kinds of weight associated with encapsulating ensembles: the material weight and the ensemble weight. Material weight is the weight per unit area (e.g., oz/sq yd) of the primary material. The ensemble weight is the total weight of the fabricated garment. In some cases, the ensemble weight reported by vendors includes only the garment and detachable gloves and boots, while in other cases ensemble weight includes breathing apparatus and cooling.

The NFPA standards require a minimum battery of physical property tests (a portion of those listed above) to be performed on the garment, glove, visor, and footwear materials of the chemical protective suit or garment. These tests complement chemical resistance testing such as for permeation and penetration resistance. This battery of physical tests standardizes the methods used for measuring each physical property so that results can be compared from one product to another, and compared against a minimum performance requirement. These physical properties are listed below:

Suit Material

- Abrasion resistance (followed by chemical resistance testing)
- Flex fatigue resistance (followed by chemical resistance testing)
- Cold temperature stiffness
- Burst strength
- Puncture propagation tear resistance

Visor Material

- Clarity
- Cold temperature stiffness
- Burst strength
- Puncture propagation tear resistance

Glove Material

- Abrasion resistance (followed by chemical resistance testing)
- Flex fatigue resistance (followed by chemical resistance testing)
- Cold temperature stiffness
- Cut resistance
- Puncture resistance

Footwear Material

- Abrasion resistance (followed by chemical resistance testing)
- Flex fatigue resistance (followed by chemical resistance testing)
- Cold temperature stiffness
- Cut resistance (upper)
- Puncture resistance (upper and sole)
- Impact and compression resistance (toe)
- Coefficient of friction (sole)

Design Features

Table 6 lists the advantages and disadvantages for many of the design alternatives discussed earlier. Chemical resistance issues are not included in the table as they are discussed elsewhere in this chapter. Ensemble design and component alternatives are characterized in a qualitative manner since there are no generally accepted quantitative criteria by which they can be assessed.

Table 6:
Advantages and Disadvantages of Full-Body Chemical Protective Clothing Components and Designs

Component	Advantages	Disadvantages	Comments
Garment Material			
Composition	Wide variety available		Chemical resistance varies from material and from vendor to vendor; Flame resistance and physical properties should be also considered as needed in the application
Supporting Fabric	Adds strength	Increases weight; Reduces flexibility; Edges must be sealed to prevent wicking	
Single layer	Material provides all needed properties	Difficult to achieve comprehensive chemical resistance, flame resistance, and high strength in single layer; material likely to be expensive	Some NFPA 1991 suits are single layer material products
Multi-layer	Combination of dissimilar materials capable of achieving desired performance in cost-effective way	Adds bulk; Both materials must be worn in together to provide intended protection	Some NFPA 1991 suits use over covers in conjunction with an inner suit
Elastomeric	Better flexibility; Good physical strength/durability	Poorer chemical resistance	Properties vary dramatically with elastomer and material construction
Plastic	Better chemical resistance; Lower cost	Less flexibility; Poorer physical durability	Properties vary dramatically with plastic and material construction

Table 6 *(continued)*

Component	Advantages	Disadvantages	Comments
Garment Material			
Wear Indicator	Indicates reduced chemical and physical protection		Preferred means of indicating wear
Increased Thickness/Weight	Increases strength Increases protection	Increases weight Reduces flexibility	
Seams			
Stitched	Applicable to all materials; stronger	Requires strapping or sealants be tested for chemical resistance penetration through pinholes; More labor intensive and expensive	Lapped and double-stitched constructions preferred
Cemented	Works well with rubber materials	Adhesive chemical resistance may differ from garment material; More difficult to achieve uniform seam	
Welded	Chemical resistance may be equivalent to primary material; Less bulky; Less susceptible to snagging	Rubber materials cannot be welded	

Table 6 *(continued)*

Component	Advantages	Disadvantages	Comments
Visors			
Rigid	Sturdy; Impact resistant; Resistant to flex damage in storage	Heavy; Many require gasket for sealing	
Flexible	Light weight	Easily damaged; Little impact resistance	
Integral	Factory sealed and tested	Replacement requires return to factory	
Detachable	Visors easily replaced	Sealing is user dependent; Chemical resistance of gasket may be a concern	
Removable Protector Films	Increase protection and service life	May interfere with vision	
Gloves			
Single	Wide range of materials available Low cracking pressures provide better mobility	Difficult to achieve comprehensive chemical resistance, flame resistance, and physical durability Low cracking pressures may not prevent inward leakage	
Combination	Increased chemical resistance can be achieved	Decrease hand function (dexterity and tactility); Difficult to redon once hand is pulled out within suit	NFPA 1991 requirements can only be met currently with glove combinations
Over Gloves	Increased protection	Decrease hand function (dexterity and tactility)	

Table 6 *(continued)*

Component	Advantages	Disadvantages	Comments
Footwear			
Standard Boot	Increased protection to foot; Physical durability; Available in several sizes	When on encapsulating suits, requires gas-tight interface	Few boots meet all NFPA 1991 requirements and are generally not used with compliant suits; Steel toe and good traction sole recommended
Bootie of Garment Material	Chemical resistance consistent with garment material; Regular seaming used for interface	No physical protection to foot without interface; Normal sizing usually not available	Over boot required
Over Boot (rigid)	Provides physical protection to inner boot or bootie	May be cumbersome; Oversized boots must be used	Steel toe and good traction sole recommended
Over Boot (soft) – worn over standard shoe or boot	Provides chemical protection to standard footwear Lessens bulk of footwear	May be subject to physical hazards	Steel toe and good traction sole recommended
Glove/Sleeve & Boot/Pant Cuff Interface			
Integral	Factory sealed and tested	Difficult to replace, may require return to factory; Not interchangeable for fit or chemical resistance	
Detachable	Easy to replace; Can be optimized for fit and chemical resistance	Seal is made in field and is not readily tested	
· Concentric Rings	No mechanical parts	Relies on compression seal; No clamp or lock; Difficult to inspect	
· Ring/Clamp	Simple; Easily inspected; Mechanical seal	May require two persons; Corrosion of clamp	

Table 6 *(continued)*

Component	Advantages	Disadvantages	Comments
Glove/Sleeve & Boot/Pant Cuff Interface			
· Connect Rings	Ease of use; Mechanical seal	Complexity; Repair may require return to factory	
· Elastic Bands	Simple; Easy glove removal	May not provide adequate seal; Glove may become loosed or dislodged in use	
Closures			
Conventional Plastic or Metal Teeth Zippers	Good physical strength	No barrier protection	Can only be used in conjunction with other closures for gas-tight or liquid-tight suit integrity
Snaps	Positive locking	No barrier protection	Can only be used in conjunction with other closures for gas-tight or liquid-tight suit integrity
Hook and Loop Tape	Good for securing flaps; Easily opens and closed	No barrier protection; May be degraded by some chemicals	Can only be used in conjunction with other closures for gas-tight or liquid-tight suit integrity
Adhesive Tapes	Good for securing flaps	Little barrier protection; Difficult to apply uniform seal	Can only be used in conjunction with other closures for gas-tight or liquid-tight suit integrity
Pressure Sealing Zipper	Simple Gas-tight	Available materials have limited chemical resistance	Stainless steel teeth and slider available; Reverse pull and Viton tape also available
Extruded Track Closure	Protects inner zipper from environment; May be used for liquid-tight suits or garments	Sealing closure takes a long time; Some extruded materials become stiff or inoperable in cold weather	

Table 6 *(continued)*

Component	Advantages	Disadvantages	Comments
Closures			
Thermally Welded Cofferdam	Chemical resistance equivalent to that of primary material	Requires heat sealing equipment; Suit donning and doffing cannot be done without assistance; Number of suit uses limited	
Front of Side Orientation	May allow donning without assistance; Better for emergency doffing	More difficult to get into and out of suit; More susceptible to chemical exposure (liquid splashes)	Ease of donning depends on closure length
Back Orientation	Less exposure to chemical splashes; Easier to put on and take off suit	Requires assistance in donning	Ease of donning depends on closure length
Cover or Flap	Provides additional splash protection to closure	Increases donning time	Flaps may be used in conjunction with hook and loop tape, snaps, or adhesive tapes
Exhaust Valves			
Number (More than one)	Less chance of overall suit failure (ballooning) from valve blockage; Greater capacity to vent exhaust air	Increased risk of inward leakage	Position of valves is important to prevent block. Sitting in a vehicle may block rear shoulder- mounted valves
Flapper	Simple; Inexpensive	Cracking pressure cannot be adjusted; Chemical resistance of diaphragm	Valves same or similar to those used as exhalation valves for respirators

Table 6 *(continued)*

Component	Advantages	Disadvantages	Comments
Air Supply Systems			
Combination Airline/SCBA	Versatile for extended operations; Supply switches to bottle if airline interrupted	Requires penetration of and fitting on suit	
Air-Purifying Respirator	Lightweight; Simple	No warning if canister overloaded; Requires knowledge of contamination levels; Lower protection factor than SCBA or airline	Not permitted for EPA Levels A and B
Communications Sets			
Voice Amplification	Simple; Inexpensive	Functional only over short distances; Available only in some SCBA or respirator facemasks	
Hard-Wired System	Extends range of operation	Requires all persons be connected by wire, limits mobility	
Wireless	Voice/hearing quality high; Different microphone options available; Allows wide range of operation	Battery life limits operation period; Some training required	User must decided on simplex, duplex, or half-duplex modes of operation

Table 6 *(continued)*

Component	Advantages	Disadvantages	Comments
Air Supply Systems			
Combination Airline/SCBA	Versatile for extended operations; Supply switches to bottle if airline interrupted	Requires penetration of and fitting on suit	
Air-Purifying Respirator	Lightweight; Simple	No warning if canister overloaded; Requires knowledge of contamination levels; Lower protection factor than SCBA or airline	Not permitted for EPA Levels A and B
Communications Sets			
Voice Amplification	Simple; Inexpensive	Functional only over short distances; Available only in some SCBA or respirator facemasks	
Hard-Wired System	Extends range of operation	Requires all persons be connected by wire, limits mobility	
Wireless	Voice/hearing quality high; Different microphone options available; Allows wide range of operation	Battery life limits operation period; Some training required	User must decided on simplex, duplex, or half-duplex modes of operation

Table 6 *(continued)*

Component	Advantages	Disadvantages	Comments
Communications Sets			
Continuous	Hands free operation	All conversation transmitted, desired or not; Quickly runs batteries down; Requires communications discipline	
Push-to-Talk	Conserves battery life; Communication occurs only when desired	Difficult to activate switch inside suit; Requires stopping activity to communicate	
Voice-Activated	Conserves battery life; Hands free operation	First word often not heard; Requires communications discipline	
Cooling Systems			
Passive Systems	Simple; Inexpensive	No control over cooling	
Powered Fluid Circulating Systems	Cooling rate controllable	Bulky; Limited service life; Requires replenishment; Hard to integrate with some suits	
Umbilical Air Cooling Systems	Simple	Noisy; Mobility restricted	
Powered Cooling Compressor Systems	Highly efficient	Heavy; Expensive; Not proven practical	

Ensemble Comfort, Function and Fit

Manned ensemble testing is often performed to determine the range of activities that a user can perform while wearing the ensemble. These may include different types of exercises or tasks that simulate the end application of the suit. Results from these tests are generally subjective regarding the design, comfort, and fit of the garment. Measurement of the wearer's physiological responses (e.g., core temperature, skin temperature, heart rate, and weight loss) during this testing can help quantify the results when compared to tests without the ensemble or with other ensembles. Procedures useful in assessing these suit or garment qualities are established in ASTM F 1154, *Standard Practices for the Qualitative Evaluation of Comfort, Function, Fit, and Integrity of Chemical Protective Suits and Ensembles*.

Manufacturers may offer one size or several sizes of their respective garments. With the exception of disposable coveralls (ANSI/ISEA 103), there are no standards for sizing these garments; consequently, one manufacturer's large may not fit the same group of people as another manufacturer's large. Generally, the more sizes offered by a manufacturer, the greater the range of people who will comfortably (and functionally) be fit by a given garment design. The only way to gauge the fit of a garment is to actually try on each of the available sizes offered by the manufacturer. It is important that suit fit and functionality be judged when wearing all of the necessary personal protective equipment such as helmets, breathing apparatus, cooling systems, and other items.

Quality Control and Assurance

Manufacturers employ varying levels of quality assurance in their fabrication of chemical protective suits. Quality control can be applied to several areas of the manufacturing process:

- in the receipt of materials and components,
- testing of subassemblies (e.g., seams, closures, valves), and
- full garment testing.

For example, some manufacturers may test each individual totally encapsulating chemical protective suit for gas-tight integrity. For NFPA 1991-compliant suits, these test results must be provided with each suit.

Only NFPA standards require some form of minimum quality control. The NFPA standards require that manufacturers establish a quality assurance program that is audited periodically by an independent certification organization. The newest editions of NFPA 1991 and 1992 further require that the manufacturer be registered to ISO 9001, *Quality Systems—Model for Quality Assurance in Design, Development, Production, Installation and Servicing*. This is one part of the certification process. The other part requires that independent testing be conducted to determine compliance with the standard, both initially and for continued compliance.

Protection from Chemical/Biological Terrorism Incidents

NFPA 1994 is a new standard released in August 2001 that sets performance requirements for protective clothing used at chemical and biological terrorism incidents. It is unique in that it defines three classes of ensembles based on the perceived threat at the emergency scene. Differences between the three classes are based on:

- The ability of the ensemble design to resist the inward leakage of chemical and biological contaminants.
- The resistance of the materials used in the construction of the ensembles to chemical warfare agents and toxic industrial chemicals.
- The strength and durability of these materials.

All NFPA 1994 ensembles are designed for a single exposure. Ensembles must consist of garments, gloves, and footwear. The requirements are very similar to those for NFPA 1991 and 1992. Table 7 describes the requirements that apply to each of the three ensemble classes. Table 8 provides a list of the chemicals involved in testing materials performance.

Table 7:
Overview of Key NFPA 1994 Performance Criteria

Property	Test Method/ Measurement	NFPA 1994 Class 1	NFPA 1994 Class 2	NFPA 1994 Class 3
Gas-tight integrity	ASTM F 1052	Passes test after simulated use		
Inward gas leakage	NFPA Test using sulfur hexafluoride	≤ 0.02%	≤ 2.0%	
Liquid-tight integrity (suits)	ASTM F 1359		No leakage after 20 minutes	No leakage after 20 minutes
Liquid-tight integrity (gloves and footwear)	ASTM D 5151 (modified)		No leakage	No leakage
Permeation resistance of primary materials and seams (industrial chemicals)	ASTM F 739 – see Table 8 for list of chemicals	Breakthrough times ≥ 1 hour (all chemicals tested at full strength)	Breakthrough times ≥ 1 hour (all chemicals tested diluted with closed top)	Breakthrough times ≥ 1 hour (liquid chemicals tested diluted with open top)
Permeation resistance of primary materials and seams (chemical warfare agents)	ASTM F 739 (modified) – see Table 8 for list of chemicals	≤ 4.0 or 1.25 μg/cm^2 (chemicals tested at full strength)	≤ 4.0 or 1.25 μg/cm^2 (chemicals tested diluted with closed top)	≤ 4.0 or 1.25 μg/cm^2 (chemicals tested diluted with open top)
Biopenetration resistance of primary materials and seams	ASTM F 1671		None after flexing and abrasion	None after flexing and abrasion
Suit and visor burst strength	ASTM D 751	≥ 200 N (45 lb_f)	≥ 156 N (35 lb_f)	≥ 134 N (30 lb_f)
Suit and visor puncture propagation tear resistance	ASTM D 2582	≥ 49 N (11 lb_f)	≥ 31 N (7 lb_f)	≥ 25 N (5.6 lb_f)
Suit and visor seam or closure strength	ASTM D 751	≥ 2.63 kN/m (30 lb_f/2 in.)	≥ 1.31 kN/m (15 lb_f/2 in.)	≥ 1.31 kN/m (15 lb_f2 in.)
Glove dexterity	ASTM F 2101	Same as NFPA 1991 (7.1.3.5)	Same as NFPA 1992 (7.2.3.6)	≤ 200% (7.3.3.7)
Cut resistance (gloves)	ASTM F 1790	≥ 25 mm (1 in.) under force of 90 grams	≥ 25 mm (1 in.) under force of 75 grams	≥ 25 mm (1 in.) under force of 60 grams
Cut resistance (footwear upper)	ASTM F 1790	≥ 25 mm (1 in.) under force of 800 grams	≥ 25 mm (1 in.) under force of 600 grams	≥ 25 mm (1 in.) under force of 400 grams
Puncture resistance (gloves)	ASTM F 1342	≥ 22 N (5lb_f)	≥ 22 N (5lb_f)	≥ 22 N (5lb_f)
Puncture resistance (footwear upper)	ASTM F 1342	≥ 36 N (8lb_f)	≥ 36 N (8lb_f)	≥ 36 N (8lb_f)
Abrasion resistance (footwear sole)	ASTM D 1630	Abrasion rating ≥ 65	Abrasion rating ≥ 65	Abrasion rating ≥ 65

Table 7 *(continued)*

Property	Test Method/ Measurement	NFPA 1994 Class 1	NFPA 1994 Class 2	NFPA 1994 Class 3
Slip resistance (footwear sole)	ASTM F 489	Coefficent of friction ≥ 0.75	Coefficent of friction ≥ 0.75	Coefficent of friction ≥ .75
Suit and visor puncture propagation tear resistance	ASTM D 2582	≥ 49 N (11 lb_f)	≥ 31 N (7 lb_f)	≥ 25 N (5.6 lb_f)
Suit and visor seam or closure strength	ASTM D 751	≥ 2.63 kN/m (30 lb_f/2 in.)	≥ 1.31 kN/m (15 lb_f/2 in.)	≥ 1.31 kN/m (15 lb_f/2 in.)
Glove dexterity	ASTM F 2101	Same as NFPA 1991 (7.1.3.5)	Same as NFPA 1992 (7.2.3.6)	≤ 200% (7.3.3.7)
Cut resistance (gloves)	ASTM F 1790	≥ 25 mm (1 in.) under force of 90 grams	≥ 25 mm (1 in.) under force of 75 grams	≥ 25 mm (1 in.) under force of 60 grams
Cut resistance (footwear upper)	ASTM F 1790	≥ 25 mm (1 in.) under force of 800 grams	≥ 25 mm (1 in.) under force of 600 grams	≥ 25 mm (1 in.) under force of 400 grams
Puncture resistance (gloves)	ASTM F 1342	≥ 22 N (5lb_f)	≥ 22 N (5llb_f)	≥ 22 N (5lb_f)
Puncture resistance (footwear upper)	ASTM F 1342	≥ 36 N (8lb_f)	≥ 36 N (8lb_f)	≥ 36 N (8llb_f)
Abrasion resistance (footwear sole)	ASTM D 1630	Abrasion rating ≥ 65	Abrasion rating ≥ 65	Abrasion rating ≥ 65
Slip resistance (footwear sole)	ASTM F 489	Coefficent of friction ≥ 0.75	Coefficent of friction ≥ 0.75	Coefficent of friction ≥ 0.75

Table 8:
Chemical Used in Permeation Testing for NFPA 1994

Chemical	State	Criteria Applied		
		Breakthrough time > 60 minutes	Cumulative permeation < 4.0 μg/cm^2 (in one hour)	Cumulative permeation < 1.25 μg/cm^2 (in one hour)
Ammonia	Gas	✔		
Carbonyl Chloride (CG)	Gas	✔		
Chlorine	Gas	✔		
Cyanogen Chloride (CK)	Gas	✔		
Dimethyl Sulfate (DMA)	Liquid	✔		
Distilled Sulfur Mustard (HD)	Liquid		✔	
Hydrogen Cyanide (HC)	Gas	✔		
Lewisite (L)	Liquid		✔	
Sarin (GB)	Liquid			✔
V-agent (VX)	Liquid			✔

Class 1 Ensembles

Class 1 ensembles offer the highest level of protection. They are required to:

- Be gas-tight by passing an inflation test (with the exhaust valves plugged).
- Show very low levels (less than 0.02 percent) for penetration of surrogate gas during an inward leakage test involving a human test subject.
- Use materials (garment, visor, glove, footwear, and seams) that provide the highest level of permeation resistance against chemical warfare agents as well as toxic industrial chemical liquids and gases.
- Use garment, glove, and footwear materials that possess relatively high levels of physical hazard resistance (to abrasion, tearing, punctures, and cuts).

Class 1 ensembles are intended for the worst-case circumstances, where the substance involved creates an immediate threat, is unidentified and of unknown concentration. Such situations would occur where there is still an ongoing release with likely gas/vapor exposure, the responder is close to the point of release, and most victims in the area appear to be unconscious or dead from exposure. Stay times in the hazard zone are likely to be short and limited to the breathing air available from the SCBA.

Class 2 Ensembles

Class 2 ensembles offer an intermediate level of protection. These ensembles have the following characteristics:

- The ensembles are not gas-tight, but still must show no more than 2.0% leakage of the outside environment into the ensemble as demonstrated by the same surrogate gas test described for Class 1 above.
- The ensemble must also pass a "shower" test that shows that the suit will not allow any penetration of liquid when sprayed from several directions.
- The materials used in the Class 2 ensembles, gloves, and

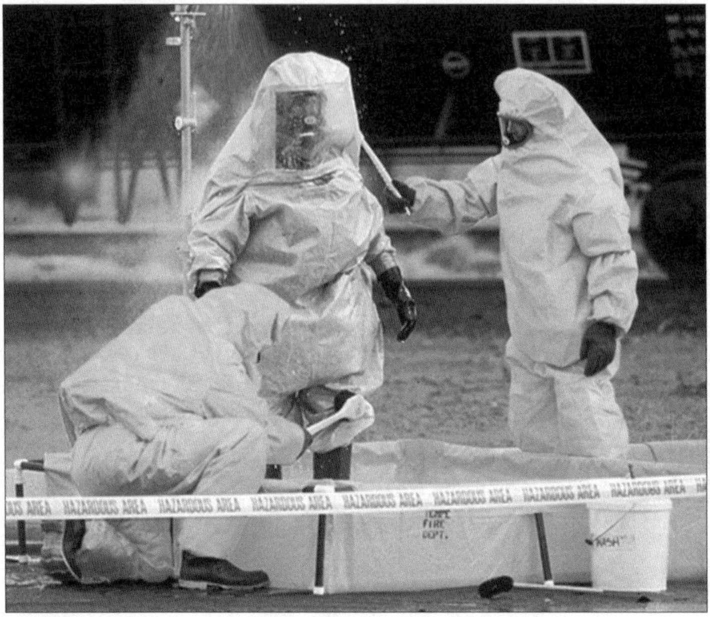

Figure 12.
Example of a proposed non-gas-tight encapsulating suit for use as Class 2 NFPA 1994-compliant ensemble. (Photo reproduced with the permission of W.L. Gore and Associates, Inc.)

footwear must demonstrate permeation resistance to chemical warfare agents and liquid/gaseous toxic industrial chemicals. However lower concentrations of these chemicals are used in testing as compared to Class 1 (for example, gases are tested at a concentration of 0.1% in air).

- Class 2 garment, glove, and footwear materials must still be rugged, but at lower physical property levels as compared to Class 1 ensemble.

An example of a Class 2 ensemble is shown in Figure 12.

Class 2 ensembles are intended for circumstances where the agent or threat has generally been identified and where the actual release has subsided. Conditions of exposure include possible contact with residual vapor or gas and highly contaminated surfaces at the emergency scene. Victims in the area may still be showing signs of movement. Wearing time will still be limited by the breathing air from the SCBA.

Class 3 Ensembles

Class 3 ensembles offer the lowest level of protection. These ensembles have the following characteristics:

- The ensembles are not gas-tight and are not required to show any resistance to leakage of vapor or gas from the outside environment. However, the suits must be liquid-tight as demonstrated by passing a short-duration "shower" test.
- Class 3 ensemble materials are not tested against gases, but still must provide permeation resistance to low concentrations of liquid chemical warfare agents and liquid toxic industrial chemicals where the liquid is permitted to evaporate during the test.

Class 3 ensembles are intended for use well after the release has occurred or in the peripheral zone of the release scene for such functions as decontamination, patient care, crowd control, perimeter control, traffic control, and clean-up. Class 3 ensembles should only be used when there is essentially no (but not necessarily none) vapor or gas exposure, exposure to liquids is expected to be incidental through contact with contaminated surfaces, and patients or self-evacuating victims. Class 3 ensembles must cover the individual and it is preferred that this clothing also cover the wearer's respirator to limit its potential for contamination. Because these ensembles are intended for longer wearing periods, the use of air-purifying respirators with these suits is likely.

Summary

The selection and use of full-body chemical protective clothing must account for a complete ensemble of clothing and equipment. While the National Fire Protection Association has prepared standards covering both vapor and liquid splash protective clothing (and more recently ensembles for domestic preparedness involving chemical agents), organizations must still consider a large number of clothing design and performance factors. As with any selection of personal protective equipment, a risk assessment must be performed to identify the hazards and specific risks while protective clothing is matched to provide protection against those hazards.

Selection of Chemical Protective Clothing

By Jeffrey O. Stull

7

Introduction

The lack of comprehensive chemical protective clothing (CPC) selection standards results in little guidance in many areas of CPC selection. This chapter discusses factors that should be considered in the selection of CPC and recommends an approach for conducting a hazard assessment that emphasizes protection from dermal exposures. In addition, comprehensive decision logic for selection and use of CPC is also included. This information addresses a significant need within the user community and can help end users comply with OSHA (and other) regulations related to the use and selection of CPC.

Past and Existing Selection Guidelines

OSHA's General Industry standards on Personal Protective Equipment (PPE) (29 CFR Subpart I) require that the employer conduct an assessment to identify hazards and then base protective clothing (and equipment) selections on the information determined during the assessment. A number of selection decision guidelines have been developed to assist in the identification of appropriate CPC. The National Institute for Occupational Safety and Health (NIOSH) created selection logic for respirators allowing for the choice of a specific respirator type based on specific inputs, such as the application and level of exposure.(1) NIOSH also developed a relatively detailed selection guide for personal protective equipment use at hazardous materials incidents.(2) This guide includes a number of decision inputs that result in recommendations of specific types of both respirators and chemical protective suits. At the time, it provided relatively detailed suggestions for an emerging area of Bio-Chemical Protective Clothing (BCPC) use. More recently, both NIOSH and the International Safety Equipment Association (ISEA) developed guides for selecting and using particulate respirators.[3,4] In addition, ISEA has proposed a new standard on the classification of chemical protective clothing that defines the types of CPC based on their integrity and material chemical resistance performance. The proposed standard, shown in Figure 1, includes a decision tree for selecting types of BCPC. Several manufacturers have also created selection guides for either clothing or gloves, but these guides are specific to the products offered by that manufacturer. Currently, a comprehensive selection guide that covers the full range of CPC is not available.

Selection Factors

With the exception of cost, the primary factors affecting appropriate CPC selection were considered in the development of the decision logic presented in the Appendix. These factors include:

- **Clothing integrity –** A characterization of CPC in terms of its overall ability to prevent inward leakage of contaminants (gases/vapors, liquids, or particulates); primarily applies to full-body clothing, but may also be used to characterize gloves and footwear.
- **Material barrier properties –** The barrier effectiveness of the materials and their seams used in the construction of CPC. Divided among different forms of chemical resistance.
- **Physical ruggedness and durability –** The ability of materials to sustain barrier performance following care and use, and their resistance to physical hazards, such as abrasion, cutting, punctures, and tears.
- **Ease of decontamination –** The ability to remove chemical contaminants from CPC to acceptable levels without degradation of CPC performance characteristics.

Figure 1:
ISEA Proposed Chemical Protective Clothing Decision Logic

KEY

Type 1 – gas–tight encapsulating suits

Type 2 – non–gas–tight encapsulating suits

Type 3 – liquid “jet” protective clothing

Type 4 – liquid “splash” protective clothing

Type 5 – particulate protective clothing

Type 6 – light liquid “splash” protective clothing

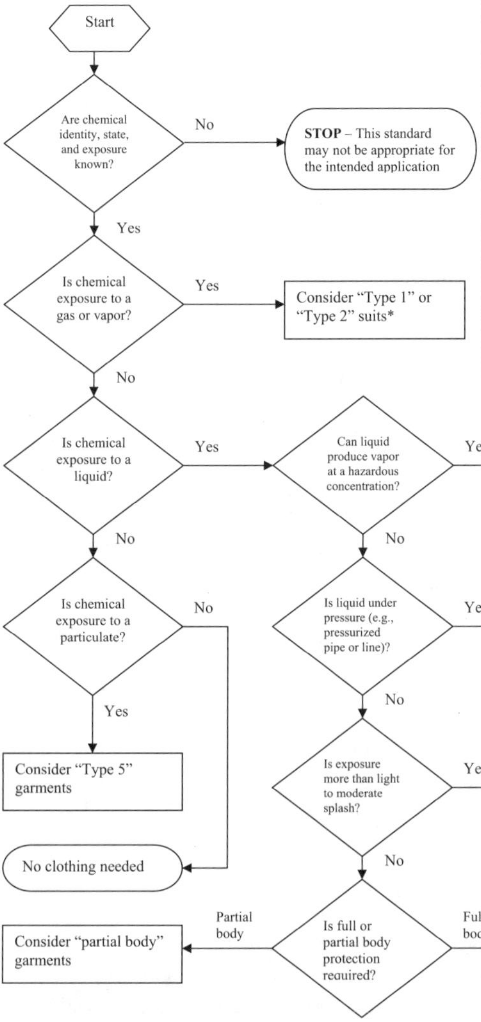

* Type 1 and 2 suits can be used in similar situations under certain conditions. Factors such as the type of respirator required, compatibility of respirator components and chemical environment, size restrictions (i.e., confined space entry), etc. should be considered when selecting between these garments.

- **Protection from other hazards** – Other hazards that CPC might be expected to protect against alone, or in conjunction with using other equipment; examples include visibility for roadside work, fall protection, drowning for work near water.
- **Integration with other equipment** – The suitability of CPC to interface properly with other necessary items of CPC or other equipment (e.g., respirators, head protection, hearing protection, etc.).
- **Impact on wearer** – The effects of the CPC on user function, mobility, vision, communication, dexterity, grip, tactility, traction, and comfort.

These factors are described in more detail in Chapters 5 and 6.

Hazard Assessment

The OSHA standard covering PPE (Title 29 CFR 1910.132) requires that employers use a hazard assessment to determine the need for and then to select PPE, including CPC. Appendix B to Subpart I of the General Industry OSHA standards provides non-mandatory guidelines for conducting PPE hazard assessment and for selecting PPE. However, it provides only a very general approach for identifying chemical hazards. The approach described here can be used for identifying hazards and determining their associated risk. The results of this approach provides information that serves as the input for the CPC selection decision logic.

Step-by-Step Approach

One method for identifying chemical hazards includes conducting a hazard assessment of the workplace and determining the risk of exposure. This information is used for priority ranking of protection needs. A suggested rating system for this approach is shown in Table 1. The five steps of this process include:

Table 1:
Rating Schemes for Likelihood and Consequences of Exposure

Rating	Likelihood of Exposure
0	Exposure cannot occur
1	Exposure very unlikely
2	Exposure possible, but unlikely
3	Exposure likely
4	Multiple exposures likely
5	Continuous exposure likely
Rating	**Consequences of Exposure**
0	No Effect
1	Temporary effect on employee (such as discomfort) with no long term consequences
2	Exposure results in temporary, treatable injury
3	Exposure results in serious injury with loss of work time
4	Exposure results in permanent debilitating injury
5	Exposure results in likely death

1. **Define each workplace and the tasks to be evaluated.** The defined workplace should include an area encompassing the range of hazards that may be encountered. Examples of a workplace include specific work locations for a particular employee, a laboratory, or part of a production process. Tasks should be defined as those worker activities that involve unique hazards and can be accomplished by a single individual or group of individuals within a given period of time.

   The principal source of information for a hazard assessment results from an inspection of the workplace. In particular, observation of specific worker tasks can provide much of the desired information. Additional details can be obtained through employee interviews. Examples of the information includes:

   - The types and frequency of hazards encountered in the task,

- Specific instances in which hazards have been encountered in the past, and
- The past effectiveness or ineffectiveness of any PPE used in the task.

A review of OSHA Form 300 (Log of Work Related Injuries and Illnesses), or an equivalent log and summary of all occupational illnesses and injuries at the workplace, is another potential source of information. As part of the review, details from any accidents or exposures should be evaluated to determine if PPE was being used at the time of the incident. If not, then the PPE needs should be assessed to determine if it should be added as a prevention mechanism. If PPE was being used at the time of the incident, the appropriateness of the clothing and equipment should be evaluated to determine if changes are warranted.

The task must be clearly defined prior to selecting clothing. The extent of the task should be documented to ensure that the type of clothing and equipment selected will protect workers during all facets of the activity. Key information from the initial evaluation used as part of this step is listed below.

Job Classification or Task

- Identify/characterize the task accurately, including both standard and non-standard activities.

Summary of Process or Task

- Summarize the operation, providing specific details, including the operation length, and location (i.e., indoor vs. outdoor).
- Determine the type of operations where CPC will be used (e.g., dipping, spraying, mixing, moving of components or parts, assembly or disassembly work, repair, demolition, emergency response).
- Note any other tasks that should be evaluated because of their impact on CPC performance of the task.

- Note the normal variation in the workplace throughout different workshifts and years (i.e., seasonal variability).

2. **Identify the hazards associated with each work task.** General hazard categories that should be considered include biological, chemical, physical, environmental, thermal, electrical, radiation, person-position, and person-equipment hazards.

   Workers can be exposed to several types of hazards. Since the focus of this book is on chemical protective clothing, the discussion below focuses on chemical or other hazards that affect the protection provided by the clothing. To identify chemical hazards, consider the following aspects of the work place.

   Potential or Actual Chemical Hazards

   - Identify the specific chemicals involved in the task.
   - For overlapping tasks, identify the chemicals and their effects.
   - Note the temperature of chemicals with which contact may occur.

   Physical Properties of Chemicals

   - Identify the chemical as a solid, liquid, or gas.
   - List the pertinent chemical properties, such as vapor pressure, boiling point, specific gravity, etc.
   - For chemical mixtures, list the properties of the major components.

   Chemical Contact Periods

   - Determine if contact with the chemical(s) will be with a solid, liquid, vapor, or gas.
   - Note the length of direct contact between the CPC and the chemicals (seconds, minutes, hours, days). Also indicate if CPC will be reused.

- Determine if the contact is routine, intermittent, or infrequent/unplanned.
- Note if the contact occurs typically results from a liquid splash.
- Determine if the wearer can change CPC quickly following exposure.
- Note if the CPC will be used for emergency response.

Type of Potential Contact

- Determine the type of chemical contact expected from both routine and emergency conditions (e.g., routine splash, pressurized spray accident).
- Note the physical hazards that will impact the integrity of the clothing, altering the ability to keep chemicals out (tearing, cut, puncture, or abrasion risks).

Other Hazards

- Physical hazards (e.g., flying debris, slippery surfaces)
- Environmental hazards (e.g., extremes of cold or heat, noise, lighting)
- Biological hazards (e.g., airborne or liquid-borne pathogens)
- Thermal hazards (e.g., contact with hot surfaces, radiant heat)
- Electrical hazards
- Radiation hazards

If a complete, well-written Material Safety Data Sheet (MSDS) is available, it can be attached to the analysis with the pertinent sections noted.

Depending on the location of the worker, additional hazards may need to be evaluated. For example, if a person is working near water, they will need a personal flotation device to prevent risk of drowning, especially if the person is encumbered with clothing and equipment. Similarly, if the work is carried out on an elevated platform, fall protection will be needed.

In some situations, the protective clothing and equipment itself may create or lead to the introduction of hazards. Some examples include:

- Creation of particles or static electricity that can damage sensitive environments,
- Allergic or other skin reactions from non-biocompatible materials,
- Retention of contamination for continued exposure,
- Reduced mobility and hand function, or impaired communications and vision,
- Lack of ankle or back support increasing the opportunity for strains and sprains, and
- Heat stress from use in warm environments.

3. **Determine each affected body area.**
   For each hazard, determine which portion of the body could be affected if exposure occurs. Exposures can be isolated to a single area, such as the hand in the case of a splash, or extend to the entire body when exposed to a gas. General body areas and systems typically affected by workplace hazards include: the entire body, trunk/torso, head, eyes/face, arms, hands, legs, feet, respiratory system, or hearing.

4. **Estimate the likelihood of employee exposure to identified hazards.**
   An estimate of the likelihood of exposure is necessary for each identified hazard as it relates to each affected portion of the body. A proposed rating scale based on the frequency of exposure is shown in Table 1. Using this scale, the ratings range from 0 — an indication that the exposure cannot occur, to a score of 5 — which indicates that continuous exposure is likely.

The mere presence of chemicals does not mean that an exposure will occur. However, in some functions, exposure is "expected" as part of the job, such as when washing parts with a solvent. In other situations, exposure might not occur unless there is a serious incident. The analysis conducted when characterizing the workplace and the type of chemicals present, forms the basis for determining the likelihood of exposure.

5. **Estimate the possible consequences of exposure to identified hazards.**

   For each identified hazard affecting a specific portion of the body, an estimate of the consequences of exposure should be made. A rating scale for the consequences of exposure is shown in Table 1. With this scale, a rating of 0 indicates that no effect results from the exposure, while a rating of 5 indicates that the exposure is likely to result in death. Regardless of the rating system used, all estimates should be based on the "worst case" scenario.

   A very important parameter that must be determined carefully is the toxicity of the chemical against which the CPC will provide protection. However, this is not a simple task, because no published standards exist for skin exposure to toxic chemicals.

   The American Conference of Governmental Industrial Hygienists (ACGIH) identifies chemicals with a "Skin" notation in their *Threshold Limit Values® and Biological Exposure Indices* booklet.[5] The "Skin" notation refers to a potential contribution to the overall exposure by the cutaneous route, which includes mucous membranes and eyes, and occurs either by airborne exposure or by direct contact with the substance. A review of the consistency that exists between chemicals with a "Skin" notation found large variations in the toxic properties of these chemicals and their effects on humans.[6] Another way to express skin hazard is:

   Skin Hazard ∞ (Toxicity * Bioavailability).[7]

Therefore, the evaluation process must use information on the toxicity and bioavailability of the chemical(s) in question. If acute oral toxicity data is used, certain derived classifications and rankings are available, but these should be used only as initial screening tools. Table 2 lists toxicity rankings for chemicals based on the LD_{50} (dose of a substance lethal to 50% of the exposed test animals) and LC_{50} (concentration in air lethal to 50% of an exposed animal population). Table 3 provides an example using actual acute toxicity data by listing the LD_{50} of some common toxic materials. A skin absorption estimate technique based on animal toxicology data is also available for certain applications. Table 4 shows an extension of this estimation process to humans. However, all of these tables are qualitative at best and therefore must be used carefully.

Currently, industry has no simple method to summarize bioavailability, so this part of the evaluation must take into account such things as water solubility and specific metabolic routes to provide an estimate of bioavailability.

As part of a proposed dermal hazard rating scheme, Eiser identified a number of factors that should be considered in relation to effects on the worker or the test animal.[(8)] The chemicals analyzed in that dermal scheme were selected from the 1987-88 ACGIH Threshhold Limit Values (TLVs®). The categories of factors listed below should be considered in the evaluation of the toxicity of chemicals for which CPC will be used.

"Skin" Designation Chemicals

- Potential for systemic effects that result from skin absorption alone
- Anticipated severity of systemic effects
- Reversible effects versus permanent damage
- Speed of action on the body
- Warning properties (odor and/or irritancy at low concentrations)
- Absorption (skin permeation) rate for the substance
- Known significant toxic dose

Table 2:
Toxicity Classes

Toxicity rating	Descriptive term	LD_{50}-wt/kg single oral dose (rats)	4-h inhalation LC_{50}-ppm (rats)
1	Relatively harmless	15 g or more	>100,000
2	Practically non-toxic	5-15 g	10,000 – 100,000
3	Slightly toxic	0.5-5 g	1000 – 10,000
4	Moderately toxic	50-500 mg	100 – 1,000
5	Highly toxic	1-50 mg	10 – 100
6	Extremely toxic	1 mg or less	<10

Table 3:
Approximate Acute Ingestion LD_{50}s of a Selected Variety of Chemical Agents

Agent	LD_{50} (mg/kg)
Ethyl alcohol	10,000
Sodium chloride	4,000
Ferrous sulfate	1,500
Morphine sulfate	900
Phenobarbital sodium	150
DDT	100
Picrotoxin	5
Strychnine sulfate	2
Nicotine	1
d-Tubocurarine	0.5
Hemicholinium-3	0.2
Tetrodoxin	0.1
Dioxin (TCDD)	0.001
Botulinus toxin	0.00001

Table 4:
Toxicity Rating Chart

	Probable oral lethal dose for humans	
Toxicity rating or class	**Dose**	**For average adult**
1. Practically non-toxic	>15 g/kg	More than 1 quart
2. Slightly toxic	5-15 g/kg	Between 1 pint and 1 quart
3. Moderately toxic	0.5-5 g/kg	Between 1 ounce and 1 pint
4. Very toxic	50-500 mg/kg	Between 1 teaspoonful and 1 ounce
5. Extremely toxic	5-50 mg/kg	Between 7 drops and teaspoonful
6. Supertoxic	<5 mg/kg	A taste (less than 7 drops)

- Immediately Dangerous to Life or Health (IDLH) potential from skin contact alone
- Chronic hazard potential, including carcinogenic effects
- Reported fatalities with use

Corrosives or Irritants (Not Assigned a "Skin" Notation)

- Necrotic skin effect versus transient irritation
- Systemic effects in addition to the local effects
- Draize Eye Irritancy Test scores, related refinements, and alternate test data
- Warning properties
- Permanent versus temporary injury or disability
- History of industrial or laboratory use

Allergic Sensitizers

- Potent versus mild sensitizers
- Anaphylactic shock responses reported with use

- Acute toxic or corrosive effects with skin contact in addition to sensitizing potential
- Chronic toxic effects in addition to sensitizing potential

Similar schemes can be employed to determine the severity of other non-chemical hazards, if present, by simply rating what would happen to an individual if exposure did occur.

Determination of Risk

Risk is a product of the likelihood and consequences of exposure. The determination of risk permits decisions on what hazards pose the greatest risk and indicate the hazards that should be considered important when selecting CPC. A completed hazard assessment provides a list of hazards, the parts of the worker's body affected, how likely exposure will occur, and the probable consequences of exposure. Using the rating scales established above, a quantitative score for risk can be calculated. Scores can range from 1 to 25, with 25 representing the highest risk to the worker. Following this approach for each of the hazards can define the relative risk to each hazard and each part of the body.

Decision Logic

A decision logic is a series of steps that can be undertaken to reach a specific conclusion or recommendation based on the information available to the user. In the case of CPC, the decision logic provides a tool for the selection of CPC. Permitting the end user to apply information about their situation (as related to the nature of the chemical exposures determined in the hazard assessment) assists in the determination of the specific types of CPC and related protective equipment that must be worn.

Scope

The field of CPC applications is especially broad. In order to provide and apply a decision logic, it is necessary to carefully define the scope of the decision logic. For the purposes of this publication, the decision logic focuses on chemical exposure as the principal potential hazard to the worker. However, cleanroom applications are not included, since the key consideration is protection of the environment, rather than protection of the worker. A decision logic can address both emergency and non-emergency situations, though a separate decision logic is used for each type of application.

Decision Logic Outputs

Ultimately, the decision logic must yield the type of protective clothing to be worn, but also, to some degree, must differentiate between overall design, design features, and material performance, allowing the end user to identify and select specific "appropriate" CPC. In this process, trade-offs exist between detailed choices and simplicity of use for the decision logic.

For CPC used in emergencies, several National Fire Protection Association (NFPA) standards specify the types of CPC for relevant situations. In these cases, it is simply enough to indicate compliance with the appropriate NFPA standard.

In the absence of standards like those available in other areas of PPE (e.g., respirators, eye/face protection, and foot protection), it is necessary to establish a system of CPC classification that meets the following objectives:

1. Define the types of non-emergency CPC, thereby allowing meaningful differentiation of the product types by end users.

2. Provide a logical basis for showing the different levels of protection offered by CPC.

3. Direct the correspondence, where possible, and when appropriate, to the types of clothing dictated by industry practice.

Since the focus of this selection guide is chemical protection, the classification system is based on the two selection factors associated with barrier protection: product integrity and material barrier performance. It is also necessary to establish different classes of chemical hazards based on the differences in exposure.

For each protection level, acceptable product designs that span the different types of CPC to be used (i.e., full-body garments, partial body garments, gloves, footwear, and eye/face protection) must be identified. A description of the general types of CPC in terms of design, design features, materials, and sizing is available in *Chemical Protective Clothing: Types of CPC*. Tables 5 and 6 list the specific types of emergency and non-emergency chemical protective clothing, which are the possible output or "choices" of the decision logic.

Definition of Necessary Decision Logic Inputs

The outcomes of the completed hazard assessment should include a list of hazards present in the workplace, an identification of body parts that may be affected, the likelihood of exposure, and the consequences of exposure. Although these outcomes serve as the primary input to the decision logic, some information must be relegated as secondary decisions for choosing CPC.

The input information required by the decision logic is structured to be readily determined by individuals responsible for making clothing selection and use decisions. Some of this information may require research (i.e., obtaining data about a chemical substance), but all can be obtained within a reasonable amount of time.

Using the hazard assessment outcomes, the inputs shown in Table 7 were used to prepare the decision logic.

Table 5:
Emergency Protective Clothing Choices

Type	General Description/ Clothing Items	Integrity	Material Barrier
E1	Vapor protective suit with attached gloves; footwear, covers both wearer and breathing apparatus (meets NFPA 1991)	Gas/vapor (passes inflation test) Liquid (passes sustained shower test)	Permeation-resistant to 21-chemical battery (garment, visor, gloves, footwear, seams)
E1a	Type E1 suit also providing flash fire escape protection	Same as above	Same as above
E1b	Type E1 suit also providing liquefied gas protection	Same as above	Same as above; also against 6 liquefied gases in liquid form
E1c	Type E1 suit with providing both flash fire escape and liquefied gas protection	Same as above	Same as above
E2	Encapsulating liquid splash-protective suit, with gloves, footwear (meets NFPA 1992)	Liquid (passes shower test)	Penetration-resistant to 7-chemical battery (garment, visor, gloves, footwear, seams)
E2a	Type E2 non encapsulating suit, separate gloves and footwear	Same as above	Same as above
E2b	Type E2 suit also providing flash fire escape protection	Same as above	Same as above
E3(1)	Protective ensemble consisting of encapsulating suit with attached gloves, also footwear (meets NFPA 1994, Class 1)	Gas/vapor (passes inflation test and inward leakage at 0.02%)	Permeation resistant to 9 liquid and gas chemical battery (high exposure level)
E3(2)	Protective ensemble consisting of encapsulating suit with attached gloves, also footwear (meets NFPA 1994, Class 2)	Gas/vapor (passes inward leakage at 0.02%) Liquid (passes shower test)	Permeation resistant to 9 liquid and gas chemical battery (moderate exposure level) Viral penetration-resistant
E3(3)	Full body garment, may or may not cover wearer respirator; separate gloves and footwear (meets NFPA 1994, Class 3)	Liquid (passes short duration shower test)	Permeation resistant to 5 liquid chemical battery (low exposure level) Viral penetration-resistant
E4a	Full-body or partial body garments (meets NFPA 1999)	Liquid (passes short duration shower test)	Viral penetration-resistant
E4b	Gloves, disposable (meets NFPA 1999)	Liquid (does not leak when filled with liquid)	Viral penetration-resistant
E4c	Face protection devices (meets NFPA 1999)	Liquid (spray impact)	Viral penetration-resistant

Table 6:
Chemical Protective Clothing Choices

Clothing Item	Type	General Description	Integrity	Material Barrier
Garments	C1	Total encapsulating suit, covers both wearer and breathing apparatus	Gas/vapor (passes inflation test)	Permeation-resistant
	C2	Total encapsulating suit, covers both wearer and breathing apparatus	Gas/vapor (cannot be inflation tested, but passes inward leakage test)	Permeation-resistant
	C2v	Total encapsulating suit, covers both wearer and breathing apparatus	Gas/vapor (cannot be inflation tested, but passes inward leakage test)	Vapor-penetration resistant
	C3	Hooded coverall, multi-piece splash suit (e.g., hooded jacket with pants or bib pants)	Liquid (passes sustained shower test)	Permeation-resistant
	C3p	Partial body clothing item – apron, jacket, hood, pants, overalls, sleeve protector, shoe/boot cover	Liquid (for body area covered by item)	Permeation-resistant
	C4	Hooded coverall, multi-piece splash suit (e.g., hooded jacket with pants or bib pants)	Liquid (passes shower test)	Liquid penetration-resistant
	C4p	Partial body clothing item – apron, jacket, hood, pants, overalls, sleeve protector, shoe/boot cover	Liquid (for body area covered by item)	Liquid penetration-resistant
	C5	Hooded coverall or encapsulating suit, may not cover breathing apparatus	Particulate (provides acceptable particle hold out)	Particulate penetration-resistant
	C6	Coverall, jacket/pant sets	Covers body area of interest	Liquid-repellent
	C6p	Partial body clothing item – apron, jacket, hood, pants, overalls, sleeve protector, shoe/boot cover	Covers body area of interest	Liquid-repellent

Table 6 (continued)
Chemical Protective Clothing Choices

Clothing Item	Typ-e	General Description	Integrity	Material Barrier
Gloves	CG1	Unsupported, supported, or flat-film glove	Gas/vapor (can be air tested)	Permeation-resistant
	CG2	Unsupported, supported, or flat-film glove	Liquid (will not leak when filled with liquid)	Liquid penetration-resistant
	CG3	Supported glove	Coves hand/wrist area	Liquid penetration-resistant
Footwear	CF1	Boot, overboot, or boot/shoe cover	Gas/vapor (can be air-tested)	Permeation-resistant
	CF2	Boot, overboot, or boot/shoe cover	Liquid (will not leak when filled with liquid)	Liquid penetration-resistant
	CF3	Boot/shoe cover	Covers foot/ankle area	Liquid-repellent
Eye/face protection	CE1	Hood with visor or respirator full-facepiece	Gas/vapor, particulate (resists inward leakage)	Permeation-resistant
	CE2	Hood with visor or respirator full-facepiece	Gas/vapor, particulate(resists inward leakage)	Penetration-resistant
	CE3	Faceshield or faceshield with goggles	Liquid splash-resistant	Penetration-resistant
	CE4	Faceshield	Covers face area	Penetration-resistant

Table 7: BCPC Design Logic Inputs

1. The general class of protection
   - Emergency
   - Biological substance
   - Chemical substance
2. The physical state of the hazard
   - Gas or vapor
   - Liquid
   - Solid particulate
3. The type of exposure to the biological agent or chemical and its consequences
   - For liquidborne (or bloodborne) pathogens
     - Relative pathogencity – low, high, ultrahigh
     - Length of exposure – incidental, splash or spray, continuous
     - Volume of exposure – small, large
     - Conditions of exposure – low pressure, high pressure
   - For airborne pathogens
     - Relative pathogencity – low, high, ultrahigh
     - Location of particles or aerosol – settled, suspended
   - For chemical particulates
     - Reactivity – non-reactive, flammable, explosive
     - Consequence of exposure – nuisance, affect skin, affect respiratory system
   - For chemical liquids
     - Reactivity – non-reactive, flammable, explosive
     - Volatility – non-volatile, volatile, highly volatile
     - Length of exposure – incidental, splash or spray, continuous
     - Volume of exposure – small, large
     - Conditions of exposure – low pressure, high pressure
   - For chemical gases and vapors
     - Reactivity – non-reactive, flammable, explosive
     - Skin effect – irritating, corrosive, skin-reactive, skin toxic
     - Concentration – below exposure level, above exposure level

Establishing Decision Logic Inputs

The risk of exposure is determined by using the hazard assessment; protection needs are ranked by the relative risk. For determining relative risk, a risk assessment form can be used to list hazards and identify associated risks including:

- Hazards identified for the specific workplace/task,
- Body areas or body systems affected by the respective hazard,
- The rating associated with the likelihood or frequency of the respective hazard, and
- The rating associated with the severity or consequences of the respective hazard.

The risk assessment then can be completed using each row to indicate a separate hazard with the associated information and ratings. For each identified hazard, the likelihood (or frequency) rating of the hazard is multiplied by the consequences (or severity) rating of the hazard to determine a risk rating of 0 to 25. Using risk scores based on information from the hazard assessment provides a method for ranking hazards associated with the workplace/task. Hazards with the highest risk score should be assigned higher priority for prevention or minimization. Hazards with a rating of 0 or a low risk score should be assigned a lower priority for prevention or minimization. Therefore, the CPC selections will be made based on the priority hazards.

In hazard and risk assessment, it is important to examine possible engineering or administrative controls for hazards with the highest risk. Engineering controls can encompass changes in the task or process, or incorporate other measures designed to eliminate or reduce possible exposure to hazards. Administrative controls can include changes in tasks or work practices to limit or eliminate employee exposure time to a hazard. If engineering or administrative controls are not possible, then the focus should be placed on identifying different types of CPC for elimination or reduction of exposure to hazards.

From information about affected body area or body systems, decisions about the type of CPC (and other PPE) necessary to eliminate or minimize the hazard can be made. In some cases, the appropriate CPC will be obvious and limited to a single general type (i.e., for example, using chemical protective gloves in handling contaminated glassware). In other cases, there may be several types of PPE that can be used to provide the needed protection.

Creation of Decision Logic Hierarchy

The decision logic design is based on a simplified flowchart format. This format identifies boxes for operations or information gathering steps, diamonds for decisions, and rounded rectangles for output. Since the decision logic covers a wide range of information, decisions, and CPC types, the process is illustrated on multiple pages to allow for branching that is difficult to organize in a simple flowchart. It also includes tables when multiple factors need to be considered in selecting CPC. As the primary decision logic focuses on CPC integrity and barrier properties, other CPC factors are accounted for separately in supplemental checklists to further provide discrimination of CPC selection decisions.

A critical part in the design of the decision logic is the provision of explanatory information and definitions. For each step of the process, relevant notes and definitions are provided on or near the specific page where the information is needed. Descriptive information is also provided as part if the decision logic.

Use of the Decision Logic

The following sections provide a narrative description of the selections steps embodied in the decision logic. A multi-page decision logic is presented in the Appendix at the end of this chapter.

General Hazard Classification

The first decision is the determination that the situation is either an "emergency" or a "non-emergency," since separate decision logic is followed for emergency and non-emergency situations. The differences exist primarily because specific CPC standards exist for emergency situations, which help define the type of CPC. Another key difference is that emergencies generally involve less characterized physical surroundings and environmental conditions.

Emergency Response Decision Logic

Specific Decisions. As with the start of all of the selection guides, the decision logic begins with information gained from the hazard assessment. Because the circumstances involving an emergency response can be uncertain, the first branch establishes identification of the hazard. If the hazards cannot be identified, then a question is posed whether site entry is needed. Usually site entry would only be required for rescue, or if a specific intervention or mitigation task is required in the "hot" zone.

- If an entry is not warranted, then no site entry is made and CPC is not needed.
- If an entry is required, then the highest level of performance, an NFPA 1991 compliant vapor-protective suit is suggested. Preferably, this suit should be equipped with flash fire escape protection.

Separate decisions are needed to determine if there is a reactive hazard (explosion) and if the situation presents a toxic threat. Determination of reactive hazards is based on a measurement of the Lower Explosive Limit (LEL). A toxic threat involves chemicals that are present at IDLH concentrations or are skin toxic chemicals. If toxic chemicals are not involved, it is possible to respond in regular firefighter protective clothing (compliant with NFPA 1971 or NFPA 1976).

Additional assessments are made when the incident involves either chemical or biological terrorism agents or gas/vapor hazards.

- If chemical or biological terrorism agents are involved, a number of considerations are provided in tabular form within the decision logic to choose the appropriate class of NFPA 1994-compliant ensembles.
- If gas/vapor hazards are involved, a separate determination is made to identify if the vapor is toxic, flammable, or a liquefied gas. The answers to these questions determine the specific type of NFPA 1991 vapor-protective ensemble to use. (NFPA 1991 includes different options for the certification of vapor-protective ensembles).

If none of the above characterizations apply, the decision logic focuses on liquid hazards. Specific determinations are made with regard to medical hazards (i.e., bloodborne pathogens), warranting the use of NFPA 1999 protective clothing, and flammable liquids, warranting NFPA 1992 clothing also certified for flash fire escape protection. Liquid exposures are characterized as "direct" or "indirect," where:

- Direct exposures involve multiple splashes, large liquid volumes, or liquid contact under pressure, and
- Indirect exposures involve a few splashes and incidental contact with no pressure.

All selection decisions end with either "no response" or use of individual CPC or an ensemble that is compliant with any relevant standards. The minimum response clothing is an NFPA 1992 compliant set of protective clothing, equivalent to the Level B protection as defined by OSHA in 29 CFR 1910.120. This recommendation is consistent with these regulations. A description of NFPA standards is provided in Chapter 6.

Garment Choices. In addition, the basic types of clothing variants for each type are defined based on options or different classes within the respective clothing standards. Chemical vapor protection during emergencies is based on four different options:

- Type E1 is an ensemble, including a totally encapsulating suit, gloves, and footwear that is worn together with a self-contained breathing apparatus, based on NFPA 1991, Standard on Vapor-Protective Ensembles for Hazardous Materials Emergencies (baseline requirements in the standard)

- Type E1a is an E1 ensemble that also is certified as meeting the optional flash fire protection.
- Type E1b is an E1 ensemble that also is certified as meeting the liquefied gas protection.
- Type E1c is an E1 ensemble that also is certified as meeting both the optional flash fire protection and the liquefied gas protection.

Chemical liquid protection during emergencies is based on three different options:

- Type E2 is an ensemble, including an encapsulating full-body suit, gloves, and footwear based on NFPA 1991, Standard on Vapor-Protective Ensembles for Hazardous Materials Emergencies (baseline requirements in the standard).
- Type E2a is an E2 ensemble except that the suit is non-encapsulating (does not fully cover the wearer) and has separate gloves and footwear.
- Type E2b is an E2 ensemble that also is certified for meeting the optional flash fire protection.

Appropriate CPC for responding to chemical and biological terrorism incidents is defined in NFPA 1994, Standard on Protective Ensemble for Chemical/Biological Terrorism Incidents. This standard defines three classes for clothing ensembles:

- Type E3(1) is a protective ensemble consisting of an encapsulating suit with attached gloves and footwear that meets NFPA 1994, Class 1 requirements for the highest threats (i.e., high concentration gases, aerosols, liquids, and particulates, including airborne and liquidborne biological agents).
- Type E3(2) is a protective ensemble consisting of an encapsulating suit with attached gloves and footwear

that meets NFPA 1994, Class 1 requirements for the intermediate threats (i.e., dilute gases, aerosols, and particulates and moderate liquid contact, including airborne and liquidborne biological agents).

- Type E3(3) is a protective ensemble consisting of encapsulating suit with attached gloves and footwear that meets NFPA 1994, Class 1 requirements for minimum threats (i.e., light liquid contact and liquidborne biological agents).

Recommendations for biological protective clothing are based primarily on NFPA 1999, Standard on Protective Clothing for Emergency Medical Operations. Defined clothing types are based on the part of the body covered:

- Type E4a is a full-body or partial body garment.
- Type E4b are medical examination gloves.
- Type E4c is a face protection device such as goggles, a faceshield, or a respirator facepiece.

Non-Emergency Chemical Hazards Decision Logic

Garment Hierarchy. Selection of CPC for personal protection uses a more complex decision logic because of the larger variety of hazards that may be encountered. For these situations, a modified form of the type classification system proposed by ISEA is proposed. This scheme provides six different types of CPC, as defined in Table 5, using different levels of garment integrity and material barrier performance. Included within this scheme are the following types of CPC for chemical protection:

- Type C1 – Encapsulating vapor-protective suits that can be inflation tested and use permeation-resistant materials.

- Type C2 – Encapsulating vapor-protective suits that offer inward leakage resistance (cannot be inflation tested) and use permeation-resistant materials.

- Type C2v – A variation of Type 2 that uses vapor penetration-resistant materials.
- Type C3 – Clothing that offers liquid protection, demonstrated through a sustained shower test and uses permeation-resistant materials.
- Type C4 – Clothing that offers liquid protection based on a shower test and liquid-penetration resistant materials.
- Type C5 – Clothing that provides resistance to particulate inward leakage with materials that also resist particulate penetration.
- Type C6 – Clothing that offers limited liquid protection based on body coverage and liquid repellent materials.

Variations for Types 3, 4, and 6 involve partial body clothing items where the specific CPC item provides the same integrity and material barrier performance, but only for the portion of the body covered by the item (e.g., a sleeve protector provides liquid protection for the arm area only).

Along with the definition of garment types are a similar array of CPC types for gloves, footwear, and eye/face protection. In general, there are fewer types of these items. They are matched with garment types in the decision logic so that the end user can choose items needed to provide the appropriate coverage for the desired level of body protection. This approach reinforces the concept that chemical protection often requires an ensemble of CPC and related equipment.

Clothing Selection Decisions. When selecting CPC for chemical protection, the initial decisions are based on the principal hazard presented by the chemical(s) in terms of the physical state encountered. Consequently, chemical hazards are classified into one of the following categories:

- Vapor hazards,
- Liquid hazards, and
- Particulate hazards.

If vapor hazards are present, the CPC selected is based

on exposure level and the presence of liquids. For high exposure levels to vapors (as determined by the relative concentration of the chemical vapor as compared to an established exposure limit), the most protective CPC level – Type 1 is suggested. For lower exposure levels, Type 2 clothing may be acceptable and the variant level, Type 2v, may be used when there is no potential for liquid exposure. This is because Type 2v materials are based on adsorbents that work well for capturing some chemicals but quickly become saturated during a liquid exposure. At the current time, this type of clothing is not common in industry, though it is used extensively in the military and may have application for agricultural pesticide operations.

If exposure to liquids is the principal hazard, then a decision logic similar to the CPC selection for liquidborne pathogens is used. In this approach, the respective types of CPC (Types 3, 4, and 6) are selected based on the volume of liquid present during the exposure, and whether or not the liquid exposure involves pressure. A separate decision involves whether full-body or partial-body protection is desired. The variants of the three types pertaining to partial body protective clothing are specified only if partial-body protection is needed.

Particulate protection is provided by a type of outfit specifically defined for environments where exposure to the particulate is harmful. This includes particulates that primarily present respiratory hazards, since these should be kept off the wearer's skin as a matter of hygiene and prevention of exposure. Also included are particulates that affect the skin, either through dermal absorption or reactivity concerns. Nuisance materials are not considered to be hazardous chemicals, so no distinction is made for exposures to these materials.

Glove, Footwear, and Eye/Face Protection Types. Parallel decisions are made with respect to gloves, footwear, and eye/face protection whether the items are used in conjunction with garments or individually, as defined in hazard assessment. However, fewer individual types are defined because the differences in these types of clothing items are substantially less in terms of material chemical resistance and item integrity.

Three types of gloves are defined, creating a hierarchy of protection:

- Type CG1 is an unsupported, supported, or flat-film glove, which has gas/vapor integrity and uses materials that are permeation-resistant against the chemical(s) of interest.
- Type CG2 is an unsupported, supported, or flat-film glove, which has liquid integrity and uses materials that are penetration-resistant against the chemical(s) of interest.
- Type CG3 is a supported glove that covers the hand and wrist area and uses materials that are penetration-resistant against the chemical(s) of interest.

Likewise, footwear is classified into three types. However, the third type simply covers other footwear.

- Type CF1 is a boot, overboot, or boot/shoe cover, which has gas/vapor integrity and uses materials that are permeation-resistant against the chemical(s) of interest.
- Type CF2 is a boot, overboot, or boot/shoe cover, which has liquid integrity and uses materials that are penetration-resistant against the chemical(s) of interest.
- Type CF2 is a boot/shoe cover, also covering the ankle area, which is liquid repellent for the chemical(s) of interest.

Eye and face protection involves a slightly different classification system because the visor may be provided by the respirator or built into the suit, especially for vapor-protective suits. As a result, four types of eye and face protection are defined:

- Type CE1 is a hood with visor or a full face respirator, which has gas/vapor integrity and uses materials that are permeation-resistant against the chemical(s) of interest.

- Type CE2 is a hood with visor or respirator full facepiece, which has gas/vapor integrity and uses materials that are penetration-resistant against the chemical(s) of interest.
- Type CE3 is a faceshield, possibly with goggles, which has liquid splash integrity and uses materials that are penetration-resistant against the chemicals of interest.
- Type CE4 is a faceshield, which covers the face and uses materials that are penetration-resistant against the chemicals of interest.

Specific Tests for Classifying Clothing Integrity and Material Chemical Resistance

The descriptions provided for non-emergency chemical protective clothing are not tied to individual standards. Therefore, it is important to provide some way of demonstrating the level of integrity and chemical resistance in a consistent manner. Table 8 offers a recommended test and criteria for providing these demonstrations of appropriate performance justifying the type classification of the clothing item. Integrity classifications are based on the following determinations:

- Integrity against gases and vapors (passes inflation test),
- Integrity against gases and vapors (passes inward leakage test),
- Integrity against inward liquid leakage (passes shower test),
- Integrity against inward particulate leakage (passes aerosol test), and
- Coverage of the body or a specific body area (no true integrity).

While material chemical resistance is based on the following tests:

- Permeation resistance,

Table 8: Recommended Test Methods for Demonstrating Chemical Protective Clothing Performance

Property	Test Method	How Applied
Integrity	Inflation – use ASTM F1052	Use minimum pressure drop to determine acceptable suit performance ($\Delta P < 20\%$ recommended)
	Inward leakage of gases or vapors – Devise test with gas/vapor challenge (see Appendix B of OSHA 29 CFR 1910.120 for ammonia test or NFPA 1994 for sulfur hexafluoride test)	Determine acceptable levels of leakage based on chemical hazard (<2% recommended)
	Liquid (moderate duration) – ASTM F1358 with 60 minute exposure	Permit no water spotting of inner test garment
	Liquid (moderate duration) – ASTM F1358 with 20 minute exposure	Permit no water spotting of inner test garment
	Liquid (short duration) – ASTM F1358 with 4 minute exposure	Permit no water spotting of inner test garment
	Particulate – use ISO 17491, Procedure B1	Determine acceptable levels of leakage based on chemical hazard (<2% recommended)
Barrier	Permeation resistance – ASTM F739 or ASTM F1383; choose conditions representative of use	Set acceptable breakthrough time and permeation rate based on conditions of BCPC use and hazards associated with specific chemicals (BT > 60 minutes, recommended)
	Penetration resistance – ASTM F903; choose exposure conditions representative of use or use Procedure B (most common in industry)	Require "pass" results (no penetration) for specific chemicals of interest
	Repellency – ASTM F2370 or ISO 6530	Set minimum runoff of chemical for acceptable performance (Penetration index of <1, repellency index of >90, recommended
	Particulate hold out – ASTM F1215, choose particle size representative of particulate hazard	Base filtration efficiency on particulate hazards (efficiency > 99%, recommended)

- Vapor-penetration resistance,
- Liquid-penetration resistance,
- Liquid repellency, and
- Particulate-penetration resistance.

While the results of integrity tests are generally easy to interpret, specific guidelines are needed for applying chemical resistance test information. Descriptions of these tests are provided in Chapter 5.

Applying Permeation Tests. The set-up for permeation testing involves a number of choices including the type of contact (i.e., continuous or intermittent), length of exposure, and test temperature. Ideally, conditions should be chosen that mimic the worst-case exposure conditions expected for the specific application. However, the majority of published chemical permeation data are generated under relatively the same conditions: continuous exposure for 8 hours at room temperature (25±2 °C). Since more end users rely on manufacturer data, these conditions become the de facto parameters for permeation testing. Nevertheless, if the work conditions are substantially different, the following factors should be considered in evaluating the application of existing permeation data:

- Continuous contact generally results in shorter breakthrough times than intermittent contact. However, intermittent contact with some chemicals results in a wetting of the CPC surface, which leads to the same permeation as continuous contact.
- Longer tests will generally provide more information about CPC performance over a full work shift and can be more representative of reuse, even though gloves are decontaminated following initial use.
- Higher temperatures result in faster permeation and shorter breakthrough times. The effect of temperature is significant, especially for volatile chemicals, though no generalized conclusion can be provided for the change in breakthrough time that results from an increase in temperature.

The best source of permeation data is from manufacturer literature, because the data are specific to the product. There are also compilations of test data that are found in numerous sources.

- Forsberg and Keith present similar data in Permeation and Degradation Compendium.[9] This information is available both as a book and as an expert system.
- Another aid to quick selection of CPC in the field is the Quick Selection Guide to Chemical Protective Clothing by Forsberg and Mansdorf.[10] This pocket-sized booklet provides an easy-to-use color-coded selection system for many hazardous chemicals against typical (generic) CPC materials. In combination with a short training class, this guide could also be provided to aid workers with their own CPC selection process.

Applying Penetration Tests. Chemical penetration resistance data are less available than permeation data. The pass or fail reporting nature of the test results make decisions on clothing use relatively easy. Yet, there are different formats for how the test may be run, which can lead to differences in test results among manufacturers. Two general approaches are used: the first approach involves testing the material over a 1-hour period with a one-minute of the exposure at 13.8 kPa (2 psi). In the second approach, the testing is completed at ambient pressure. The pressure used in the first approach represents a more rigorous test and better assessment of material barrier performance to more severe liquid exposures.

Consideration of Other Material Properties

Depending on the other hazards present, additional properties should be considered in the evaluation and selection of each chemical protective clothing type. Other than emergency CPC, there is no consensus on what tests to use or the respective levels for these tests. Chapter 5 provides a comprehensive review of different test methods that can apply to garments, gloves, footwear, and eye/face protection. As a general guideline, the following property evaluations are recommended:

- Garments – tensile strength, tear resistance, burst strength, seam strength.
- Gloves – classification of physical properties according to ANSI/ISEA 105.
- Footwear – toe impact/compression resistance and sole puncture according to ANSI Z41; cut and puncture resistance of footwear upper; abrasion resistance of footwear sole.
- Eye/Face Protection – Classification and conformance to ANSI Z87.1.

Table 9 provides a list of other properties that can be applied to the selection of different types of chemical protective clothing.

Table 9:
List of Other Key Performance Properties and Standard Test Methods

Physical Property	Type of Test/Application	Available Test Methods
Physical Properties		
Breaking strength	Textiles	ASTM D5034/ASTM D5035
	Leather	ASTM D2208/ASTM D2009
	Rubber/coated fabrics	ASTM D412/ASTM D751
	Plastic	ASTM D638/ASTM D882
	Seam strength	ASTM D751/ASTM D1683
Burst Strength	Textiles	ASTM D3786/D3787 ASTM D3940
	Coated fabrics	ASTM D751
	Leather	ASTM D2207
Impact resistance	Footwear	ANSI Z41
	Headwear	ANSI Z89.1
	Eye and facewear	ANSI Z87.1
Projectile resistance	Bullet resistance – garments (vests)	Department of Justice Standards
	Small projectile – eye and facewear	ANSI Z87.1
Tear or Snag Resistance	Textiles	ASTM D1424/ASTM D2261 ASTM D3939/ASTM D5362
	Plastics	ASTM D1004/ASTM D1922
	Leather	ASTM D4704
	Puncture propagation tear resistance	ASTM D2582
Abrasion resistance	Textiles	ASTM D3884/D4966 ASTM D4157/D4158
	Rubber and coated fabrics	ASTM D3389
	Plastics (lenses)	ASTM D1044
	Rubber footwear soles	ASTM D1630
Cut resistance	Gloves and footwear	ASTM F1790
	Chainsaw method	ASTM F1414 ASTM F1458

Table 9: (continued)
List of Other Key Performance Properties and Standard Test Methods

Physical Property	Type of Test/Application	Available Test Methods
Physical Properties (continued)		
Puncture resistance	Gloves and footwear upper materials	ASTM F1342
	Footwear soles	ANSI Z41
Flex fatigue	Flat material	ASTM F392
Slip resistance	Footwear soles	ASTM F489/F609
Shock absorption	Cushioning materials	ASTM D4168
	Headwear	ASTM F429
	Gloves	ISO 10819
PPE Resistance to Environmental Conditions		
Water repellency and absorption resistance	Surface wetting spray method – garments	AATCC 22
	Spray impact method – garments	AATCC 42
	Water absorption/ capacity test – garments	ASTM D 5802
Water penetration resistance	Hydrostatic method	ASTM F903/AATCC 127
Salt spray/corrosion resistance	PPE hardware Face/eyewear	ASTM B117/ASTM 27 ANSI Z87.1
Stiffness/resistance to cold temperatures	Low temperature stiffness – polymers	ASTM D 1053
	Low temperature bending – coated fabrics/rubber	ASTM D2136
	Impact – plastics/rubber	ASTM D1790
UV-light resistance	Textiles	AATCC 16/ASTM G23
	Rubber	ASTM D1148
Ozone resistance	Rubber and coated fabrics	ASTM D518/ASTM D1149
Noise attenuation	Ear protectors	OSHA 20 CFR 1910.95 Appendix B
Light attenuation	Eyewear	ANSI Z87.1
Visibility	Fluorescence method	ANSI/ISEA 107
	Retroreflective method	ANSI/ISEA 107

Table 9: (continued)
List of Other Key Performance Properties and Standard Test Methods

Physical Property	Type of Test/Application	Available Test Methods
Flame and Heat Resistance		
Flame resistance	Flame spread, 45 degree, surface exposure, butane	*16 CFR Part 1610* ASTM D 1230
	Burn time, 90 degree, edge, methane	ASTM D 3659/ASTMD6413
	Burn time, 90 degree, edge, propane	FTMS 191A,5905.1
	Burn time, 90 degree, folded edge, methane	ASTM F1358
	Burn time, 180 degree, edge, propane	ASTM D635
Heat resistance	Heat effects/thermal shrinkage – textiles, other PPE	ISO 17493
	Blocking – coated fabrics	ASTM D751
Conductive heat resistance	Clothing materials	ASTM D1518/ASTM F1060
Radiant heat resistance	Clothing materials	ASTM F1939
Thermal protective performance	Clothing and glove materials	ASTM D4108 NFPA 1971
	Whole clothing items	ASTM F1930
Molten metal resistance	Splash method	ASTM F955
	Small splash (droplet) method	ISO 9150
Electrical Properties		
Electrical insulation	Gloves	ASTM D120
	Footwear	ASTM F1116 ANSI Z41.1
	Headwear	ANSI Z89.1
Conductivity and electrical resistivity	Textiles	AATCC 76
	Footwear	ANSI Z41.1
Static charge accumulation resistance	Clothing materials	FTMS 101C,4046 INDA IST 40.2
Electrical arc protective performance	Clothing materials	ASTM F1958
	Overall clothing	ASTM F1959

Table 9: (continued)
List of Other Key Performance Properties and Standard Test Methods

Physical Property	Type of Test/Application	Available Test Methods
Human Factors		
Material biocompatibility	Medical device (gloves, gowns) biocompatibility tests	ISO 10993
	Natural rubber latex proteins	ASTM D5712
Thermal insulation and breathability	Air permeability – materials	ASTM D 737
	Moisture vapor transmission – materials	ASTM E 96
	Thermal resistance – materials	ASTM D1518
	Thermal resistance – garments	ASTM F1291
	Total heat loss – materials	ASTM F1868
Mobility and range of motion	Overall product evaluation – chemical protective clothing	ASTM F1154
Hand function	Gross dexterity – gloves	ASTM F2010
	Fine dexterity – gloves	EN 420 NFPA 1999
	Grip strength – gloves	NFPA 1971
Ankle support	Footwear	Field studies
Back support	Garments or accessories	Field studies
Clarity and field of vision	Eyewear – prismatic power	ANSI Z87.1
	Eyewear – refractive power, resolving power, astigmatism	ANSI Z87.1
	Eyewear – haze	ANSI Z87.1 ASTM D1003
	Eyewear – transmittance	ANSI Z87.1
Ease of communications	Hearing protectors	ANSI S3.2
Sizing and fit (adjustment)	Garment sizing/dynamic fit test	ASTM F 1731 ANSI/ISEA 101
Donning and doffing ease	Full body suits	NFPA 1991

Notes (Sources of Standards):

AATCC – American Association of Textile Chemist and Colorists (Research Triangle Park, NC), www.aatcc.org

ANSI – American National Standards Institute (New York, NY), www.ansi.org

ASTM – American Society for Testing and Materials (West Conshohocken, PA), www.astm.org

CAN/CGSB – Canadian General Standards Board (Hull, Quebec), www.cgsb.ca

EN – European Norm from Committee on European Standardization (Brussels, Belgium), www.cenorm.be

FTMS – Federal test method standard from U.S. Government Printing Office (Washington, DC), www.gpo.gov

IER – Institute of Environmental Sciences and Technology (Mount Prospect, IL), www.iest.org

INDA – Association of the Nonwovens Fabric Industry (Cary, NC), www.inda.org

ISO – International Standards Organization (Geneva, Switzerland), www.iso.ch

MIL – Military standard from U.S. Government Printing Office (Washington, DC), www.gpo.gov

NFPA – National Fire Protection Association (Quincy, MA), www.nfpa.org

Summary

The selection of CPC is an assignment that must be carried out carefully and diligently. All selection decisions begin with a thorough and detailed hazard assessment and determination of relative risk. The hazard assessment process entails an analysis of the workplace and decisions on the likelihood of exposure and consideration of the consequences of exposure. When this analysis focuses on the portions of the body likely to be affected, the results yield the protection priorities that can then be used in a decision logic for selecting specific types of chemical protective clothing.

A decision logic that addresses both emergencies and non-emergencies should be used. Emergency situations dictate use of CPC that conforms to different NFPA standards depending on the circumstances. For non-emergency situations, a CPC type classification system has been proposed for garments, gloves, footwear, and eye/face protection to offer appropriate clothing to match the specific risks and exposure conditions.

No decision logic is perfect for every situation. One that attempts to cover every possible situation becomes complicated and unusable by the ordinary end user. But, a decision logic that is too simple will omit some of the key attributes that should be considered in making CPC selections. Summary

References

1. Bollinger, N.J. and R.H. Schutz, *NIOSH Guide to Industrial Respiratory Protection*, DHHS (NIOSH) Publication No. 87-116, U. S. Department of Health and Human Services, National Institute for Occupational Safety and Health, Morgantown, WV, September 1987.

2. Ronk, R., M.K White, and H. Linn, *Personal Protective Equipment for Hazardous Materials Incidents: A Selection Guide*, DHHS (NIOSH) Publication No. 84-114, U.S. Department of Health and Human Services, National Institute for Occupational Safety and Health, Morgantown, WV, September 1987.

3. *NIOSH Guide to the Selection and Use of Particulate Respirators Certified under 42 CFR 84*, DHHS (NIOSH) Publication No. 86-101, U.S. Department of Health and Human Services, National Institute for Occupational Safety and Health, Cincinnati, OH, January 1996.

4. *Use and Selection Guide for Non-Powered Air-Purifying Particulate Respirators*, Industrial Safety Equipment Association, Arlington, VA, 1996.

5. American Conference of Governmental Industrial Hygienists (ACGIH), *Threshold Limit Values® and Biological Exposure Indices for 2003*, American Conference of Governmental Industrial Hygienists, Cincinnati, OH (2003).

6. Grandjean, P., "Preventing Percutaneous Absorption of Industrial Chemicals: The 'Skin' Notation," *Am. J. Ind. Med.*, 14, 97 (1988).

7. Brown, U.K., *Acute Toxicity in Theory and Practice*, John Wiley and Sons, Inc., New York (1986), p.3.

8. Eiser D.N., AT&T Microelectronics, Reading, PA, private communication.

9. Forsberg, K, and L. H. Keith, Chemical Protective Clothing: Permeation and Degradation Compendium, Lewis Publishers, Boca Raton, FL, (1995).

10. Forsberg, K., and S.Z. Mansdorf, 4th Edition, Quick Selection Guide to Chemical Protective Clothing, John Wiley and Sons, New York, NY, (2003).

APPENDIX

Decision Logic for the Determination of General Hazard Classification

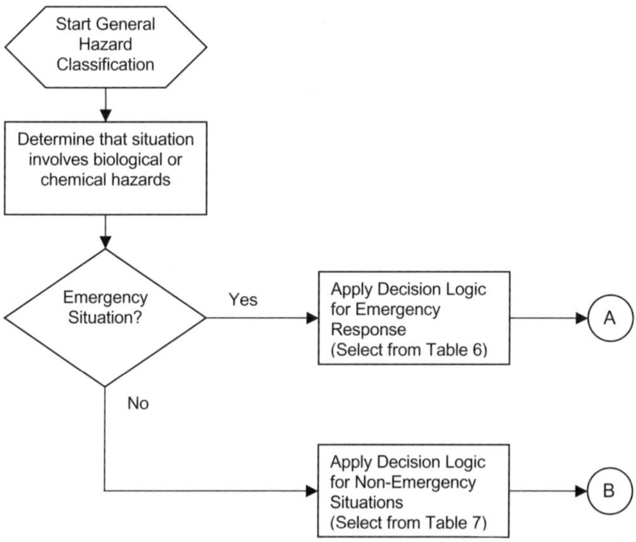

Emergencies
Emergencies are unplanned events involving either chemical or biological hazards. Normally, specific individuals or groups will be designated to act in the event of an emergency.

Both Biological and Chemical Hazards Present
If both biological and chemical hazards are present, go through each decision logic and choose the most protective of the two sets of BCPC recommended.

Decision Logic for the Emergency Response Scenarios

Step A – Emergency Response: Part 1 (Initial Assessments)

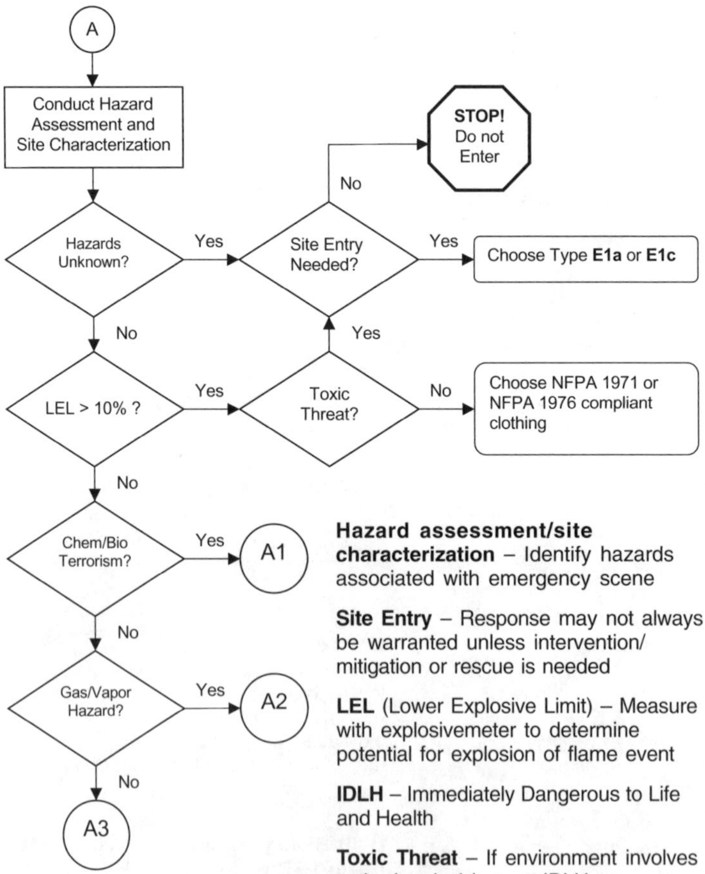

Hazard assessment/site characterization – Identify hazards associated with emergency scene

Site Entry – Response may not always be warranted unless intervention/ mitigation or rescue is needed

LEL (Lower Explosive Limit) – Measure with explosivemeter to determine potential for explosion of flame event

IDLH – Immediately Dangerous to Life and Health

Toxic Threat – If environment involves toxic chemical (e.g., at IDLH concentration)

Chem/Bio Terrorism – Chemical or biological release associated with terrorism event

Gas/Vapor Hazard – Chemical present as gas or liquid evolves vapor

Step A1 – Select Appropriate Ensemble for Chem/Bio Terrorism Incident

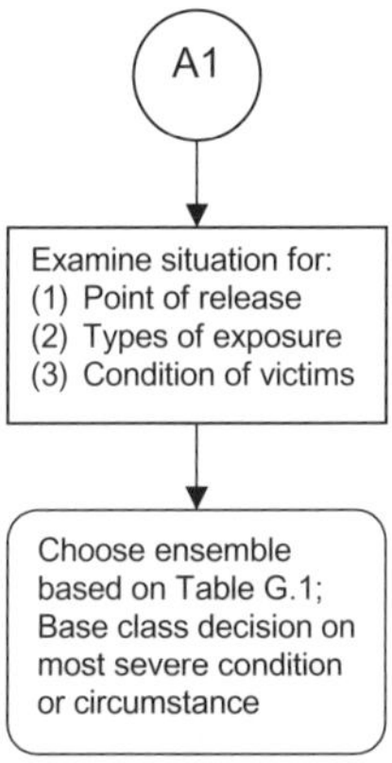

Class	Point of Release	Types of Exposure	Conditions of Victims
1 Type E3(1)	Responders close to point of release in both time and distance	• Vapors and gases (high concentrations) • Aerosols, liquid dispersions • Liquid puddles • Liquid containing pathogens	Most of victims dead or showing serious signs and symptoms of chemical agent exposure
2 Type E3(2)	Responders separated from point of release by either time, or distance	• Vapors and gases (low concentrations) • Aerosols, liquid dispersions (low concentrations) • Liquid residue • Liquid containing pathogens	Majority of victims survive, but non ambulatory
3 Type E3(3)	Responders separated from point of release by both time and distance	• Liquid residue • Liquid containing pathogens	Mostly ambulatory

Step A2 – Select Appropriate Ensemble for Vapor Protection

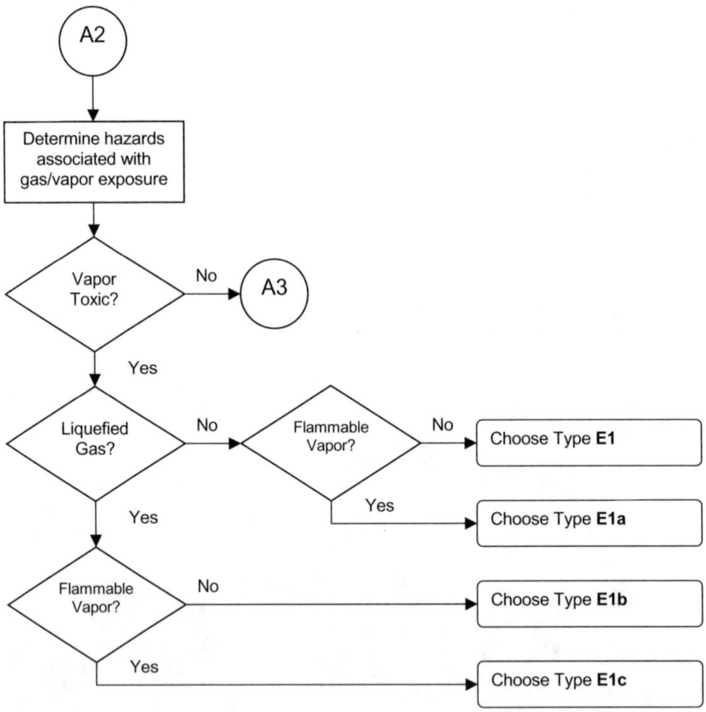

Vapor Toxic – Vapor presents toxic (or other adverse effects) on skin; also may be at high concentrations above exposure limit

Liquefied Gas – A gas that under its charged pressure is partially liquid at 21°C (70°F)

Flammable Vapor – Vapor is rated as having flammability potential by DOT placard, NFPA flammability rating or other indication

Other Vapor Protection Ensemble Considerations:

(1) Permeation data for specific chemicals

(2) Reusable versus disposable suit

(3) Location of closure system

(4) Type of glove system

(5) Type of footwear system

Step A3 – Select Appropriate Ensemble for Liquid Protection

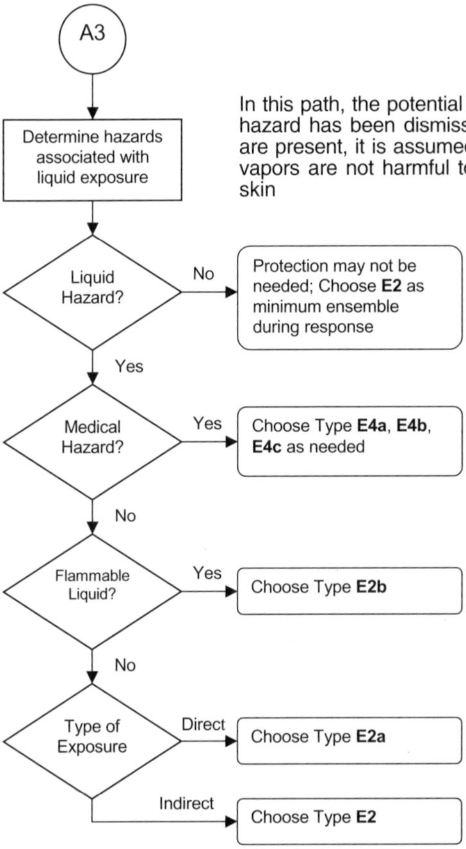

In this path, the potential for a vapor hazard has been dismissed; if vapors are present, it is assumed that the vapors are not harmful to the wearer's skin

Liquid Hazard – substance is determined to be liquid; may also be particulate

Medical Hazard – liquidborne or bloodborne pathogen, usually in emergency medical situation

Flammable Liquid – A liquid that is rated as having flammability potential by DOT placard, NFPA flammability rating or other indication

Type of Exposure – Refers to type of liquid exposure:

Direct exposure involves multiple splashes, possibly with liquid under pressure

Indirect exposure involves 1-2 splashes with no pressure

Decision Logic for the Non-Emergency Situations

Step B – Select Chemical Protective Clothing

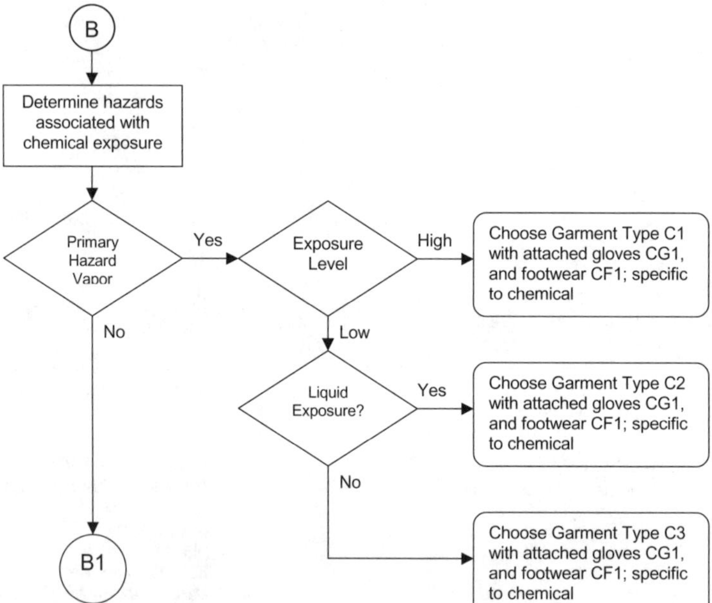

Primary Vapor Hazard – If principal hazards associated with chemical exposure occur via gas, vapor, or vapor associated with liquid

Exposure Level – Judge exposure level based on concentration of gas or vapor in air at work site; compare to exposure limits set using those permitted by OSHA, threshold limit values or short term exposure levels recommended by ACGIH, or other action levels appropriate to organization; also applies to chemicals known to be toxic through skin absorption

Liquid Exposure – If liquid is also present and if potential also exists for direct exposure to liquid; this is significant because liquid may saturate capability of resistant garment in preventing vapor penetration

Step B1 – Select Full-Body Chemical Protective Clothing for Liquid and

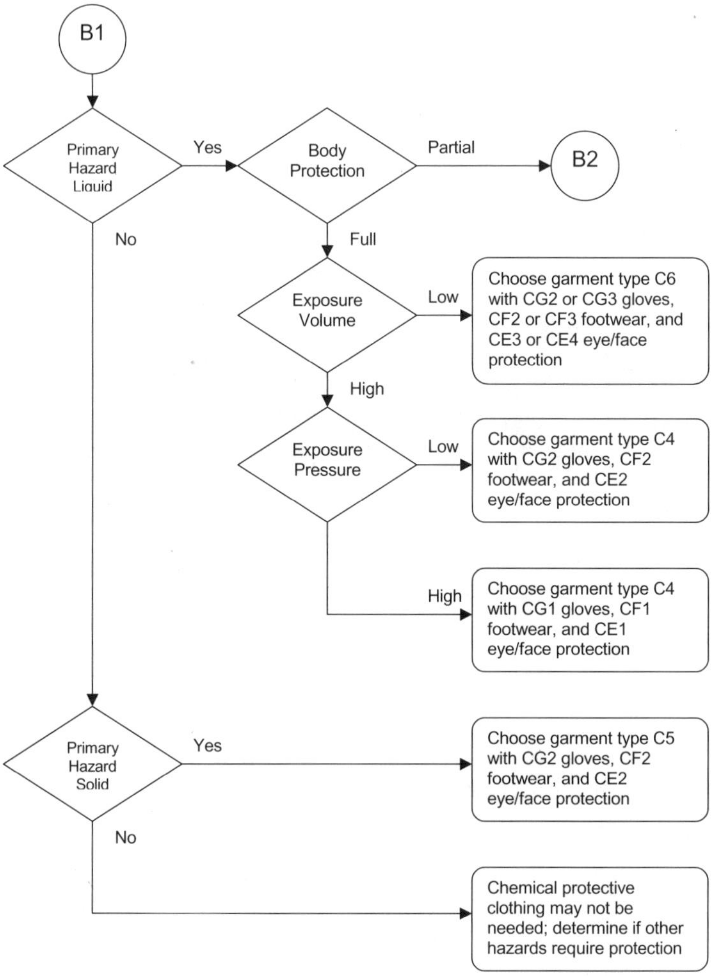

Step B1 – NOTES for Selection of Liquid and Particulate Chemical Protective Clothing

Primary Liquid Hazard – If principal hazard associated with chemical is through liquid contact, may be continuous, or via splashes; does not pertain to complete immersion, except for portions of hands and feet.

Body Protection – Distinguish between full-body protection, requiring ensemble of clothing and equipment, or partial-body protection, such as protecting the hands, feet, or front torso, where partial body garments would be used.

Exposure Volume – Based on relative volume of liquid chemical. Some organizations have suggested that 500 mL can be difference between high and low volume exposures. However, the decision between a high and low volume should be based relative to the chemical of concern. Exposure volume may occur in single exposure or over multiple exposures in single wearing of BCPC item.

Exposure Pressure – Pressure in work task that may be accompanied by kneeling or leaning in chemical, or as might occur with release of chemical, such as from burst pipe. Tasks involving these exposures would be considered high pressure. Low pressure involves minor contact with little or no force against chemicals.

Principal Particulate Hazard – Principal hazard is from particulates, that either pose hazard by direct contact or reaction with the skin, or by their rerelease to the atmosphere and exposure to the wearer's respiratory system. Does not include nuisance particles such as dust, dirt, or debris.

Step B2 – Select Partial-Body Chemical Protective Clothing for Liquid

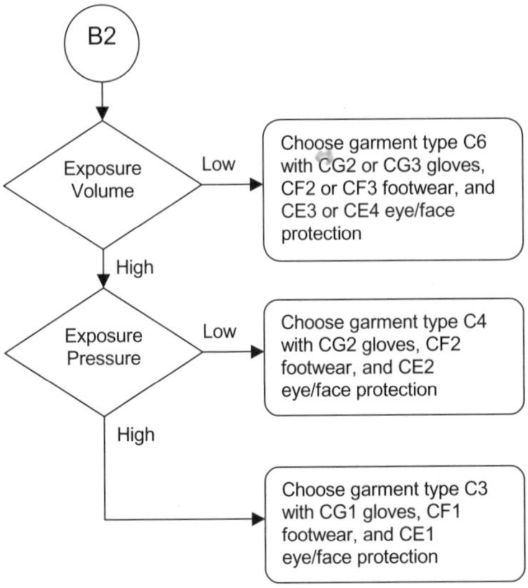

Exposure Volume – Based on relative volume of liquid chemical. Some organizations have suggested that 500 mL can be difference between high and low volume exposures. However, the decision between a high and low volume should be based relative to the chemical of concern. Exposure volume may occur in single exposure or over multiple exposures in single wearing of BCPC item.

Exposure Pressure – Pressure in work task that may be accompanied by kneeling or leaning in chemical, or as might occur with release of chemical, such as from burst pipe. Tasks involving these exposures would be considered high pressure. Low pressure involves minor contact with little or no force against chemicals.

8

Decontamination of CPC

By Michael H. Ziskin, Jeff Behar, and S. Zack Mansdorf

Introduction

Effective decontamination of chemical protective clothing (CPC) is critical for all clothing regardless if it's worn once and discarded or meant for multiple use. More importantly, CPC users should be aware that they also may become contaminated and therefore may require some level of decontamination. To prevent or reduce the risk of personal contamination, CPC users must be trained to properly remove all contaminated garments. All clothing items require cleaning or decontamination, and multiple use items require inspection before reuse. In addition, safe and proper disposal of chemically contaminated clothing is required by federal and state environmental regulations.

For the purpose of this chapter, decontamination is defined as the process of removing, isolating or reducing contaminants that are known or suspected to be present on CPC. Decontamination can be employed to (1) protect the wearer from becoming exposed to the contaminants of concern, (2) minimize the effect contaminants may have on chemical protective clothing (chemical attack), (3) avoid cross-contamination, and (4) return CPC to service in a condition that will provide the same protection as prior to the first use.

Selection of a decontamination method depends on many variables including, but not limited to:

- The type of contamination (solid, liquid or gas/vapor),
- The physical, chemical and toxicological properties of the contaminant,
- The permeability of the contaminant through the skin,

- The delivery dose,
- The location and area of CPC affected,
- The number of CPC users requiring decontamination, and
- The physical location or facility where decontamination is to be performed.

In specific situations, especially for outdoor locations, environmental conditions will have a significant influence on the decontamination process.

For most CPC users, decontamination that results in a "zero" exposure level is usually the only acceptable choice. Contamination of CPC can degrade the performance of the clothing and decrease the subsequent resistance to breakthrough performance and increase the permeation rate upon continued exposure. For some contaminants, improper decontamination can lead to immediate effects, such as chemical burns.

The following case study illustrates possible outcomes when proper decontamination procedures are not established.

> *A first-shift worker in an aluminum can manufacturing plant was preparing a solution of hydrofluoric acid. This solution was used to "polish" beverage cans prior to painting. The solution was prepared by diluting a 37% stock of hydrofluoric acid to approximately a 3% bath solution. The job required the use of long gloves, a face shield, and an apron to protect against the corrosive effects of the hydrofluoric acid. (Hydrofluoric acid in dilute concentrations, that is generally below 50% v: v, can cause delayed burns occurring hours after initial contact.)*
>
> *The worker apparently damaged the gloves (pinhole leaks) that he found on top of the 55-gallon drum of the 37% hydrofluoric acid used to make up the dilute solution. He proceeded to make up the solution, not knowing that the gloves he was wearing were admitting hydrofluoric acid.*

Upon task completion, the first shift worker replaced the gloves, he thought were resistant to the permeation of hydrofluoric acid, on top of the drum where he had found them. This practice of leaving the gloves on top of the drum for the next worker was routine. The second-shift worker completed the same task, using the same pair of gloves that the first shift worker had used. The third-shift worker also followed this practice.

Approximately six hours after the first worker had completed his work shift, he noticed a severe burning sensation in his fingers. He knew immediately he had been burned by the hydrofluoric acid and went to the local emergency room for calcium gluconate injections (a complexing agent) under the fingernails.

By the time the first-shift worker had completed treatment, the second-shift worker, now at home, noticed a burning sensation as well. The third-shift worker was just completing his shift at this time. The first shift worker called the plant and the third-shift worker went immediately for treatment. In this example, three separate workers suffered the effects caused by one pair of damaged, contaminated gloves because the gloves were not inspected or decontaminated prior to each use.

This chapter presents information on decontamination processes and methods, planning considerations, types of contamination and the effects on CPC, management of decontamination operations, and special considerations for certain exposure scenarios. Since some forms of CPC have been tested and proven protective against other contaminant types, discussion of biological decontamination, and to a limited extent, radiological decontamination is included in some sections.

Although decontamination of respiratory protective equipment and tools, vehicles and other equipment is not addressed specifically in this chapter, many of the considerations contained herein would apply.

Decontamination Plan

A decontamination plan for CPC should be developed as part of an overall Personal Protective Equipment Program or site-specific Health and Safety Plan. Development of the decontamination plan should occur before personnel enter into areas where exposure to contaminants may occur. The plan must include a hazard and risk assessment and address the jobs and duties performed during decontamination. In many instances, the decontamination plan may be developed as part of the CPC selection process.

The first step in the development of any decontamination plan is to evaluate the contaminants of concern. A competent individual familiar with the contaminants that may be present should perform this task.

In addition to the type of contaminant, the level of the contaminant that may be encountered and the time length of the task to be performed must be evaluated. The first step should be to consult the Material Safety Data Sheet (MSDS) for the hazardous substance of concern. The MSDS can provide important information such as:

- Supplier Information – Including the name, address and phone number of the chemical manufacturer.
- Hazard Ingredients/Identity Information – This information is important if the contamination is due to a trade product that contains many chemicals that could react or effect the health of personnel adversely. This information would prove helpful with identification of proper disposal methods.
- OSHA's Permissible Exposure Limit (PEL) and/or the American Conference of Governmental Industrial Hygienists' Threshold Limit Value® (TLV®) for safe exposure – This provides a reference value to compare the air monitoring results of the decontamination facility, personnel and surroundings.
- Physical/Chemical Characteristics – Physical/chemical characteristics (such a vapor pressure, vapor density,

evaporation rate, solubility in water, specific gravity and flash rate) define contaminant behavior and help to determine how to handle disposal of decontamination waste.

- Special firefighting procedures, unusual fire and explosion hazards, reactivity data, stability and incompatibilities – This may assist in determining the appropriate level of protection necessary for decontamination workers, as well as the equipment to use, and the procedures to implement.
- Health hazards (acute and/or chronic), signs and symptoms of exposure, emergency first aid procedures – This information should be known prior to handling the contaminant waste streams so that (1) personnel are aware of the early signs and symptoms of exposure, and (2) medical personnel are aware of signs and symptoms of exposure.
- Precautions for Safe Handling and Use/Control Measures – Although some MSDSs do not provide specific information regarding personal protection (such as glove materials, etc.), this section may provide important information, such as the requirement to use supplied air, vs. an air filtering face piece.
- Ecological information/waste disposal procedures – Ecological information may include the effects of the chemical (or waste) if released into the environment, as well as the chemical's environmental fate. Some MSDSs do not provide specific information regarding proper waste disposal requirements, since many states may have more aggressive state requirements, but the MSDS can assist with identifying federal requirements regarding disposal of waste generated during decontamination.

The decontamination plan should be developed in conjunction with the effect of the decontamination method on the CPC material. Some materials are easily cleaned, while others, including porous materials, may require more significant steps as part of the decontamination process. The process in itself may render the equipment or protective clothing reusable or, in many instances, not reusable. CPC manufacturers may also have recommendations regarding the decontamination of their product and this information should be considered in planning for decontamination. When CPC is not reusable, it must be disposed of properly. Waste characterization on CPC must be performed and disposal arrangements must follow federal and state requirements.

Record keeping and reporting of decontamination practices should also be established as part of the program. Various logs may be created to document the CPC exposure record, decontamination performed and inspection results. This information may prove useful in determining the "use" life projection of the CPC.

The decontamination plan should include specific methods for each of the known or suspected contaminants. Any health or safety hazards that may be introduced as a result of decontamination methods should also be addressed. The number and placement of decontamination stations should be identified, along with the general position of the decontamination work areas, relative to those activities adjacent to these work areas. The necessary decontamination equipment and methods should be clearly identified. Checklists can be used for quality control during the implementation of decontamination procedures.

During removal of contaminated CPC, personnel must be protected from cross-contamination. Standard Operating Procedures (SOPs) to prevent cross-contamination and to minimize worker contact with contaminants during removal of CPC should be part of the decontamination plan. This requires that personnel using CPC as well as those assisting in the decontamination operations follow procedures similar to the "universal precautions" used for avoiding contamination by blood borne pathogens. When these procedures are followed, there is increased likelihood that cross-contamination will be minimized. Even with no visible CPC barrier damaged and no expected permeation of the contaminant, cross-contamination is always a concern in the decontamination process.

The complete decontamination plan should be reviewed periodically and the plan should be updated and revised anytime workplace conditions or site-specific hazards change.

Prevention of Contamination

The best control option for prevention of CPC contamination is to avoid contaminant contact. Although obvious, this tenet should be continuously re-emphasized when implementing any CPC use.

Avoiding contamination can be accomplished by a number of methods. Establish work practice controls or procedures that restrict the CPC user from contact with the contaminants. For example, not walking through areas of obvious contamination or not touching surfaces unnecessarily can be very effective when enforced. Other ways to practice contamination avoidance include (1) reducing the number of CPC users with the potential for contact, (2) using remote handling tools rather than direct contact, (3) replacing more expensive or reusable items with less expensive or disposable items, and (4) encasing the source of contaminants with isolation materials, such as plastic sheeting.

As a part of avoidance, consider methods to minimize the potential for contaminants to bypass CPC. This can be accomplished in a number of ways. One way is to don the CPC prior to entering contaminated areas. Another way is to inspect CPC prior to each use for signs of physical damage from cut, puncture, tear or abrasion. Gloves and boots could be tucked under the sleeves and legs of outer clothing, while hoods could be worn outside the collar. CPC can be designed with appropriate closures protecting zippers, buttons, snaps and interfaces of protective equipment items. The CPC ensemble should prevent solid or liquid contaminants from entering at any closure or at interfaces between pieces of personal protective equipment.

To minimize the risk to personnel selected to wear CPC, users should be free of injuries to the skin's surface. Cuts and scratches may enhance the potential for contaminants or infectious agents to penetrate into the body when contact occurs.

CPC Contamination

Contaminants may collect on the surface of CPC then penetrate and/or permeate through CPC materials. Although surface contamination may be easy to detect and remove, this type of contamination may permeate the CPC and may be difficult to detect without testing a sample of the affected CPC. Decontamination efforts may be effective to permit safe removal of the CPC, but not adequate for the CPC to be reused. In order to determine the limitations of the decontamination, it is important to understand how the contaminants can be removed, based on the deposition of contaminants and their effects on CPC.

Many factors affect the extent of CPC contamination. For instance, the longer a contaminant is in contact with CPC, the greater the probability and extent that permeation might occur. In this case, contamination avoidance would be effective. For other factors, the following relationships may help decide how to minimize contamination:

- Concentration – With increased concentration of contaminants an increase in permeation is likely. Remote handling methods should be implemented in high concentration areas.
- Temperature – Permeation rates tend to increase as the temperature increases.
- Size of the contaminant molecule and pore space – Permeation increases as the contaminant molecule becomes smaller, and as the pore space of the material to be permeated increases.
- Physical state of the contaminant – Gases, vapors and low-viscosity liquids tend to permeate more readily than high-viscosity liquids or solids.

Surface Contamination

When surface contamination of a CPC garment occurs, the contaminant does not enter into the pores or molecular matrix of the CPC; only surface adhesion occurs. An example of a surface contaminant is dry pigment dust. Washing or brushing often easily removes surface contaminants. Water-soluble contaminants are more efficiently dissolved with emulsifiers or wetting agents (e.g., soaps and detergents). Some surface contaminants may require neutralization or complexing for effective decontamination. Inorganic acids that do not permeate the surface of the barrier may still remain on the outside surface. Effective neutralization may include the use of a weak acid or base, or a buffered solution (e.g., citric acid for bases and baking soda solution for acids). Some acids, such as hydrofluoric acid, may require complexing through the use of an agent that can bind to the fluoride ion.

Matrix Contamination

Matrix contamination occurs when the contaminant actually permeates the molecular matrix of the protective barrier. Matrix contamination is basically an artifact of permeation, and therefore the most difficult to manage.

An illustrative example of matrix contamination is the diffusion of ethylene oxide from rubber catheters after being sterilized. The gaseous ethylene oxide permeates the rubber of the catheter in the sterilizer and then continues to outgas small quantities of ethylene oxide after removal from the source. Aeration is required to "rinse" the ethylene oxide from the rubber matrix. Application of an aeration process depends on a number of factors including:

- Solubility parameters of the barrier and the contaminant,
- Type of fillers, such as clays,
- Enhancers used, such as carbon blacks, and
- Additives to polymers, such as butyl rubber or PVC plastic, and compounding ingredients such as anti-oxidants.

NIOSH research demonstrated that the time it took for chemical breakthrough through two different samples of the same generic glove material was significantly different statistically, suggesting that the glove formulations were different. Additives tend to vary from manufacturer to manufacturer. Changes and enhancements in formulations can create variability within a single manufacturers product line. Regardless of the specific polymer formulations, if permeation of the barriers can occur the potential for matrix contamination exists.

For many types of CPC, cleaning and decontaminating are synonymous since the intent is to remove unwanted residues. In practical terms, "cleaning" or laundering may be required before reuse of protective clothing such as gloves or suits that become "dirty" either from non-hazardous materials (e.g., dirt, dust, grease, etc.) or from normal body emissions (e.g., perspiration). For reusable items this type of decontamination is typically achieved by standard laundry techniques.

When contamination is possible, the decontamination technique and level of efficiency should be pre-determined before the CPC is actually used. Advanced preparation must incorporate training for the anticipated users on how to minimize contamination potential and risks, and how to perform the prescribed decontamination procedures.

Effects of Contamination on CPC Materials

Surface contamination generally does not damage the barrier as significantly as matrix contamination. Nevertheless, surface contamination can result in a physical deterioration of the barrier. Discoloration, pitting, cracking or other physical changes may indicate this deterioration. Matrix contamination can wash out additives such as the plasticizers, causing changes to the physical properties. Some rubbers may become brittle after the contamination and decontamination cycle. Re-exposing damaged materials is the primary concern, because the protective properties may be much less than expected. Tests have demonstrated that if the matrix is contaminated, shortened breakthrough times may result. Additionally, a greater permeation flow rate of the contaminant to the inside of the barrier may occur.

Laundering Considerations

In general, routine work clothing that has become chemically contaminated should be laundered separately from other clothing. Advance notice to outside laundering facilities should include information about the potential health hazards associated with the contaminants. The effectiveness of chemical contaminant removal by laundering varies greatly. Documentation of the contaminant removal process and assessment of the risk of worker exposure to any residual amount of contamination should be maintained.

Since a laundering process may distribute the contaminant throughout the CPC, most CPC should not be laundered. For example, laundering gloves may contribute to contamination inside of the glove. Since most CPC is not designed for laundering, the process may ultimately damage the clothing. CPC should be placed on a use schedule and be disposed of properly.

Decontamination Methods

Decontamination methods should control how contaminated CPC, equipment, and tools, etc. are "left behind" in contaminated areas. These controls may include removal methods, administrative procedures and engineered systems, such as work zone delineation. The selection of a decontamination method may depend upon factors including the type of contamination, cost of CPC, availability of decontamination resources, and the probability of successful decontamination. CPC decontamination methods must be effective for the contaminant of concern and the method itself should not pose any health and safety hazards to personnel, the public, or the environment. Administrative procedures and engineered systems for decontamination are discussed later in the chapter.

Contamination removal methods can be an effective means to protect:

- The worker from the contaminated CPC, while removing it,
- Other workers from becoming contaminated or exposed to workers wearing contaminated CPC,
- "Clean," non-contaminated work areas from becoming contaminated, and
- The worker that reuses previously contaminated CPC from becoming contaminated upon donning CPC.

Typically, contamination removal methods are employed when the contaminants can be (1) physically removed, (2) rendered harmless by chemical detoxification or disinfection/sterilization, or (3) removed by a combination of methods involving both physical and chemical means.

Physical Removal

In some instances, contamination may be removed by dislodging, displacement, rinsing, wiping or through evaporation. These removal methods may utilize hand tools (brushes, towels, wipes, etc.) or mechanized devices such as portable hand-pressurized pump sprayers or spray/shower-type systems, some of which can also be heated. Caution should be exercised when using mechanized devices since they can spread contamination, destroy CPC, or even expose/burn the CPC wearer.

Loose contaminants such as dusts and vapors can adhere to CPC surfaces and become entrapped in zippers, or interface areas around glove assemblies or hood/respirator areas. These types of contaminants can usually be removed by using either a dry mechanical method or a wet method using water or a water/surfactant mixture. Anti-static solutions can also be applied if contaminants are electrostatically charged.

Glues, cements, resins, muds, or other contaminants that adhere to CPC may resist removal by spray alone and may require additional techniques such as scraping, solidification, freezing, absorption or adsorption. Caution should be taken not to introduce new hazards by tearing or scraping the CPC or by causing the user to inhale powered adsorbant spill cleanup products.

Volatile liquid contaminants may simply be removed by allowing enough time for the contaminant to evaporate, assuming the time is available and the weather conditions are favorable. Caution should also be taken to ensure that the wearer does not inhale vaporized contaminants.

Gross Decontamination

Gross decontamination is analogous to kicking mud from boots before rubbing them on a doormat, or rinsing sand off at the beach before showering inside. The concept is simply to remove as much contaminant as possible before the cleaning or decontamination processes. This can extend the life of decontamination equipment and washing/rinsing agents.

Dry Decontamination

In some instances dry decontamination may be necessary. Situations where dry decontamination methods can be used include, but are not limited to:

- Removal of water reactive contaminants,
- The inability to contain liquid decontamination run off,
- The lack of available wet decontamination substances or,
- Weather conditions that cause water soluble decontamination liquids to freeze.

Dry decontamination may be performed using brushing, wiping, blotting, or vacuuming techniques, or with dry powders, such as soap, detergents, earth or flour. A hazard and risk analysis should be performed to determine if dry decontamination is adequate. Dry decontamination may be acceptable without further removal methods, or as an interim measure that can be employed immediately and followed by additional removal methods.

When wet decontamination methods are needed but are not immediately available, dry decontamination may be used as an interim step. An example would be emergency response and rescue operations involving contaminated victims and hazardous waste operations where ground surface contamination levels are varied. Dry decontamination may be practiced in work areas to minimize contaminant transfer to other less contaminated areas.

Dry decontamination may be followed by wet decontamination when conditions allow. In some instances, as in the case of water reactive materials, water may not be used due to the potential for contaminant reaction. CPC wearer safety may be of concern since direct exposure may occur during the doffing process. In these instances, where residual CPC contaminant levels may still exist even after dry decontamination is performed, isolation methods may be used. Oils, such as mineral oil may be used to coat the CPC surface, isolating the residual contaminants.

Chemical Removal

If the physical removal of contaminants was not effective, additional removal methods would be required. Washing and rinsing methods involving water or other chemicals may be used to:

- Dissolve contaminants,
- Reduce adhesion forces between contaminants and the CPC surface being cleaned,
- Solidify liquid or gel contaminants,
- Rinse and dilute, and
- Disinfect/sterilize.

Each chemical removal method has specific applications, limitations, and health and safety risks to both the CPC wearer and any personnel assisting in the decontamination process.

Chemical removal by dissolving contaminants generally implies using solvents that are compatible with the contami-

Table 1:
General Guide to Solubility of Contaminants in Four Solvent Types

Solvent	Soluble Contaminant
Water	Low-chain hydrocarbons Inorganic compounds Salts Some organic acids and other polar compounds
Dilute Acids	Basic (caustic) compounds Amines Hydrazines
Dilute Bases (detergent, soap, etc.)	Acidic compounds Phenols Thiols Some nitro and sulfonic comounds
Organic Solvents (alcohols, ethers, ketones, aromatics, straight-chain alkanes (e.g., hexane) common peteroleum products (e.g., fuel oil, kerosene)	Nonpolar compounds (e.g., some organic compounds).

nated CPC. For example, CPC constructed of organic materials could be damaged or dissolved by organic solvents. From a health and safety perspective, organic solvents could represent hazards to both the personnel wearing CPC and those assisting in the decontamination process. In addition, fire codes and environmental regulations may govern the use, storage, transportation and disposal of waste materials generated in the decontamination process. Organic solvents include alcohols, ethers, ketones, aromatics, straight-chain alkanes, and common petroleum products.

Halogenated solvents are generally incompatible with CPC materials and more toxic than organic solvents. These solvents should only be considered for use in decontamination of equipment and building surfaces when other cleaning agents will not remove the contaminants. Table 1 provides a general guide to solubility of contaminants in four solvent types.

The use of surfactants is perhaps the most common of all decontamination methods. Examples of commonly used surfactants include household detergents, soaps, etc. Surfactants are used because they tend to be:

- Compatible with physical and other chemical removal methods,
- Readily accessible,
- Fairly inexpensive,
- Safe to handle and use,
- Easy to store, and

In addition, surfactants can prevent the redeposit of contaminants on the surface of decontaminated CPC.

Solidification agents completely solidify or gel contaminants, making the physical removal process easier. Most solidification agents remove the moisture using absorbents such as grounded clay or powdered lime, polymerize liquids through chemical reactions via polymerization catalysts and chemical reagents, or freeze materials using ice water. Solidification agents are more suited for equipment and surfaces than for CPC.

Rinsing contaminated CPC with water, surfactants or other liquids is performed when dilution or solubilization is identified as the chemical removal method. Multiple rinses of CPC with clean solutions can be more effective in removing contaminants than a single rinse. Continuous rinsing can be even more effective than multiple rinsing, since increasing the rinse time and volume of rinse liquids remove more contaminants. However, increasing the volume of rinse liquids may increase costs associated with disposal of the contaminated rinse liquid.

Health and Safety Hazards

Health and safety hazards must be evaluated when selecting contaminant removal methods. Hazards may result from incompatibility between chemicals used in the removal method and contaminants of concern. In some cases the chemical used in the removal method may react with contaminants to produce an explosion, heat or toxic product. The incompatibility also could be between the chemical used in the removal method and the CPC material itself. In this type of situation, the solvent may permeate and/or degrade the CPC itself. In other instances, the reactions may produce toxic or flammable vapors. Figure 1 presents a decision aid for health and safety risk assessment of decontamination methods.

Selection of Decontamination Methods by Type of Contamination

Surface contamination can be removed by physical means including brushing, spraying, washing, and scraping. If the contaminant is water soluble, washing is effective. If the contaminant is marginally water soluble, detergents and wetting agents can enhance removal (e.g., use of an emulsifier for oils). Some contaminants may require neutralization or complexing.

Protective clothing that has *pore contamination* usually cannot be decontaminated and may need to be handled as hazardous waste.

Permeated CPC materials must be decontaminated if feasible. If decontamination is not possible, the CPC should be discarded. In research conducted by Schlatter, Berardinelli, Perkins, and Stull, aeration was determined to be as effective or more effective than traditional water washing, dry cleaning, Freon cleaning, and other methods. "Hot air" aeration is effective when temperatures are above 100°F (38°C). Research performed by Perkins and others demonstrated the effectiveness of this technique for the removal of various volatile solvents from butyl rubber using 50° C air temperatures.

For certain chemicals, the use of hot water, detergents, and other additives can increase the efficiency of decontamination. Nevertheless, multiple washings may be necessary.

For some non-polar and non-volatile contaminants, decontamination options are limited. One approach is to use a

Figure 1:
Decision Aid for Health and Safety Risk Assessment of Decontamination Methods

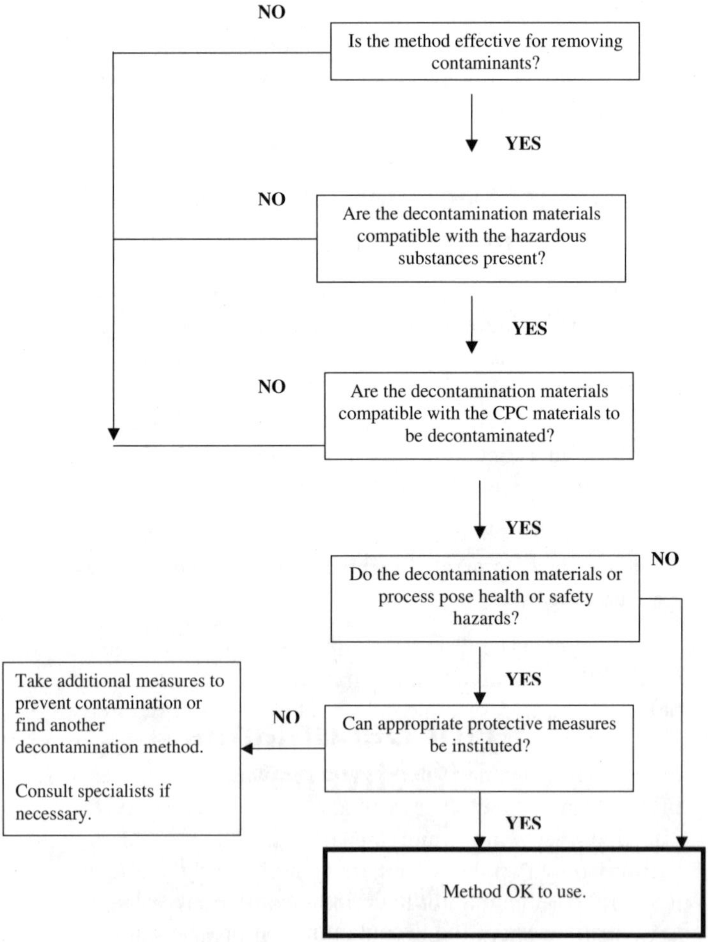

"wash" solvent with a high affinity for the contaminant. The solvent used in this approach may damage the protective barrier by removing essential plasticizers or other compounding ingredients from the CPC material. The decontaminating solvent may simply replace the contaminant in the barrier, resulting in a barrier contaminated with the cleaning solvent.

Dry cleaning using perchloroethylene is of limited value, as reported by Schlatter, since the dry cleaning method could degrade the barriers. Studies of Freon 113® washing of butyl, neoprene, and Viton® suits contaminated with nitrobenzene showed efficiencies of 45%, 67%, and 83% respectively, for an initial treatment. These efficiencies were increased upon subsequent washing.

Research clearly demonstrates that the standard boot rinse or suit wash used by many hazardous waste site workers and emergency responders is simply not effective except in rare circumstances.

Removal of Low Hazard Contaminants

In some instances, CPC is used for protection from low hazard contaminants. Decontamination of low hazard contaminants can be as simple as disposing of CPC when doffed, or wet wiping with warm water and a mild surfactant CPC prior to removal.

Testing for Decontamination Effectiveness

The effectiveness of CPC decontamination can be measured. Guidelines or procedures should be established to ensure that the desired outcomes of the decontamination are obtained. If the effectiveness of the decontamination process cannot be determined, then the process should be modified. CPC decontamination can involve non-destructive testing including visual observation, air monitoring, wipe sampling and chemical solution analysis. CPC decontamination should not be determined by visual inspection alone. The effectiveness of a decontamination program can be assessed in a variety of ways, such as the collection of wipe samples on decontaminated equipment and on surfaces in clean areas, analyzing the final decontamination rinse water for the presence of contaminants, or visual inspection of PPE for signs of leakage or failure.

Non-destructive testing is feasible for most contaminants with vapor pressures above 10 mm Hg. The basic principle is to capture and measure the evolving contaminant from the barrier. For example, photoionization detection may be used to scan the protective item. By evaluating the atmosphere with an appropriate chemical detector tube, the tester is able to determine if the protective item has off-gassed the potential contaminant.

Tests for contamination can range from simple visual observations to relatively sophisticated chemical analyses. Visual observation is appropriate if the contaminant does not permeate the barrier and can be seen with the naked eye (e.g., a dry pigment). Other types of visual observation may be conducted using natural or ultraviolet light. Natural light may show discolorations, stains, corrosive effects, dirt or alterations in the CPC fabric. Ultraviolet light may cause certain contaminants to phosphoresce such as polycyclic aromatic hydrocarbons Safety precautions need to be exercised when using UV light to test for contamination.

For other materials that do not permeate the barrier, simple "swipe" testing of the barrier surface is appropriate. For example, pH paper can be used to test for contamination by acids and bases.

Techniques and media for collection of wipe samples from surfaces vary with the agent and purpose of the sample. Classic wipe sampling techniques involve wiping a surface with a filter, which is then submitted for chemical analysis. For a limited number of chemicals, direct-reading colorimetric wipe sampling procedures are available for qualitative or semi-quantitative detection of surface contaminants such as acids, bases, isocyanates, aromatic amines, organic solvents (not solvent specific), lead, platinum salts, explosives and hydrazine. Colorimetric wipes can be sent for laboratory analysis to identify the chemical solvent or compound of concern.

Immunoassay kits can provide qualitative or semi-quantitative information on-site, and within about an hour for a variety of pesticides and other toxic chemicals.

Although wipe sampling may also be used to evaluate the effectiveness of decontamination procedures, it is not used to assess the health risk resulting from contamination. Only a few surface contamination concentration guidelines have been published. Typically, the employer must establish concentration limits for a specific task. These limits should be based on sufficient initial sampling to determine a "normal" range of contamination that can be expected after utilizing prescribed cleaning procedures. It would be appropriate to include documentation for the limits and their purpose in the worksite Safety and Health Program.

Gloves and other personal protective equipment (PPE) can be evaluated for effectiveness in actual field use conditions with Permea-Tec Sensors or similar analyzers. Permea-Tec Sensors include both direct-reading colorimetric detectors and a micro-encapsulated solvent indicator system. A color change on the Permea-Tec Sensor indicates chemical penetration through PPE. Permea-Tec Sensors can also be sent for laboratory analysis to identify the chemical solvent or other compound penetrating the gloves or garment.

Analyzing the contaminants collected in the cleaning solution may assess the effectiveness of CPC decontamination efforts. In this manner, it may be possible to determine if additional decontamination is necessary.

Destructive testing may be required for detecting CPC contamination, particularly when matrix contamination has occurred or when low-volatility contaminants are encountered. Selection of the appropriate analytical method must be based on the nature of the potential contaminant. Examples of analytical methods include permeation testing, digestion, and analysis. To avoid the cost of destroying expensive multiple-use protective equipment, swatches of the barrier material can be obtained from the manufacturer to facilitate pre-testing. Another approach is to affix the swatch to the protective clothing and test after exposure. For gloves and other limited use items, the actual item of protective clothing should be tested.

Decontamination for Reuse

Evaluation of the effectiveness of a decontamination method is challenging. Reuse of CPC will depend on a variety of factors including:

- An understanding of how contamination occurs,
- Effective removal of the chemical during cleaning as demonstrated by analysis for residual contamination,
- Continued integrity of CPC, and
- No impact of the cleaning on CPC performance.

Since it is usually difficult to establish safe levels for the residual contamination on CPC, most "used" CPC is considered exposed and is not reusable. Measurement technology may be limited and samples taken from the decontaminated CPC would most likely destroy the integrity of the CPC. However, some benchmarking of exposure scenarios with subsequent laboratory analysis of exposed and decontaminated CPC may provide reference for future decontamination decisions. The decision to reuse CPC should be based on exposure scenario data.

For radioactive contamination, measurement of "safe" levels of residual contamination may be determined by use of direct reading instrumentation and/or wipe sampling. In this case, if the CPC has not been physically damaged or the integrity of the CPC compromised, the CPC may be reused.

Work Zone Control

OSHA's work zone control requirements for CPC decontamination can be found in 29 CFR 1910.120 and 1926.65, the Hazardous Waste Operations and Emergency Response (HAZWOPER) standard. These requirements clearly address the need to establish controlled work zones for CPC decontamination. These work zone control concepts have been in place since 1986 and have been used successfully for a variety of CPC decontamination activities associated with routine industrial operations, hazardous waste site activities and during emergency response to man-made disasters, hazmat and terrorism incidents. Work zone control requirements are performance-based concepts. These concepts involve containment or isolation of contaminants and control of exposures through the use of risk reduction measures that protect adjacent workers, the public and the environment. Each CPC decontamination activity should employ work zone control concepts.

When establishing and maintaining work zones, there should be written procedures that incorporate the changing site operations. The decontamination work zone should be configured with safety as the key factor. The decontamination work zone should "buffer" clean areas, support zones, from contaminated areas, exclusion zones. Clean areas should not be immediately adjacent to a contaminated area or in a downwind location where the contamination could spread. Contaminated areas should remain as small as possible.

Where contaminants may translocate from one work zone to another through airborne transfer, liquid run-off, pedestrian or vehicular movement, the establishment of a buffer zone or contamination reduction zone (CRZ) becomes important in order to decontaminate CPC while preventing the transfer of contaminant to clean areas. The CRZ should be positioned between the exclusion zone and the support zone. Surveillance of the boundaries of the CRZ should be monitored visually and by use of direct reading instruments to assure that the decontamination process does not release airborne hazards into other work zones.

All decontamination activities should occur in an area demarcated as a CRZ. The CRZ should be chosen based on the potential for exposure associated with the decontamination task. The zones should not be removed until the area has been inspected by a competent safety professional that can confirm that the area is clean. The boundary between the CRZ and clean support zone must also be clearly demarcated. All equipment and personnel to be decontaminated must pass through the established decontamination corridor to minimize the potential for cross-contamination.

The size of the area needed to adequately decontaminate workers and equipment determines the size of the CRZ and equipment and the distance necessary to prevent physical transfer of hazardous substances into clean areas. Within all work zones, hazard levels should be measured and documented on an ongoing basis to ensure the effective establishment of boundaries. A waiting area should be present at the decontamination zone entry point so that contaminated personnel and equipment can be staged without spreading contamination further.

Hazardous Waste Site Operations and Emergency Response

Decontamination procedures for protective clothing and equipment used in Hazardous Waste Site Operations and Emergency Response must be clearly identified as part of a Site Safety and Health Plan, as required by 29 CFR 1910.120 and 1926.65 of the HAZWOPER standard. Most of the information presented in this chapter is consistent with HAZWOPER requirements.

Hazardous waste sites and emergency responses may be characterized in many ways, and the site boundaries depend upon many factors. For the purposes of this chapter, CPC decontamination at hazardous waste site operations and for emergency responses is characterized as area-specific, worker-specific and task-specific. The contaminants may be either single agents or a mixture of multiple constituents. The waste constituents can range from contaminated virgin chemical product to soil and/or groundwater with relatively low levels (parts per billion) of contamination. Even construction debris may be classified as "contaminated."

Hazardous waste sites and emergency responses are generally transitional; they will most likely be cleaned, consolidated and/or stabilized. CPC decontamination will most likely be temporary until the work is completed. The range of site activities makes it very likely that CPC will only be a part of the overall worker protection strategy. Additional personal protection may include hard hats, welding aprons/gloves/ helmets, harnesses, personal flotation devices, etc. Unfortunately, most of this equipment may not be decontaminated for reuse in "clean" work environments but could possibly be reused in contaminated work areas. Reusable contaminated personal protective equipment should be evaluated to ensure that it will function as originally designed and pose no additional threat to the user. The possibility of residual contamination levels stressing CPC material, seams, etc. from cut, puncture, tear and abrasion will also need to be evaluated.

Decontamination for emergency response personnel can be complex, since responders are often at risk of exposure before the contaminants and their respective concentration levels are known. Respondents to hazardous material incidents must be prepared to decontaminate their CPC at various locations, whether it is in the field at the incident site, or back at the emergency responder's base of operation. The level of decontamination necessary will depend on a variety of factors, including the type of contaminant encountered, availability of resources, and the level of protection worn. The authority having jurisdiction over the scene (i.e., fire chief, plant manager, etc.) should delegate expertise to research CPC protection levels and decontamination methods.

For both hazardous waste operations and emergency response, decontamination stations may be as simple as large wading pools to hold wash and rinsate solutions or as sophisticated as multi-stage decontamination shower trailers. Factors affecting the type of systems employed may include:

- Contamination type,
- Level and extent of contamination,
- Locality of response actions,
- Length of response action,
- Visibility of response action, politics, and
- Required specifications.

Decontamination stations for Hazardous Waste Site Operations should always be located upwind from the Exclusion Zone and within the CRZ.

CPC decontamination for hazardous waste operations and emergency response will most likely be more difficult to establish than the more predictable, routine workplace activities, since more information is initially available on these activities, increasing the likelihood that the contaminant is known and the work practices are well defined.

For some CPC decontamination operations involving low hazard contaminants a detergent and water can be used. In other instances, such as emergency response to Anthrax, OSHA had recommended that a 0.5% hypochlorite solution be used. In this case however, cross contamination with bleach may lead to skin and eye irritation. In addition, bleach may damage some types of CPC and other equipment including respirator face pieces, SCBA components, etc. Personnel involved at hazardous waste sites and in emergency response will require CPC protection when they are working in the contaminated area or the CRZ, and therefore will require CPC decontamination. After decontamination and doffing CPC, workers should practice good personal hygiene including washing with soap and water.

Decontamination of Chemical Warfare Agents

Chemical warfare agent (CWA) decontamination can pose unique challenges. It is important to identify the substance's physical properties so that appropriate counter-agents for neutralization and decontamination can be chosen. Once identification of the toxic substance has been accomplished, technical information about its physical properties must be obtained. It should then be determined whether the agent is a persistent or non-persistent agent. Many non-persistent agents, such as chemical or biological agents, dissipate or degrade fairly readily following contact with wind, water, sunshine, or other natural elements. They can also be cleaned from CPC and equipment with soap and water. The most commonly referenced universal liquid decontamination agent for CWAs is 0.5% hypochlorite solution. This solution should be used when further information regarding the warfare agent is not known.

With the more persistent oil-based nerve agents, such as Tabun, Sarin, Soman, and VX, an organophosphate-like compound, the decontamination procedure is likely to be more difficult. Decontamination of persistent agents may require the use of alcohol, acetone, or other solvents. Multiple stage decontamination may also be required. Decontamination rinsate containing mixtures of contaminants and alcohol, acetone, or other solvents used in the decontamination process must be properly disposed in labeled drums or containers that can be properly destroyed.

Dry decontaminants are available to decontaminate CPC and equipment from specific CWAs. The military uses a universal dry decontaminant known as M291 resin, which is available as pads packaged in small individual packets. M291 resin is a dry black carbonaceous material that decontaminates by absorption and physical removal of the CWA from the victim. M291 resin is used for spot decontamination of skin exposed to CWAs. Flour, followed by wiping with wet tissue paper is reported to be effective against nerve agents soman, VX and mustard gas contamination. In some cases, the nature or the extent of contamination may result in the need to destroy all CPC and equipment.

Decontamination of Radioactive Agents

Some chemical contaminants may also exhibit radioactivity and pose:

- A contact hazard by direct deposit on the skin,
- A body absorption hazard by exposure to penetrating radiation, and
- An inhalation hazard to radioactive particulates.

Most CPC contaminant removal methods discussed in this chapter will remove radiological contamination. At a minimum, the physical methods of contaminant removal would apply. Typical CPC decontamination involves the use of brushes and a water/trisodium solution followed by spraying copious amounts of low-pressure water. This solution must be containerized for appropriate disposal. Personnel in the decontamination process must wear appropriate protective clothing and respiratory protection. Worker exposure to radiation should be measured. Workers wearing CPC, as well as the CPC that has been decontaminated, should be surveyed for residual radiation levels. Radioactive and/or chemically contaminated CPC will be subject to the federal regulations for transportation and disposal.

Decontamination of Biological Agents

Biological agents pose a contact hazard by direct deposit on the skin and by indirect transfer from a contaminated surface to the skin, mouth, nose, or respiratory tract. If exposure is suspected, decontamination should occur immediately to reduce the risk of infection and spreading of the agent.

The critical point to understand about biological agent contamination is that although the decontamination process physically removes the agent from the skin or equipment, the agent may still be viable within the victim.

In the case of most biological agents, a diluted solution of common household bleach may be the most effective means for decontamination. Sodium hypochlorite solutions work through physical removal and oxidation and/or hydrolysis of the agent. Water also does this, but at a much slower rate.

Antiseptics and disinfectants appropriate for the individual bio-toxin can also be used. When faced with suspected virulent biological agents, full strength sodium hypochlorite may be used to assist in preventing a further spread of the contaminant. However, not all types of biological agents will be immediately killed by the application of bleach. Sodium hypochlorite solutions may degrade CPC and damage equipment. When using concentrated sodium hypochlorite, appropriate PPE and respiratory protection should be worn. Wash water and waste should be properly disposed in labeled drums or other containers that can be destroyed. Personnel involved in the decontamination process should shower using germicidal soap after removing any contaminated protective clothing.

In some cases, the nature or the extent of the etiologic contamination may make it prudent to dispose of all protective clothing and equipment, rather than attempt to decontaminate.

Cross-Contamination Issues

Cross-contamination is the transfer of harmful chemicals or other agents from one item to the next by human hands, equipment, or through the air. Cross-contaminating equipment and/or clothing can result in potential unanticipated exposure risks. Cross-contamination results primarily from improper decontamination techniques or can also occur because of an improper decontamination set up. Wind can facilitate the transfer of contaminants from the Hot Zone to the Support Zone after wind shifts or because of improper location of the support zone.

Proper decontamination techniques and proper upwind support zone locations are essential to minimize the risk of cross-contamination. All personnel involved in decontamination activities must be properly trained prior to the start of work. Proper training should be hands-on in relevant ensembles (such as EPA Level A, B, and C ensembles, as appropriate). Proper training should include the use of decontamination lines, donning and doffing of protective equipment to a level commensurate with the employee's anticipated job function and responsibility.

To minimize cross-contamination issues the following precautions should be taken:

- Minimize the handling of contaminated PPE and equipment.
- Personnel should be instructed on how to properly doff contaminated PPE.
- Ensure that all reusable PPE is appropriately decontaminated. Inspect all materials for discolorations, stains, corrosive effects, and substances adhering to objects, which indicate surface contamination.
- Control access/egress to the exclusion zone.

- Wipe-test the inner and outer surfaces of protective clothing and external equipment after decontamination.
- Utilize industrial hygiene monitoring and direct reading equipment to support boundary placement for the various zones.
- Consider routine area testing of the CRZ and support zone to confirm effective decontamination techniques.

Differences in Decontamination Approaches Between Hazmat and Incidents involving Weapons of Mass Destruction

There are several differences between decontamination procedures for everyday HAZMAT incidents and incidents involving weapons of mass destruction (WMD).

1. **Time** — Minimizing the duration of time that injured personnel are in the hazardous environment is critical. Since clothing and exposed skin may be contaminated with liquids, vapors or even particles, it is important to remove these from the casualty so that exposure time is minimized and the possibility of further inhalation or ingestion are both minimized.
2. **Higher toxicity** —Risk to the responders, victims, and onlookers may require increasing zone distances, frequent monitoring for symptoms and higher levels of PPE.
3. **Large number of victims** — Responders may be required to control, decontaminate, triage, and track hundreds or thousands of people at the site.
4. **Scene control** — An incident involving weapons of mass destruction may involve a larger area, and a mass casualty situation with numerous responders, as well as a huge media presence seeking information on the incident.
5. **Resource intensive** — More personnel and material will

be required than is normally available, driving the need for additional contingency planning before responders are faced with the situation.

6. **Crime scene evidence protection** — An everyday HAZMAT incident site is generally not a federal crime scene. In addition to decontaminating the area, evidence must be preserved for eventual use in apprehending and prosecuting the perpetrators.

Public Safety Considerations

Decontamination efforts should be conducted where all modes of traffic between controlled work zones can be regulated. The environmental impact from decontamination operations should be characterized. Whenever possible, decontamination should be located upwind from the contaminated area and downwind from the public. Any potential release of vapor, aerosol or liquid from the decontamination operations should be considered during the planning process.

Regulatory Considerations

CPC decontamination operations will expose, both CPC users and decontamination personnel to a variety of health and safety hazards. The public and environment also must be protected from these hazards. At a minimum, employees involved in CPC decontamination must be protected in accordance with OSHA regulations. This includes any regulatory requirements for personal protection, hazardous materials, hazardous waste operations and emergency response, walking and working surfaces, stairways and ladders, electrical, and use of powered hand tools.

From an environmental perspective, waste generated in the CPC decontamination operation must be characterized, properly managed, and disposed. CPC waste types may be hazardous or non-hazardous, regulated, mixed, infectious, and radioactive. Federal enforcement agencies such as the Environmental Protection Agency, the Nuclear Regulatory Commission, the Department of Energy, Department of Defense and the Department of Transportation, as well as many states, impose restrictions on the generation, storage, transport and disposal of these types of waste. Segregation, packaging, labeling, marking and record keeping requirements will vary depending on the types and levels of CPC contamination. Any CPC that cannot be reused, must be considered waste.

CPC decontamination operations may also be subject to further requirements should other authorities have jurisdiction over the deposition of the CPC waste. The FBI may require that CPC used in response to a terrorist attack be retained as evidence from the crime scene.

Decontamination Facility

Design Considerations

Prior to performing decontamination activities, a decontamination area layout should be developed and included as part of the decontamination plan. The decontamination operations should be isolated from all contaminated areas. When choosing the decontamination method, the level of decontamination necessary and the area in which to perform CPC decontamination, the following factors should be considered:

- Contaminant properties – physical, chemical and toxicological,
- The pathogenicity of infectious materials,
- The amount, location, and containment strategy,
- Available space to perform decontamination,
- The potential for, and location of, exposure based on assigned worker duties, activities, and functions,
- The potential for contaminants and other chemicals on site to permeate, degrade, or penetrate CPC,
- The movement of personnel and/or equipment among different zones,
- Availability of first aid, eye wash, emergency shower

equipment and other emergency equipment

- The methods available for protecting workers during decontamination operations,
- The impact of the decontamination process and related exposures on worker health and safety.
- Outdoor conditions
    - Prevailing weather conditions including precipitation, temperature, and wind direction,
    - Slope of the ground (because of runoff consideration).

For emergency response, decontamination operations the following additional factors should be considered:

- Proximity of the incident,
- Proximity of the equipment and personnel to be decontaminated,
- The need for medical care or mass casualty management,
- The availability of required resources (electrical, lighting, water, etc.), and
- The location of drains, sewers, aquifers, etc.

Procedures and Guidelines

Decontamination procedures should provide an organized process by which levels of contamination are reduced. The overall decontamination process should be operated as a series of procedures performed in a specific sequence. For example, outer, more heavily contaminated items should be decontaminated and removed first, followed by decontamination and removal of inner, less contaminated items. Each procedure should be performed at a separate station in order to prevent cross-contamination. The sequence of stations is called the decontamination line.

Stations should be separated physically to prevent cross contamination and should be arranged in order of decreasing contamination, preferably in a straight line.

Separate flow patterns and stations should be provided to isolate workers from different contamination zones containing incompatible contaminants, or representing different levels of risk. Entry and exit points should be conspicuously marked. The entry to the CRZ from the EZ should be separate from the entry to the EZ from the CRZ. Dressing stations for entry to the CRZ should be separate from redressing areas for exit from the CRZ. Personal who wish to enter clean areas of the decontamination operations, such as locker rooms, should be completely decontaminated. Examples of a decontamination layout are provided in Figures 2 and 3.

Decontamination should always emphasize thoroughness, not speed, unless a life threatening situation exists. In this case, both issues should be considered and the decontamination should be as thorough as it is practical.

Decontamination activities should be confined to a designated area within the CRZ. All personnel, clothing equipment, and samples leaving the contaminated area of a site must be decontaminated to remove any harmful chemicals or infectious organisms.

Specific decontamination procedures and guidelines shall be included as part of the site-specific health and safety program or plan.

Decontamination Equipment Selection

The selection of decontamination equipment is dependent upon several factors. These factors may include cost, available space, available resources (such as water, electricity, etc.), contaminants of concern, number of personnel, environmental conditions, and duration of tasks just to name a few.

Basic Decontamination Equipment (partial list)

- Plastic drop cloths for storing heavily contaminated equipment and outer chemical protective clothing

Figure 2:
Contamination Reduction Zone Decontamination Layout

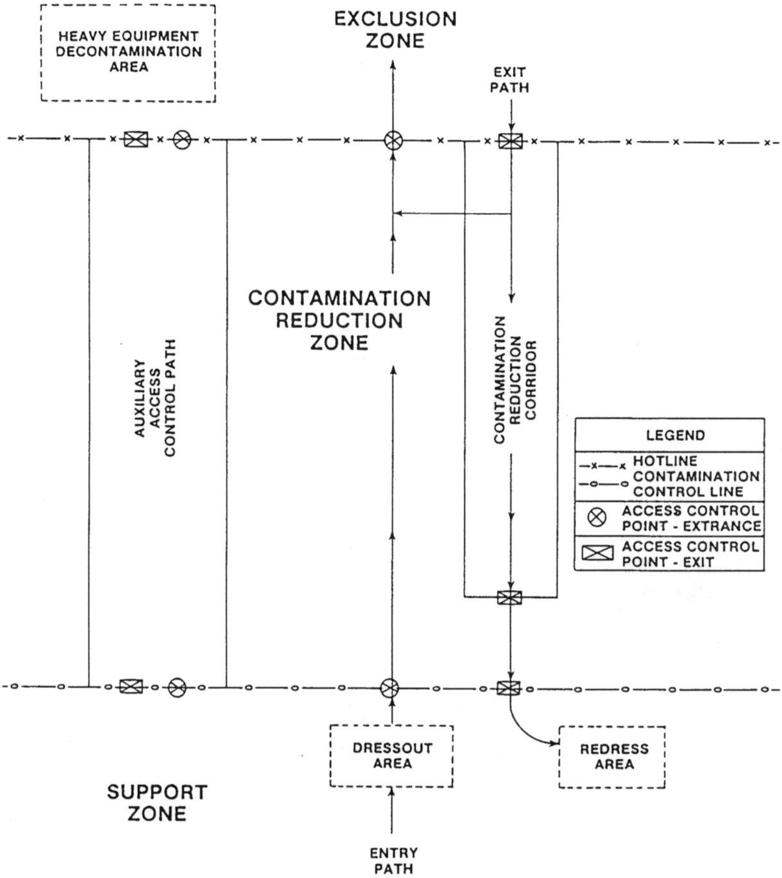

Figure 3:
Contamination Reduction Zone and Support Zone Layout

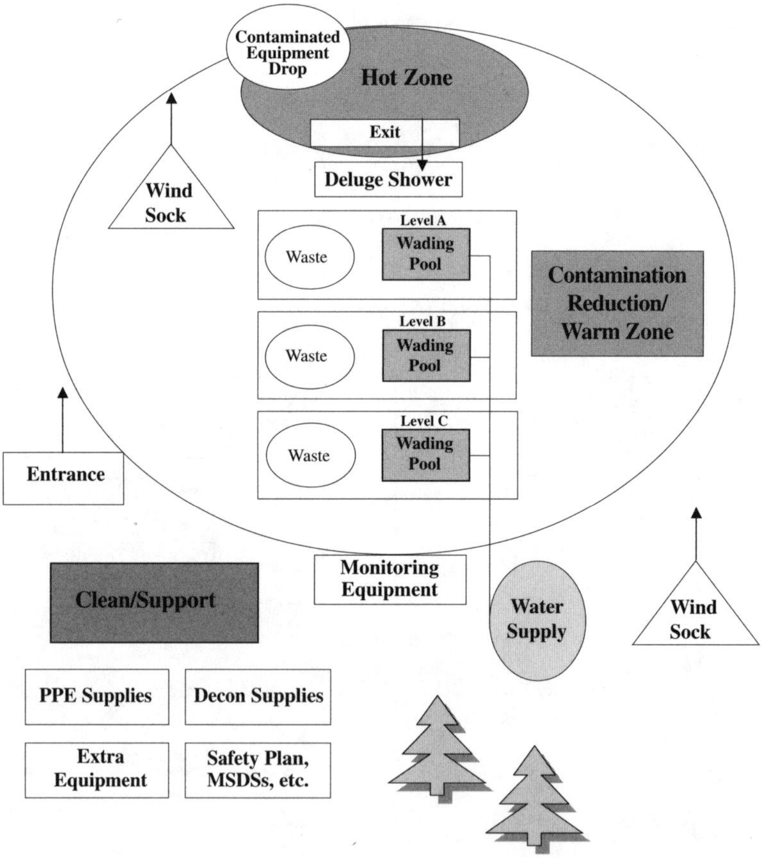

- Washing and rinsing solutions selected to reduce contamination and the hazards associated with contaminants
- Drums with removable lids or suitably lined trash cans for storing disposable clothing and heavily contaminated PPE that must be discarded
- Metal or plastic storage containers for storing contaminated solutions
- Hazardous waste labels for identifying contaminated materials and solutions
- Lined boxes with absorbents for rinsing off solid or liquid contaminants
- Long-handled, soft-bristled brushes to help wash and rinse off contaminants
- Large galvanized tubs, or children's wading pools to hold wash and rinsate solutions. Galvanized tubs/ wading pools must be large enough for personnel to place both booted feet in, and should have either no drain or be connected to a collection tank or appropriate treatment system.
- Plastic sheeting (6 mil)
- Soap for surface contamination/decontamination (e.g., trisodium phosphate, etc.)
- Labeled hazard bags (to contain contaminated materials)
- Black garbage bags (to contain non-contaminated materials)
- Appropriate personal protective equipment (gloves, respirator, etc.)
- Absorbent materials to limit spread of contamination
- Duct tape
- Wind sock to indicate wind direction
- Power source (e.g., generator)

- Ladder
- Bound log book (to document decontamination tasks, entry through decon, etc.)
- Tools (wrench, hammer, etc.)
- High efficiency HEPA filter vacuum (collection of gross particulate contaminated debris)
- Pump and flex hose (to pump out contained rinsate to storage drums)
- Equipment for containing and collecting of contaminated wash and rinse solutions spilled during decontamination
- Equipment for containing and collection of contaminated wash and rinse solutions generated during decontamination
- Paper or cloth towels for drying protective clothing and equipment
- Plastic bags for contained and cleaned equipment and CPC
- Lockers and cabinets for storage of decontaminated clothing and equipment
- Shower facilities for full body wash (or at a minimum, personal wash sinks)
- Collection tank or appropriate treatment system to collect rinsate that can be treated

Waste Disposal

Before disposing of waste, it should be appropriately characterized. All decontamination waste, including disposable PPE and rinsate should be disposed in accordance with all local, state and federal regulations. Requirements that should be considered include waste labeling, packaging treatment and manifesting.

Emergency Decontamination

Decontamination activities in support of emergencies must be established for both responders and victims. Personnel, equipment and victims may receive similar decontamination treatments based on the threat hazardous materials or agents represent, as well as the circumstances. Contaminant removal from personnel (responders and victims) and equipment would most likely involve different procedures yet, incorporate the same performance criteria to:

- Reduce the transfer of contaminants from contaminated areas into clean zones,
- Reduce contamination of and subsequent damage to personal protective equipment,
- Prevent the transfer of contaminant from the contaminated areas into clean zones,
- Prevent the spread of the contaminant from the incident site into other locations in the community, and
- Decontaminate to remove PPE safely without incurring any additional exposure to contaminates than received while wearing PPE.

For emergency responders, contamination avoidance is important. All considerations must be made to prevent contamination and control site operations by establishing work zones, security and communication systems. Prevention techniques for responders could involve parking in clean areas, positioning themselves and their equipment upwind and upgrade of an incident (vehicles should face away from the incident for rapid evacuation if necessary) and staying out of smoke/particulate or visible vapor/gas plumes and not to stepping into liquid or powder. Personnel should be instructed on "hands-off" behavior and not touch, taste or smell any suspected contaminated surface or object. Responders will be performing a variety of tasks requiring different levels of protection and decontamination.

Decontamination for each category of responders must be established.

- Responders performing entry into the incident location(s) to rescue victims, should collect evidence, mitigating the route cause of the incident, etc.
- Responders stationed at various personnel and equipment decontamination stations, should assist entry personnel in the decontamination of their PPE and equipment.
- Responders decontaminating victims and performing triage should provide medical care and transporting victims to health care facilities.
- Responders performing victim interviews, maintaining security and initiating evacuation of people from vulnerable areas.

Site Control Measures

Security measures should be established to restrict personnel in zones where contamination is likely. These measures should assure the following:

- PPE for each category of responder needs to be assigned prior to any responders taking offensive or defensive measures.
- Chemical and physical decontamination removal methods must be established. Consideration should be given to the chemical, biological and radiological properties of the contaminant. These methods must be established prior to responder entry into contaminated areas.
- Victims must be transported through a designated route out of the contaminated areas to decontamination stations.
- Only response personnel with appropriate PPE should be allowed into contaminated work zones.
- Response personnel entering contaminated areas should not be allowed to enter by themselves or be left alone in contaminated areas.

- Communication must be maintained between entry personnel and other responders, including those at decontamination stations.
- Complete control of personnel from the contaminated work zones through the decontamination stations must be maintained.

Self-Decontamination

Self-decontamination is required when any responder has been exposed to a contaminant through physical contact with a contaminated victim, a contaminated surface or object. This is particularly important for those responders not in appropriate PPE who become contaminated. Self-decontamination should be performed immediately.

For gas/vapor or liquid contamination, the following steps may be necessary:

1. Evacuate the area.
2. Get upwind into a clean area.
3. Communicate condition of exposed responder to other response personnel.
4. Wait until medical support arrives or proceed to designated location as directed.

For contamination by a visible liquid contaminant, the following additional steps may be necessary:

1. Scrape off contaminant with a blunt object (i.e., tongue depressor, plastic card, stick, edge of book).
2. Use absorbent material to soak up the contaminant (i.e., soap detergents, dirt, flour).
3. Remove any potentially contaminated clothing.
    a) Clean with soap and water to remove chemicals. If no soap and water is available, use physical removal method below.
    b) Flush with water for physical contaminant

removal. If no water is available, use dirt, flour or whatever absorbent material is available.

c) Don clean garments.

d) Report to the decontamination station and proceed as directed.

e) Dispose of all contaminated clothing properly.

Decontaminating Victims

Chemical warfare and/or HAZMAT victims may bypass the decontamination procedures at the disaster incident site if the severity of injuries or the lack of decontamination facilities necessitates this action. These victims could pose a threat to emergency personnel, hospital staff and other patients in emergency departments when they arrive at a health care facility. Therefore, the use of PPE by medical response, and hospital-based decontamination for hospital staff, are essential in handling of victims from incidents. When treating victims of exposure, decontamination is of primary importance, provided the victim does not require immediate life-saving interventions. Emergency personnel should be consulted to gather whatever information may have been collected regarding the exposure. In the absence of reliable information, symptoms should be carefully reviewed.

For contamination by a gas/vapor, the following steps may be necessary:

1. Relocate the victim to an upwind location.
2. Remove and discard clothing that has been contaminated by biological agents.
3. Remove all liquid contaminated clothing and discard properly. If clothing has been exposed to vapor, remove outer clothing and discard properly.
4. Wash hands and follow it by a shower and shampoo of hair.
5. Don clean garments.

6. Flushing exposed eyes with copious amounts of normal saline, water or eye solutions as determined by medical response personnel. Wounds should be immediately flushed with surgical irrigation solution, normal saline or water flush as per medical direction.

For liquid contamination, the following steps may be necessary:

1. Remove contaminated clothing.

2. Flush exposed areas of the victims' bodies with soap and water.

3. Don clean garments.

4. Approach victims only if wearing an appropriate level of PPE.

5. Begin decontamination immediately using water and then soap, as available.

For all victim decontamination instances:

1. Contain runoff, if possible.

2. Prevent victims from freezing temperatures.

3. Provide privacy for disrobing and decontamination (i.e., tents, sheet fence, anything to block view of onlookers).

For most biological contaminants, the following considerations may apply:

1. Use dilute hypochlorite solutions on the skin and soft-tissue injuries, including open lacerations.

2. Avoid the use of dilute hypocholorite solutions for penetrating abdominal wounds, in the eye, in open chest wounds, or for open brain or spinal cord injuries.

3. Irrigate skin or soft-tissue wounds with copious amounts of sterile saline solution.

Emergency Mass Casualty Decontamination

There are two levels of decontamination involved in mass casualty situations: Emergency and Technical. These combine to cover the rapid decontamination of victims and the deliberate decontamination of the responders and equipment at the scene.

Emergency decontamination is employed to save the lives of victims by first removing contaminated clothing, then removing the agent hazard from the skin by washing off or neutralizing the agent on the skin. Additional emergency decontamination setups may be required at supporting medical facilities away from the incident scene to treat "self-referrals" who left the incident before responders gained control. A follow-up decontamination procedure may also be performed at the medical facility on casualties who have already been through an emergency decontamination process. This secondary decontamination, known as definitive decon, assures all body surfaces are free of any residual contamination. Technical decontamination is performed to remove or neutralize all contamination from personnel, their equipment, and contaminated facilities in a deliberate fashion.

Preparation for Emergency Decontamination

Responder safety is paramount. All personnel who contact, or have the potential to contact exposed casualties, must wear appropriate CPC and respiratory protection. Decontamination procedures must also be established. Established decontamination work practices for response personnel should include the following:

1. Situation evaluation – Weapons of mass destruction agents were designed to be more toxic than other hazardous materials, and as such, present some unique challenges:

    a) Zone adjustment. The hot and warm zone

locations must be periodically reassessed to ensure contaminants are contained within the hot zone and that decontamination efforts are within the warm zone. Runoff patterns or changes in the wind speed and direction may require alteration of zones or relocation of decontamination corridors. The area must be large enough to handle several decontamination lanes.

b) Monitor warm zone. Monitor the site continuously for the presence of nuclear, biological and chemical agents, since a very small quantity can cause additional casualties. If the contaminants have spread, neutralize or isolate the larger area.

c) Establish the decontamination lanes in the warm zone near the hot zone, orienting the exit point to extend away from the hot zone. This will limit the traffic of victims into the hazard area and reduce their potential for additional exposure.

d) Establish the shower point upwind and up grade of the hot zone. The water runoff and any vapors from contamination will tend to flow back into the hot zone.

e) Obtain additional decontamination support. Liquid contamination can be transferred to initial responders, medical transport vehicles and crews, and maybe even the receiving hospitals.

- Other response vehicles can also transfer contamination. If a vehicle enters the hot zone, it must remain there until monitored as clean and, if necessary, decontaminated. Until proven otherwise, assume that any equipment entering the hot zone is contaminated. This may reduce resources in the short term, but it will prevent contamination transfer that could result in greater loss of resources over the long term.

- Hospital emergency rooms and clinics may

require decontamination support, since many victims may have departed the scene for the hospitals before the first responders arrived. It is possible that some victims transported from the site might not have been decontaminated.

f) Integrate procedures. Closely coordinate EMS and decontamination activities. Some lifesaving procedures may have to be performed during decontamination, or within the decontamination corridor, especially for the symptomatic and non-ambulatory victims.

g) Minimize direct contact with the victims.

h) Avoid any liquid contamination.

i) Use copious amounts of water. Physical removal is the primary method of decontamination. The more water used, the more effective the removal.

j) Control the runoff to the greatest extent feasible; prevent flow from entering into clean areas. Monitor the direction of the runoff to prevent cross contamination.

k) If possible, and if time and resources allow, confine the runoff to an isolated area. This is more critical for radiological materials, which are not neutralized by decontaminants.

l) Notify the sanitary waste station and other locations downstream if the runoff enters the sanitary sewer or storm drain systems.

- For chemical agents, some liquid agents will be hydrolyzed by the water and not present a long-term hazard downstream. Mustard gas and VX are notable exceptions, however, the large volumes of water can dilute them.

- For biological agents, downstream contamination will vary, depending on the agent and how it survives in treated water.

However, biological agents will not present a hazard in water unless the water is ingested or enters the body through cuts, abrasions or other orifices.

- For radiological materials, hot spots downstream present a radiation hazard when the materials have collected as sediment and become concentrated.

m) Responders should perform self-monitoring to ensure that they are aware of the signs and symptoms of exposure.

n) Consider each responder's CPC to be contaminated. From the moment the responder has entered the decontamination area, or come in contact with a casualty, consider both the responder and their CPC to be contaminated.

o) Provide shelter from the weather, staring bystanders, and the news media both during and after decontamination. A "cold shower" will chill the victims, even in warm weather. The TV telephoto lenses can capture embarrassing views of people following the removal of contaminated clothing. This can lessen the willingness of the asymptomatic victims to cooperate. A quick modesty cover should be erected at the undressing point and continue past the shower points. Examples of simple methods to provide cover include:

- Park ladder trucks side by side, extend the base ladder sections, and cover it to establish a walk-through lane.
- Erect tent-like covers using pike poles, ground ladders, or ropes strung between large vehicles.

2. Isolate the Casualties – This includes both the victims

and any other potentially exposed people.

a) Communicate directions to all responders and victims. Communicate the need for assistance and provide instructions.

- Use a loudspeaker or other voice amplification system.

- Post signs in the decontamination corridors and areas to reinforce direction of travel and procedure.

- Maintain communications, particularly between responders from different organizations working in the same area or on the sane task.

b) Keep victims in a comfortable location away from the hazard.

c) Segregate ambulatory from non-ambulatory, symptomatic from asymptomatic, male from female.

- Consider families, small children, handicapped, and elderly people.

- Isolate people refusing to cooperate from the rest of the victims.

d) Collect personal items. Collect clothing and personal items.

- Use plastic bags, tags, or other means of identification.

- Pack contaminated items into drums later.

e) Evidence. Some of these items may contain evidence, such as a residual agent, which can be collected from contaminated clothing or personal items.

Emergency Decontamination Triage

In many instances, the number of people that need to be decontaminated will quickly overwhelm decontamination capabilities. In this case, a triage process must be applied to do the most good for the greatest number of people.

Ambulatory. The first triage criterion is between ambulatory and non-ambulatory victims. Ambulatory victims can walk through the decontamination corridor.

a) Those exhibiting symptoms should receive decontamination first, then be referred for medical treatment.

b) Victims with no symptoms, but with signs of contamination, should be decontaminated next and placed under medical observation.

c) Victims with no symptoms or signs of contamination can be held and processed last.

Non-ambulatory. Non-ambulatory victims have to be processed through a litter-based decontamination corridor, assisted by at least two responders per victim. Those who require immediate care because of a life threatening condition should be decontaminated first. Multiple decontamination lanes should be established to process the victims as quickly as possible. Operations level responders on the first units deployed may have to start this triage before HAZMAT teams and additional units arrive to assist with emergency decontamination.

Emergency Decontamination Process

Emergency decontamination from exposure to weapons of mass destruction may involve a 4-step process:

Step 1 - Blot or wet – For chemical contamination, care must be exercised to fully blot the agent off of exposed skin immediately. Rubbing actions should be discouraged as this action may only increase the potential spread and penetration of the agent. Once they have been carefully checked for contamination, clothes may be used to blot contamination from the skin,. For contamination by biological agents or radioactive materials, wetting down the exposed surfaces first will help prevent the contaminant from re-suspending in the air.

Step 2 - Strip – Victims should be instructed to strip off all clothing down to the briefs. This will remove approximately 80% of the contamination. Nylons or pantyhose should be removed since they can hold a large amount of liquid or vapor close to the skin and they cover a significant amount of skin.

Step 3 – Flush – Flush the affected area with large amounts of water.

Step 4 – Cover – Cover for decontaminated victims should be provided to prevent hypothermia and for modesty. Blankets, towels, sheets, or tablecloths should be procured from available resources (local stores, hotels, hospitals, or other nearby facilities). Disposable ponchos, coveralls, or cut up salvage covers may be available for use as covers. Decontamination trailers may be available and designed to completely contain the decontamination process.

Emergency Decontaminants Of Choice

The availability of various surfactants can enhance the effectiveness of the decontamination process. The addition of soap when flushing will help lift the contaminant off the skin. Although decontamination with a bleach solution has been shown to be beneficial in reducing damage caused by some agents and neutralizing chemical and biological agents in runoff, the contact time of a bleach solution on casualties is usually not sufficient to neutralize the agent. Increasing the contact time may result in chemical burns from the bleach. Use of bleach solutions for emergency mass casualty decontamination is not recommended. High volume, low-pressure water showers are recommended for emergency mass casualty decontamination.

Technical and Equipment Decontamination

The need for a rapid response to any emergency is critical. Response personnel should be on the scene early to conduct reconnaissance, rescue victims, and identify the WMD. For the response personnel to perform their tasks in CPC, a technical decontamination capability must exist. The decontamination set up must be established before responders enter the hot zone. If the first responders reached the scene before the WMD agents were identified and the hot zone was designated, emergency decontamination may be needed.

The technical decontamination corridor should be established away from the emergency decontamination corridors. Responders are likely to be contaminated and the victims will not have protective clothing. Some victims awaiting emergency decontamination may become unruly or interfere with the process when they observe responders moving to the head of the line or receiving a different decontamination effort. With emergency decontamination, emphasis is on the speedy removal of WMD agents from the victims and their clothing since the victims have no protection from WMD agents. Technical decontamination concentrates more on thoroughness and neutralization of the WMD agent on the CPC.

When emergency response vehicles and equipment may become contaminated, they must be isolated and assessed. This may be done using direct-reading instruments, surface wipes, or other types of sampling that requiring subsequent laboratory analysis.

Victims' personal belongings should be bagged and tagged. A plan for the disposition of these belongings should be developed. Monitoring of bagged personal items can identify contamination problems. Agent vapors will collect in the bag, and can be checked with direct reading instrumentation, colorimetric tubes, or other monitoring devices. Procedures also must be in place for the handling of weapons belonging to law enforcement personnel.

Considerations for Hospital and other Health Care Facilities

Hospitals and other facilities used for support to the responders or victims may become contaminated. Much of this can occur from cross-contamination that results from emergency decontamination, transportation of victims, or self-evacuation of contaminated persons. These facilities must eventually be closed, monitored, decontaminated, and thoroughly inspected before being reopened to the public.

All emergency facilities included in an emergency response plan for hazardous materials incidents must meet OSHA requirements (29 CFR 1910.120[q]) for both staff training and response to hazardous materials, because they likely will be presented with a chemically exposed patient who has not been decontaminated at the scene. Under these regulations, emergency medical personnel who may decontaminate victims exposed to a hazardous substance should be trained, at a minimum, to the first-responder operational level. For response to an unknown hazard, OSHA regulations require Level B protection, which includes a positive-pressure self-contained breathing apparatus and splash-protective chemical-resistant clothing.

Off-site emergency services/health care facilities that handle chemical contamination and biological contamination incidents must be equipped with showers, washing liquids and rinse water, and a containment system for all run-off. Procedures must be written to address safe work practices, and the establishment of a decontamination zone. All staff personnel involved in handling chemical contamination triage must be appropriately trained. Hands-on training should be conducted in a simulated work environment that permits staff personnel to have experience performing tasks, making decisions, or using equipment appropriate for handling chemical contamination and biological contamination incidents.

Glossary

ACGIH – American Conference of Governmental Industrial Hygienists. Publishes Threshold Limit Values® (TLV®s) occupational exposure guidelines and BEIs (biological exposure indices), which are used as maximal exposure criteria for many substances.

air monitoring – Actions to detect and quantify airborne contaminant conditions by the collection of an air sample using personal pumps, detector tubes, or direct monitoring instrumentation to indicate the amount and type of hazards present in the workplace ambient atmosphere.

area sampling – Conducted to assist with defining the extent of contamination or to measure the effectiveness of engineering controls. During area sampling, the air sampler is placed in a fixed location in the work area or near the suspected source of the hazard.

BEI – Biological Exposure Indices.

biological agents – Include bacteria, viruses, molds and fungi, and may be present as contaminants in the air or on work surfaces.

chemical agents – Agents that may be present as contaminants in the air, in water, on work surfaces, and as bulk materials. They may be present as gases, vapors, dusts, fumes, mists, fibers, particulates, or other aerosols, as pure bulk materials or as contaminants in other bulk materials. Hazardous chemicals include carcinogens, toxic or highly toxic agents, reproductive toxicants, irritants, corrosives, sensitizers, hepatotoxins, nephrotoxins, neurotoxins, agents which act on the hematopoitic systems, and agents which damage the lungs, skin, eyes, or mucous membranes.

CWA – Chemical Warfare Agents.

competent – means possessing the skills, knowledge, experience, and judgment to perform assigned tasks or activities satisfactorily as determined by the employer.

Contamination Reduction Zone (CRZ) – a buffer zone between the exclusion zone and the support zone

demonstration – The showing by actual use of equipment or procedures.

direct-reading instruments – Allow collection of significant amounts of data and the determination of the employees' exposure profile during operations. They also provide qualitative worker exposure data. However, they may not have a high enough specificity, detection limit, or precision for compliance monitoring or exposure assessment. Direct-reading instruments shall be used at for screening purpose only to identify contamination presence.

EPA – Environmental Protection Agency.

hands-on training – Training in a simulated work environment that permits personnel to have experience performing tasks, making decisions, or using equipment appropriate to the job assignment for which the training is being conducted.

initial training – Training required prior to beginning work.

OSHA – Occupational Safety and Health Administration.

Permissible Exposure Limit (PEL) – The concentration level of an agent above which an employer may not expose his employees. The PEL is typically stated in terms of an 8-hour time-weighted average (TWA) of exposure. PEL(s) are part of the OSHA regulations and are enforceable by law.

Personal Protective Equipment (PPE) – Interim control devices worn or used while working to protect employees from exposure to hazards. PPE includes items such as gloves, respirators, safety glasses, earmuffs, etc.

proficient – Meeting a stated level of achievement.

SOPs – Standard Operating Procedures

SSAHP – Site Safety and Health Plan

Threshold Limit Value® (TLV®) – The concentration in air of a substance to which it is believed that most workers can be exposed daily without adverse effect (the threshold between safe and dangerous concentrations). These values are established (and revised annually) by the American Conference of Governmental Industrial Hygienists and are time-weighted concentrations for a 7 or 8-hour workday and a 40-hour workweek. For most substances the value may be exceeded, to a certain extent, provided there are compensatory periods of exposure below the value during the workday (or in some cases the week). For a few substances (mainly those that produce a rapid response) the limit is given as a ceiling concentration (maximum permissible concentration - designated by "C") that should never be exceeded."

WMD – Weapons of Mass Destruction

References

1. Baker, R.L. and G.C. Coletta, Eds. *Performance of Protective Clothing* American Society for Testing and Materials, West Conshohocken, PA. 1986.

2. Baird, N., E. Cook, and W. Perry. Evaluating Surface Contamination at Hazardous Waste Sites. Occupational Health & Safety Vol. 65 No. 1. January 1996.

3. Bentz, A.P. and C.B. Billing. Determination of solubility parameters of new suit materials, in *Performance of Protective Clothing*, STP 989. S.Z. Mansdorf, R. Sager and A.P. Nielsen (eds.), American Society for Testing and Materials, West Conshohocken, PA. 1988.

4. Berardinelli, S.P. and R.C. Hall. Decontamination of chemical protective clothing exhibiting matrix release, in *Performance of Protective Clothing: Second Symposium*, ASTM STP 989, S.Z. Mansdorf, R. Sager and A.P. Nielsen (eds.), American Society for Testing and Materials, West Conshohocken, PA. 1988.

5. Bolz, F., K.J. Dudonis, D.P. Schulz. The Counterterrorism Handbook. Florida: CRC Press. 2002.

6. Clayton, G.D. and F.E. Clayton. Patty's Industrial Hygiene and Toxicology. New York: John Wiley and Sons. Vol. II. 1981.

7. Coletta, G. C., S.Z. Mansdorf, and S. P. Berardinelli. Chemical protective clothing test method development: Part II, degradation test method, *American Industrial Hygiene Association Journal*. 41:26-33.

8. Cox R.D. Decontamination and management of hazardous materials exposure victims in the emergency department. Ann Emerg Med 1994 Apr; 23(4): 761-70 [Medline]

9. Federal Emergency Management Agency, US Department of Justice. Emergency Response to Terrorism Job Aid. 2000.

10. Forsberg, K. and S. Faniadis. The permeation of multi-

component liquids through new and pre-exposed glove materials. *American Industrial Hygiene Association Journal* 47:189-193. 1986.

11. Garland, C.E., and A.M. Torrence. Protective clothing materials: chemical contamination and decontamination concerns and possible solutions, in *Performance of Protective Clothing*, STP 989, S.Z. Mansdorf, R. Sager and A.P. Nielsen (eds.), American Society for Testing and Materials, West Conshohocken, PA. 1988.

12. Hansen, C.M. The three-dimensional solubility parameter-key to paint component affinities: I. Solvents, plasticizers, polymers and resins. *J. Paint Tech.* 39:104-117. 1967.

13. Maniscalco, P.M., and H.T. Christen. Understanding Terrorism and Managing the Consequences, Prentice Hall, New Jersey. 2002.

14. Mansdorf, S.Z. and T.J. Roseman. *Controlled Release Delivery Systems*, Marcel Dekker, Inc., New York. 1983.

15. Mansdorf, S.Z. Risk assessment of chemical exposure hazards in the use of chemical protective clothing-an overview, *Performance of Protective Clothing*, ASTM STP 900, R.L. 1986.

16. Mansdorf, S.Z. Chemical Protective Clothing Use in U.S. Hazardous Waste Operations – Observations and Recommendations in *Proceedings of the Third Scandinavian Symposium on Protective Clothing Against Chemicals and Other Health Risks*, Gausdal, Norway. 1989.

17. Mansdorf, S.Z.: Personal protective equipment for hazardous materials spills response. *Professional Safety*, American Society of Safety Engineers, Chicago, 37(12). 1992.

18. Mansdorf, S.Z.: Personal Protective equipment decontamination for hazardous waste operations and emergency response, *Performance of Protective Clothing*: Fourth Volume, ASTM STP 1133, J. McBriarty

and N. Henry, Eds. American Society for Testing and Materials, West Conshohocken, PA. 1992.

19. Mickelsen, R.L., and R.C. Hall: A breakthrough comparison of nitrile and neoprene glove materials produced by different glove manufacturers. *American Industrial Hygiene Association Journal* 48:941-947. 1987.

20. NFPA, (National Fire Protection Association Hazardous Material Response Handbook. Quincy, MA 02269-9101 USA. 1997.

21. NIOSH, OSHA, EPA, USCG Guidelines for Hazardous Waste Operations and Emergency Response. 1986.

22. OSHA: Regulations (Standards - 29 CFR). Hazardous Waste Operations and Emergency Response. 1910.120. Washington DC: Government Printing Office; 2002.

23. Perkins, J.L., J.S. Johnson, C.R. Sackett, et al.: Residual spilled solvents in butyl protective clothing and usefulness of decontamination procedures, *Applied Industrial Hygiene* 2:179. 1987.

24. Perkins, J.L.: Decontamination of protective clothing. *Applied Occupational and Environmental Hygiene*, 6:29-35. 1991.

25. Schlatter, C.N.: Effects of water rinsing on subsequent permeation of rubber chemical protective gloves, in *Performance of Protective Clothing*, STP 989, S.Z. Mansdorf, R. Sager and A.P. Nielsen (eds.), American Society for Testing and Materials, West Conshohocken, PA. 1988.

26. Schlatter, C. N. Decontamination of protective clothing, in *Chemical Protective Clothing*, J.S. Johnson and K. Anderson, (eds.), Vol. I, Chapter 8, AIHA Press, Fairfax, VA. 1990.

27. Sidell, F., W. Patrick, T. Dashiell, K. Alibek, and S. Layne: *Jane's Chem-Bio Handbook*, Second Edition. Jane's Information Group, Virginia. 2002.

28. Smith, I.D., and K.E. Burke: Decontamination of protective suit materials, in *Performance of Protective Clothing*, STP 1037, J.L Perkins and J.O. Stull (eds.), American Society for Testing and Materials, West Conshohocken, PA. 1996.

29. Stull, J.O., C.R. Dodgen, M.B. Connor, and R.T. McCarthy: Evaluating the effectiveness of different laundering approaches for decontaminating structural fire fighting protective clothing, in *Performance of Protective Clothing*, STP 1037, J.L Perkins and J.O. Stull (eds.), American Society for Testing and Materials, West Conshohocken, PA. 1996.

30. van Hooidonk C.: CW Agents and the skin: Penetration and decontamination. In Proceedings of the International Symposium on Protection of Chemical Warfare Agents; June 6-9, 1983; Stockholm, Sweden, National Defense Research Institute).

31. Zajtchuk R, ed: Decontamination. In: *The Textbook of Military Medicine. Part I. Warfare, Weaponry and the Casualty*. Vol 3. 1997: 351-360.

32. Ziskin, M.H., and D. Han, "Personal Protective Equipment," in *Hazardous Materials Desk Reference,* Cox, DB, ed. McGraw-Hill. 1999.

33. Ziskin, M.H., "Dermal Protection Strategy for Response to Chemical and Biological Terrorism Incidents." Presented at the American Industrial Hygiene Conference, San Diego, CA. 2002.

34. Ziskin, M. H., "Personal Protective Equipment Programs for Unknown Encounters at Unknown Incidents." Presented at the American Industrial Hygiene Conference, Toronto, Canada. 1999.

9

Managing Heat Stress

by Thomas E. Bernard and

Frank N. Dukes-Dobos

Introduction

The basic premise behind chemical protective clothing (CPC) is that the clothing item will provide a barrier between the chemical and the skin of the user. Unfor tunately, the barrier may reduce the wearer's ability to dissipate internally generated heat through sweat evaporation. Depending on the thermal environment and work demands, the clothing required may cause an excessive level of heat stress—a hazard that must be considered. Results of heat stress include reduced performance and increased risk of accidents, injuries, and heat-related disorders. Ideally, heat stress should be eliminated as a hazard, but often the selected CPC is essential to manage risk associated with exposure.

Heat stress is the confluence of environmental conditions, work demands, and clothing requirements that necessitate a thermoregulatory response from the person. Determination of the level of heat stress requires identification and assessment of these factors. Important environmental conditions are air temperature and humidity, air motion, and sources of infrared radiation. Metabolic demands are an internal source of heat generation, which must be dissipated to the environment. The higher the metabolic demands, the cooler and drier the air must be.[1] Clothing modifies the ability of the heat generated by metabolic work to move to the environment.

Heat strain is the collection of physiological responses marshaled by the body to dissipate heat and maintain thermal equilibrium. Monitoring physiological responses such as heart rate and body temperature provides a means of confirming that workers are adequately protected from heat stress.

This chapter provides general guidance for heat stress evaluation and control. More detailed information on heat stress can be found in references 2 through 8 in the reference section.

Effects of Clothing on Heat Stress

Thermal balance is achieved when heat gained from work demands and the environment is balanced by heat losses to the environment from sweat evaporation and dry heat exchange (convection and radiation).[6, 8] In principle, the rate of sweat production is regulated by the body to achieve an adequate level of evaporative cooling. The ability to evaporate sweat is influenced by air humidity and motion as well as clothing. The evaluation of heat stress, therefore, depends on the following factors:

- Required rate of evaporative cooling (by sweat evaporation) to maintain body core temperature within acceptable limits
- Maximum rate of evaporative cooling that can be achieved by the combination of environmental conditions and clothing.

As the required evaporative cooling rate approaches the maximum possible rate, maintaining thermal equilibrium becomes much more difficult. Risks of heat-related disorders increase. Once the required cooling exceeds the maximum feasible, then all control is lost and body core temperature steadily increases. If an uncontrolled body temperature increase is not interrupted, then a serious heat-related disorder is likely.

Two important clothing factors that affect the maximum rate of evaporative cooling are fabric characteristics and clothing construction and configuration. In terms of fabric characteristics, the ability of water vapor to move through the fabric influences the maximal evaporative cooling rate most significantly. Moisture vapor transmission rate (MVTR) and permeability index (i_m) are traditional, frequently reported measures of water vapor transport. MVTR is based on a bench test and is specific to a fabric. i_m is based on a bench test for a specific fabric or on a manikin or wear test for a complete ensemble. In general, the greater the MVTR and permeability index, the more likely that the fabric will permit greater rates of evaporative cooling. In practice, the important factor is the ability of air, which carries water vapor with it, to move through the fabric (air permeability),[9] but this information is less readily available.

Garment construction and configuration influence air flow. If the garment is designed to be loose fitting so that air can flow through openings and around the wearer of the clothing, then the maximal rate of evaporation will be greater. This characteristic of design is referred to as garment ventilation. Garment ventilation is enhanced by external air movement (e.g., fans) and by body movements during work (i.e., pumping effect).

The amount of body surface area covered significantly influences evaporative cooling. The less skin surface covered, the more evaporative cooling occurs. If a second layer of protective clothing is necessary, evaporative resistance increases. Covering arms or legs only, if appropriate for essential protection, is better than covering the whole body. Finally, introducing multiple layers adds significantly to the evaporative resistance, thus reducing the maximal rate of evaporative cooling.

Evaluation of Heat Stress

Heat stress is assessed through consideration of environmental conditions, work demands, and clothing requirements. WBGT (wet bulb globe temperature) is a preferred index of environmental conditions in industrial hygiene practice.[1] Because the WBGT method was developed based on ordinary work clothing, adjustments for wearing CPC are necessary, and one method is discussed in this chapter.

The goal of controlling heat stress is to prevent or control an excessive physiological response (i.e., excessive heat strain). The relative ease of assessing heat stress has overshadowed the value of demonstrating the need for heat strain controls or the evaluation of the effectiveness of control measures through the measurement of heat strain. Physiological monitoring to assess heat strain is important because

- WBGT methods may not account well for heat stress effects from some types of CPC
- Established threshold exposure limits are set low to accommodate high individual variability
- Administrative controls may require fine-tuning in the field.

Selected methods of physiological monitoring are described in a later section.

WBGT-Based Methods

The WBGT is an empirically derived blending of natural wet bulb temperature (T_{nwb}) and globe temperature (T_g) with an adjustment for dry bulb temperature (T_{db}) for conditions of direct sunlight. The American Conference of Governmental Industrial Hygienists (ACGIH) adapted the National Institute for Occupational Safety and Health (NIOSH) recommendations for evaluating heat stress.[2] In 2000, the ACGIH Physical Agents Committee published a change to the Heat Stress Threshold Limit Value® (TLV®), calling it the Heat Stress and Strain TLV.[4] The heat stress thresholds are shown in Figure 1. Figure 1 illustrates the relationship between environmental conditions and metabolic rate. But clothing also has an effect on the level of heat stress. One way of expressing quantitatively the effects of clothing on heat stress is to use a clothing adjustment factor expressed as degrees WBGT. This factor can be added to the ambient WBGT to reflect an effective WBGT experienced by the person in the CPC.[10] Clothing adjustment factors for selected ensembles are listed in Table 1.[4,10] The three steps necessary to evaluate the heat stress level are:

- Determine the effective WBGT at each work location
- Estimate the metabolic rate for each task
- Determine the hourly time-weighted averages of each.

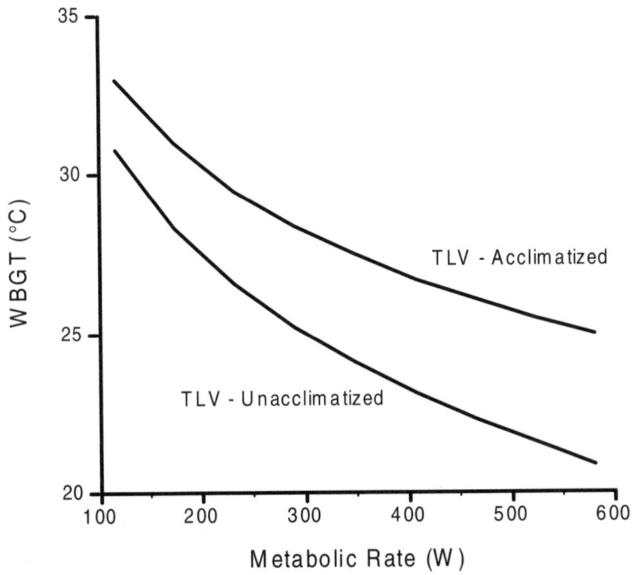

Figure 1. Heat stress TLV®-based on WBGT and metabolic rate.

To account for environmental effects on heat stress, the WBGT should be measured at each worksite location including the locations used for recovery. The data sheet in Figure 2 can be used as a model. Steps for completing the worksheet and calculating heat stress level follow.

1. Enter each task/location combination. Each combination represents a significant change in environmental conditions, clothing, or work demands.
2. Enter time spent performing the task at each location.
3. Enter dry bulb, natural wet bulb, and globe temperatures that represent environmental conditions. (If WBGT is known, enter it in the appropriate column.)
4. Note clothing requirements. (Form provides values for five common ensembles with space to enter two more that may apply—using data from Table 1 or other sources.)
5. Estimate metabolic rate for task. Ideally, estimate

Table 1:
List of Clothing Ensembles with Clothing Adjustment Factors (CAF) as °C-WBGT*[4,10]

Ensemble Description	CAF [C-WBGT]
Work Clothes Standard ensemble that includes 4-oz cotton shirt with long sleeves and 8-oz cotton pants. Represents reference ensemble for setting Clothing Adjustment Factors.	0.0
Cloth Coveralls (CC) Standard configuration of coveralls with fabric weights of about 8 to 9 oz.	0.0
SMS Polypropylene Coveralls Limited-use coveralls with hood (particle barrier only) made from a spunbonded, meltblown polypropylene.	-1.0
MB Polyethylene Coveralls Limited use coveralls (particle barrier only) made from a meltblown polyethylene (Tyvek® 1424).	1.0
Polyester Coveralls Light-weight polyester (98% Nylon) (3-oz) coveralls worn with hood of same material (ProTech2000).	0.0
Water-Barrier, Vapor-Permeable Coveralls Coverall made from laminate fabric with microporous film. (Note: There is some variability among films in this category and some may have higher adjustment factors.)	2.5
Vapor-Barrier Coveralls Coveralls and hood made of lightweight fabrics designed for limited use. Typical fabrics might be polyethylene coated spunbonded polyethylene or polyvinylchloride.	8.5
Encapsulating Suit or Turn-out Gear Total encapsulating suit (e.g., Level A); reusable, whole-body chemical protective suits; fire-fighter turnout gear.	11

Note: Fabric weights reported in ounces per square yard (oz/yd^2). Temperatures for the adjustment factors are rounded to the nearest 0.5°C-WBGT.

*Unpublished data from NIOSH 5-R01-OHO3983-01, Making Heat Stress Analysis Relevant Again, performed at the University of South Florida. Presented at the 2002 American Industrial Hygiene Conference and Exposition (AIHce) in San Diego. See www.hsc.usf.edu/`tbernard for updates.

metabolic rate to within 50 W. Suggestions for estimating metabolic rate are available from NIOSH,[2] ACGIH,[4] Bernard,[6] ISO,[11] Tayyari,[12] and Bernard.[13] Values shown and summarized in Table 2 provide rough estimates of metabolic rate to approximate degree of heat stress. (Care should be taken when using broad categories, because there is a tendency to overestimate.)

6. Compute WBGT:

   For most conditions: $WBGT = 0.7\ T_{nwb} + 0.3\ T_g$.

   In direct sunlight: $WBGT = 0.7\ T_{nwb} + 0.2\ T_g + 0.1\ T_{db}$.

7. Note clothing adjustment factor (CAF) for each location (Table 1). Use professional judgment when applying these factors or refer to similar metrics found in the literature.

8. Calculate effective WBGT ($WBGT_{eff}$), which is WBGT plus CAF.

9. Compute hourly time-weighted average (TWA) for metabolic rate (M) and effective WBGT ($WBGT_{eff}$).

10. After computing TWA-M and TWA-$WBGT_{eff}$, refer to Figure 1 to assess heat stress level. Generally, the highest combination for any hour of continuous work is applied to decision making. If the heat exposure is intermittent, up to a 2-hour TWA is acceptable.

Because this type of analysis cannot be performed in real time, it is important to understand the exposure scenarios and to use some professional judgment in selecting those that are most likely to have the highest and most representative levels of heat stress.

Physiological Evaluation

Uncompensable heat stress describes conditions in which an individual cannot maintain thermal equilibrium and the body core temperature steadily rises. When work is performed above the TLV of Figure 1, risk for uncompensable heat stress increases with higher elevations above the line.

Task	Description	Time [min]	T_{db}	T_{nwb}	T_g	Clothing (see codes below)	Met. Rate	WBGT	CAF (see below)	Effective WBGT	Time x Met Rate	Time x Eff WBGT
1												
2												
3												
4												
5												
6	Total		Sum the following columns: Time, Time x Met Rate, Time x Eff WBGT									
Time-Weighted Average		Divide Total of Time x Met Rate by Total Time and Enter under M. Divide Total of Time x Eff WBGT by Total Time and Enter under WBGT.									M	WBGT

Clothing Ensemble	Code	CAF [°C]	CAF [°F]*	Other Clothing Ensemble	Code	CAF [°C]	CAF [°F]*
Work Clothes	WC	0.0	0.0				
Cloth Coveralls	CC	0.0	0.0				
Limited-Use Particle Barrier	PB	1.0	2.0				
Water-Barrier Vapor	WBVP	2.5	4.5				
Permeable							
Limited-Use Vapor-Barrier	VBL	8.5	15.5				

*CAF [°F] rounded to nearest 0.5 °F.

Figure 2: Example of a data collection sheet for the analysis of heat stress

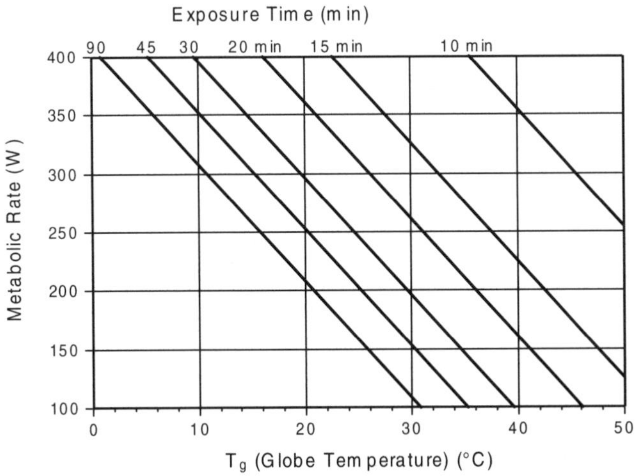

Figure 3. Relationships among metabolic rate and environmental conditions to estimate time to raise body core temperature 1°C while wearing vapor-barrier clothing.[17]

In Figure 3, some basic relationships are illustrated between metabolic rate and environmental heat reflected in globe temperature (T_g) for a person weighing 70 kg (154 lb.) who is wearing vapor-barrier clothing.[14] The time to increase core temperature 1°C is marked by the slanted lines in Figure 3, which is safe as long as the core temperature is initially less than 37.5 °C. The time to increase core temperature decreases if the metabolic rate and environmental heat are higher.

Because many working conditions place workers in uncompensable heat stress situations, physiological monitoring is important in managing exposures.[4] Control of body core temperature is the traditional goal of heat stress management.

Heart rate is useful as a measure of physiological strain because it is influenced by the cardiovascular response needed to support metabolic demands and to remove metabolic heat from the core. Both peak and sustained metabolic demands can be indicative of excessive heat stress exposure.

Body Core Temperature

Body core temperature for industrial hygiene purposes is the deep body temperature. A World Health Organization (WHO) expert panel recommended a peak value of 39°C under closely monitored conditions and 38°C for steady exposures. The ACGIH recommends 38.5°C to allow a margin of safety for peak exposures.[4] Assuming that heat stress exposures are intermittent with time for recovery allowed between exposures, a transient increase in core temperature to 38.5°C is safe. The average core temperature over the day, however, should not exceed 38°C.

Oral temperature is widely used to estimate body core temperature, with an 0.5°C offset that is added to oral temperature to estimate core temperature. An oral temperature of 38°C is very close to a core temperature of 38.5°C. A good time to assess core temperature is at the end of an exposure or a work cycle. The worker should not eat or drink anything for 15 minutes prior to the measurement. The thermometer is placed under the tongue for about 3 minutes, depending on the equilibrium time of the thermometer, and the worker must keep the mouth closed. Repeated measurements are recommended.[2]

An alternative to oral temperature is earcanal temperature. An ear temperature sensor is placed in the earcanal and insulated from the outside environment to reduce the influence of air temperature and radiant heat. A direct reading can be made as an approximation of core temperature. Using a commercially available device, the temperature can be calibrated in the same environment in which it is used by assessing oral temperature plus 1°C. The resulting core temperature-equivalent reading should be limited to 38.5°C.

Another commercially available device assesses core temperature by a thermometer placed on the skin and insulated from the environment. While there is individual variability, the alarm setting is based on an upper limit of 38.5°C for most of the users.[15]

Heart Rate Evaluation

When the cardiovascular system is overwhelmed, the body's ability to thermoregulate is limited. Heart rate is indicative of impending, if not actual, elevated core temperature. Therefore several methods for assessing heart rate are recommended.

Table 2:
Categories of Metabolic Rate with a Representative Rate*

Category Description	Rate [W]
Sedentary Relaxed, typically sitting, some hand activity	100
Light Relaxed, some walking and hand activity. Sustained indefinitely with ease.	150
Moderate Obvious physical demands. Sustained indefinitely with normal breaks every couple of hours.	275
Heavy Obvious physical demands, with marked breathing efforts. Breaks required every hour. Note: Breaks will typically reduce the overall rate to Moderate.	350
Very Heavy Obvious physical demands, with marked breathing efforts. Frequent breaks required. Note: Breaks will typically reduce the overall rate to Moderate.	500

*Note: There is a tendency to over-estimate metabolic rate when using category values.

Recovery heart rate is particularly well suited to spot check at the time an oral temperature is taken.[16] Once work is halted, the worker is asked to sit quietly and the oral temperature is taken. After the worker sits for about 1 minute, the heart rate beat count is made for 10 seconds (e.g., 55 to 65 seconds after sitting) and multiplied by 6. The result of this calculation is the recovery heart rate. A value of 110 beats per minute (bpm) or less indicates acceptable physiological strain.[4,17] If the rate is greater, there is some risk of excessive cardiovascular strain and loss of thermoregulatory capacity.

The ACGIH recommends that a sustained heart rate be limited to 180 bpm minus the age of the worker in years (i.e.,

180—Age).[4] A continuously recording heart rate monitor with an alert can be set to this value and if the alert level is sustained for several minutes, the exposure is excessive.

If heart rate is monitored for the day, the average for an 8-hour day should not exceed 115 bpm.[6]

Bernard and Kenney recommended monitoring moving time averages of heart rate over short to long intervals (5 to 90 minutes).[15] As implemented using a commercially available device, this method can track important trends in heart rate over longer periods of time. The device includes the insulated skin temperature measure of core temperature as an additional metric of exposure in the alerting process.

Comments on Physiological Monitoring

Physiological monitoring helps in safeguarding workers exposed to high heat stress, especially if they are wearing vapor-barrier clothing. Monitoring demonstrates good management of heat stress exposures. This result was intended when the Physical Agents Committee recommended the Heat Stress and Strain TLV.® One approach to physiological monitoring at a hazardous waste site (or similar operation that involves restricted entry work) is to note oral temperature and heart rate pre- and post-entry. A pattern of increasing values in either measure is a matter of concern. If post-entry values increase, then work periods should be shortened. If pre-entry values increase, then recovery periods should be extended.

Control of Heat Stress

Heat stress controls can be divided into the following categories:

- General controls
- Job-specific controls.[4,6]

General controls are implemented if there is a potential for heat stress in the workplace. Job-specific controls are directed to the distinct situations that cause excessive exposures.

General Controls

General controls include training, heat stress hygiene practices, and medical surveillance. These activities are fundamental to worker protection.

Heat stress training should be included in routine training programs for other workplace hazards. This includes preplacement and annual training as well as special alerts if extraordinary exposures might occur, as during heat waves.

Some heat stress industrial hygiene practices are the responsibility of the exposed individual. These practices reduce the risk for heat-related disorders. They include

- Self-determination
- Fluid replacement
- Healthy lifestyle maintenance
- Health status monitoring
- Performance expectations adjusted during acclimation
- Medical surveillance.

Self-determination gives the individual as much control over the pace of work as is possible and the discretion to stop an exposure causing extreme discomfort or the symptoms of heat-related disorders such as fatigue, light-headedness, nausea, dizziness, or headache.

Fluid replacement is the consumed water lost by sweating. Sweat loss may be as high as one liter (quart) per hour. Drinking water should be a habit rather than driven by thirst, because satiety is not a sufficient indicator of adequate hydration. The habit of drinking water can be developed by encouraging workers to drink small amounts frequently and by making the water readily available. The crucial beverage is water, but commercial drinks and flavored drinks are generally consumed in greater quantities than is plain water. In any case, the beverage should be readily accessible and chilled.

Healthy lifestyle maintenance means a well-balanced diet, adequate sleep, limited heat stress exposures off the job, and no drug abuse, including use of alcohol.

Health status monitoring has two features. First, anyone under continuing care should inform his/her health care provider of occupational heat exposure. Virtually every chronic disease, or the drugs provided to treat these diseases, can have an effect on heat tolerance. The health care provider can adjust treatments or provide better counsel to the individual if made aware of the heat exposure. The employer may solicit through its medical authority information on drug prescriptions. Acute illness, treated with prescription or over-the-counter drugs, may directly or through the drugs reduce heat tolerance. Employees should be particularly concerned about illness accompanied by fever, nausea, vomiting, or diarrhea.

Performance expectations adjusted during acclimation. Because acclimation (acclimatization) increases heat tolerance, gradual heat stress exposure over several consecutive days is important. During this period, workers more recently introduced to heat stress conditions should recognize that they cannot perform at the same levels as those who are acclimated. That is, the unacclimated workers and their supervisors must manage work expectations. Acclimation is completely lost after 3 weeks without heat stress exposures, and can be lost after a week of illness.[6]

Medical surveillance incorporates preplacement physicals with written opinions about the ability to tolerate heat stress and counseling of workers with personal risk factors. The health care officer may suggest routine or *ad hoc* followup, as appropriate. Health and safety personnel should monitor incidents of heat-related disorders.

Job-Specific Controls

When there is a specific exposure to heat stress that may lead to excessive physiological strain, job-specific controls should be considered. These controls are divided into the following traditional hierarchy:

- Engineering controls
- Administrative controls
- Personal protection.

Engineering controls focus on reducing or eliminating heat stress through changes in the environment, metabolic work, and evaporative resistance of the clothing. Reducing metabolic cost can be accomplished through mechanization of the work. Other effective engineering techniques include changing the clothing requirements to reduce the resistance to evaporative cooling and reducing the air humidity and temperature with spot ventilators. (In the context of heat stress, changing the CPC to reduce evaporative resistance is an engineering control. The CPC is still personal protection against the chemical.)

Administrative controls do not change the underlying level of heat stress but manage the exposures by limiting the work time and providing adequate recovery to prevent a cumulative increase in body core temperature. Classic administrative controls are work/rest cycles or worker rotation.

Personal protection usually means some sort of personal cooling system. Effective systems include liquid cooling systems that cover a substantial portion of the body; air cooling systems that circulate air under the clothing covering a large portion of the body; and water ice vests that cover at least the chest and back. Other systems are being introduced and the performance of personal cooling systems varies widely. Evaluation of new devices is appropriate during the performance of work typical of the anticipated application. The evaluation should include objective performance measures where possible. (Care should be taken when the only measure of performance is subjective.)

References

1. Dukes-Dobos, F., and A. Henschel, Development of Permissible Exposure Limits for Occupational Work. *ASHRAE Journal* 15, 57-62 (1973).
2. NIOSH, *Criteria for a Recommended Standard ... Occupational Exposure to Hot Environments*, Revised Criteria 1986, U.S. Department of Health and Human Services, NIOSH (1986).
3. OSHA, *Technical Manual (TED 1-0.15A)*, Heat Stress (Section III, Chapter 4) (see www.osha-slc.gov/dts/osta/otm/otm_iii/otm_iii_4.html).
4. American Conference of Governmental Industrial Hygienists (ACGIH), 2000, *Threshold Limit Values and Biological Exposure Indices for Chemical Substances and Physical Agents,* ACGIH, Cincinnati, OH (2000).
5. Ramsey, J.D., F.N. Dukes-Dobos, and T.E. Bernard, Evaluation and Control of Hot Working Environments: Part I – Guidelines for the Practitioner. *International Journal of Industrial Ergonomics* 14, 119-127 (1994). (Further information in T.E. Bernard, F. N. Dukes-Dobos, and J. D. Ramsey, Evaluation and Control of Hot Working Environments: Part II – Knowledge Base for Guide. *International Journal of Industrial Ergonomics* 14, 129-138 [1994].)
6. Bernard, T.E., Thermal Stress. In *Fundamentals in Industrial Hygiene*, 4th ed., edited by B. Plog. National Safety Council, Chicago (1995).
7. Ramsey, J.D., and M.Y. Beshir, Thermal Standards and Measurement Techniques. In *The Occupational Environment—Its Evaluation and Control*, chapter 25, edited by S.R. DiNardi. AIHA Press, Fairfax, VA (1997).
8. Ramsey, J.D., and T.E. Bernard, Heat Stress. In *Patty's Industrial Hygiene and Toxicology, 5th ed.*, edited by R.E. Harris. John Wiley and Sons, New York (2000).
9. Gonzalez, N.W., T.E. Bernard, N.L. Carroll, M.A. Brenner, and J.P. Zeigler, *Maximum Sustainable Work for Five Protective Clothing Ensembles and the Effects of Moisture Vapor Transmission Tate and Air Permeability* (manuscript in preparation, based on MSPH thesis of N.W. Gonzalez).
10. O'Connor, D.J., and T.E. Bernard, Continuing the Search for WBGT Clothing Adjustment Factors. *Applied Occupa-*

tional and Environmental Hygiene 14, 119-125 (1999).

11. International Organization for Standardization, *Ergonomics—Determination of Metabolic Heat Production.* ISO 8996-1990 (1990).
12. Tayyari, F., C.L. Burford, and J.D. Ramsey, Guidelines for the Use of Systematic Workload Estimation. *International Journal of Industrial Ergonomics* 4, 61-65 (1989).
13. Bernard, T.E., and B.S. Joseph, Estimation of Metabolic Rate Using Qualitative Job Descriptors. *American Industrial Hygiene Association Journal* 55, 1021-1029 (1994).
14. Bernard, T.E., and W.L. Kenney, Rationale for a Personal Monitor for Heat Strain. *American Industrial Hygiene Association Journal* 55, 505-514 (1994).
15. Fuller, F.H., and P.E. Smith, Evaluation of Heat Stress in a Hot Workshop by Physiological Measurements. *American Industrial Hygiene Association Journal* 42, 32-37 (1981).
16. NIOSH/OSHA/USCG/EPA, *Occupational Safety and Health Guidance Manual for Hazardous Waste Site Activities.* DHHS (NIOSH), Washington, DC, 85-115 (1985).
17. Bernard, T.E., Heat Stress and Protective Clothing: An Emerging Approach from the United States. *Annals of Occupational Hygiene* 43, 321-327 (1999).

10

Development of a CPC Program

By S. Zack Mansdorf

Introduction

Effective management of chemical protective clothing (CPC) is crucial to ensure user safety. Since CPC represents the "last line of defense," a failure of the program instituted to manage the CPC system can result in user injury. As with the program management for other personal protective equipment, such as fall protection and respiratory protection, a formal program should be developed. The basis for the formal program should include published standards, policies and procedures, and should emphasize a comprehensive systems approach. In fact, an integrated system that incorporates the management of all personal protective equipment is optimal.

Historically, most liquid-proof protective materials were considered to be generally "impermeable" to all chemicals. Clothing items were believed to provide an absolute barrier and to adequately safeguard workers exposed to potentially hazardous gases, liquids, and solids. However, beginning in the 1980s, published research demonstrated that protective clothing, if not appropriately selected, provided a route of entry for hazardous materials through permeation, penetration, and/or degradation. Research in standard testing methods and the performance of protective clothing has led to the development of a wide range of new products and manufacturing methods to produce CPC materials that exhibit superior chemical resistance, user comfort, and functionality.

Use of CPC is pervasive in modern industrial and maintenance activities as well as in special use situations, such as emergency response. Applications for CPC use range from light coveralls in clean room applications to the more exotic, fully-encapsulating ensembles used for work in environments that are immediately dangerous to life and health (IDLH). Since the barrier presented by CPC is the last line of defense, the worker must be adequately protected.

The Occupational Safety and Health Administration (OSHA) requires "...all personal protective equipment shall be of a safe design and construction for the work to be performed." Since no protective clothing will present an absolute barrier to all potential chemical and physical hazards, a standardized approach must be implemented to ensure proper selection, use, and maintenance.

Key to a Successful CPC Program

As with all environmental, safety, and health programs management commitment is a basic necessity for a successful CPC program. Management of the program should be guided by a competent safety and health professional with the appropriate knowledge, training, and skills related to CPC. Responsibility for the implementation of day-to-day activities should be delegated throughout the organization and include the individual CPC users.

The relative complexity of an individual CPC program will vary from company to company. In general, the program components will be determined by the level of protection required, the frequency and extent of CPC use, the risk to the users if the CPC fails, and the available staff and financial resources of the organization. The following program elements represent components typically included in a comprehensive formalized program where the major aspects of OSHA requirements are considered. However, most of the program elements described are relevant to even the casual user of CPC.

Workplace Hazard Assessment

A hazard assessment must be conducted to determine the need for protective equipment. Generally, professionals competent in recognition of safety and health hazards should perform the assessment and determine the levels of potential

risk. For compliance with the OSHA standard, written certification of the assessment is necessary.

A review of potential hazards associated with the process or position by conducting a survey of the workplace should provide the information necessary to select appropriate CPC. The most common assessment technique will include an on-site evaluation of the work task and the production process to identify the potential hazards. Considerations should include physical, chemical, and biological hazards; potential routes of exposure; types of exposure anticipated (e.g., physical state and contact type, such as a splash); maximum expected concentrations; maximum expected exposure duration; physical demands of the job; and the level of hazard associated with routine, intermittent, or potential accident scenarios resulting in exposure.

In addition to OSHA compliance, documentation of the initial hazard assessment and consideration of alternatives to CPC can provide information to assist in incident investigations following a CPC failure. Also, the documentation may assist in a worker's compensation investigation, litigation, or other third party review.

Another goal of the hazard assessment is to determine the feasibility of engineering or administrative controls that would eliminate the hazard or reduce the risk of exposure, therefore reducing the need for CPC. Methods to consider might include:

- Substitution with a less hazardous material,
- Redesign of the process or work task to reduce or eliminate the potential for exposure,
- Use of mechanical devices, tools, or robots to reduce or eliminate potential contact,
- Use of ventilation or other engineering controls to reduce emissions, and
- Administrative controls to reduce potential for exposure (e.g., rotation of workers for heat stress).

Crucial to effective management of risk is adoption of a

hierarchy of controls that requires that CPC only be considered when other means of control are not feasible or practical. Typical examples of work tasks not readily amenable to other control technologies include emergency response, spill clean up, chemical fire control, maintenance activities, and non-routine tasks.

Hazard assessments and other elements of the selection process can be documented in several ways. Documentation intended to provide certification of the hazard assessment should include identification of the workplace evaluated and the evaluator, and the date(s) of the hazard assessment. A log or other narrative can be maintained, or a standardized form can be completed. An example of a completed assessment form containing information important to the selection of CPC is provided in the Appendix.

Selection Strategies

With no reasonable alternative to CPC, the next phase is to select the optimal ensemble or single article of clothing for the task to be performed. The approach in selecting personal protective clothing must encompass an ensemble of clothing and equipment items that are easily integrated to provide both an appropriate level of protection and still allow the safe completion of work activities.

The hazard assessment, justification of need, and identification of CPC performance criteria provide the basis for selection of the level and type of CPC required for a specific task or job. The assessment may reveal alternatives to CPC use or that CPC should not be used because of the risk level, ergonomic constraints, or other problems.

A general classification scheme for acute effects of dermal exposure is valuable in the CPC selection process. In the following simplified scheme, effects of skin exposure are listed in increasing order of severity:

Category 1. Minimal irritation or other non-permanent effects

Category 2. Moderate irritation or other non-permanent effects

Category 3. Severe irritation, dermatitis, or other non-permanent effects

Category 4. Severe toxicity, burns, de-pigmentation, or other permanent effects

Category 5. Immediately dangerous to life and health

CPC selected for a Category 1 hazard, such as skin contact with isopropyl alcohol, might be any chemically resistant glove, but Category 4 or 5 hazards would require clothing that provided more assurance of protection from potential permeation, penetration, and degradation. Chemical carcinogens are not incorporated into this scheme since their effects are chronic rather than acute. Nevertheless, given the concern over even minimal contact with known carcinogens, potential cancer-causing agents are usually treated as Category 4 or 5 hazards. Physical hazards, such as thermal burns, are not considered in this scheme.

Determination of the protection and performance levels necessary to protect the worker should include the identification of appropriate CPC data and reference materials used to make the selection. If no documentation exists for physical and chemical resistance properties with the American Society for Testing and Materials Committee F23 Standard Test Methods, or an equivalent entity, laboratory testing must be conducted to determine performance data (e.g., breakthrough data, cut resistance, and thermal resistance). Identification of laboratory or field methods to ensure that the CPC selected will perform as expected and external laboratories to conduct specialized performance tests may be necessary. Additional information on laboratory and field validation can be found in Chapter 5.

When selecting CPC for specific activities, both physical work task and chemical resistance requirements must also be evaluated. For example, will the garment or accessories be required to be flame retardant, thermally protective, abrasion resistant, or cut resistant?

Once the necessary physical performance characteristics of the CPC are met, the essential chemical resistance characteristics are considered. This evaluation is primarily based on user risk if the CPC fails. If a failure would result in only minor effects (e.g., skin irritation), then the selected CPC does not have to be completely fail-safe. However, if the effects of a failure could result in serious or permanent injury, the selection requires greater diligence and may require redundancy or additional protection (e.g., double gloving).

The best protection, however, does not mean overprotection. Wearing more CPC than needed to appropriately reduce the risks of exposure can result in additional heat stress, unnecessary discomfort, and a reduced ability to perform. Overprotection also can reduce the credibility of the CPC program, which could lead to improper or reduced use of the CPC when it is needed.

For unknown hazards or for high-risk tasks where the failure of CPC and equipment might result in serious injury or death, only the highest levels of protection may be appropriate. For emergency responses to chemical spills and certain clean up operations, the U.S. Environmental Protection Agency (EPA), in cooperation with other agencies, has developed guidelines. However, these generic guidelines, which were loosely developed from a dermal hazard standpoint, are somewhat outdated.

Similarly, the current Hazardous Waste Operations and Emergency Response (HAZWOPER) rules and guides (i.e., the appendices) offer generic recommendations. The Industrial Safety Equipment Association (ISEA), through the American National Standards Institute (ANSI), developed a more comprehensive approach to the classification of protective ensembles, based on current European classification schemes (ANSI/ISEA 103-2000). Whatever the source of guidance, exposure scenarios require the utmost care in selection, use, and maintenance of the CPC.

Validation Methods

After the initial selection, job characteristics and work tasks must be carefully reviewed to ensure all the necessary performance characteristics and ergonomic constraints are satisfied. Following the selection of the CPC, the items should be "field tested." The CPC should be subjected to actual trials to verify suitability for the intended use. Additionally, worker acceptance should be ascertained.

Although medical surveillance (or monitoring), including biological monitoring, can be applied to confirm that the CPC selection is effective, in no case should it be a substitute for selection of the best protection available. Medical surveillance is based on the effects, markers, biochemical byproducts, or other artifacts of exposure (and potential damage).

Decontamination Methods

A comprehensive management system for CPC includes a standardized approach to the evaluation of the potential necessity for decontamination procedures for both disposable and reusable items. Users of disposable CPC could become contaminated during doffing or might cross-contaminate themselves if not properly trained in appropriate removal techniques. Methods for decontamination and assignment of responsibilities related to the decontamination process should be included.

Multiple-use items of CPC typically require some form of cleaning or decontamination. Cleaning is required for items that do not present a risk of retaining harmful contaminants with common cleaning techniques (e.g., water-soluble substances, biologically inert materials, and normal body emissions on the inside surfaces). For situations where contamination is likely, or decontamination is required as a precaution, the decontamination technique and efficiency should be predetermined before use of the CPC. Preparation for essential decontamination includes information for users on how to minimize the potential for contamination, the risks from contamination, and the decontamination procedures.

Training

Since proper use of the CPC is essential for protection of the user, training is a vital element in the overall program. The CPC user, as well as the immediate supervisor, should be trained. Content should include the nature and extent of the hazard, intended use, limitations of use, inspection, donning and doffing, decontamination, maintenance, emergency use, and other relevant topics. Training information should also include the minimum qualifications and experience requirements for designated trainers.

The key objective for training specifications is the user must fully understand the nature of the hazard and the proper use and limitations of the CPC. OSHA regulations contain several requirements for training users in CPC, and generally overlap requirements for training and information under the Hazard Communication rules.

Training programs for CPC users should include the following information, at a minimum:

- When CPC is necessary – Including the hazards for which the CPC is selected, and the circumstances when CPC is necessary,
- What CPC is necessary – Identification of the ensemble and each of the items of clothing that are necessary,
- How to properly don, doff, adjust, and wear the selected CPC,
- What procedures to follow in the event of a CPC failure or emergency. This may also include the symptoms and effects of exposure to help in the identification of CPC failures,
- The limitations of the CPC,
- The proper care, maintenance, useful life, storage, and disposal of the CPC, and
- Procedures for decontamination, if necessary.

Inspection Procedures

Even the best chemically resistant products will not provide adequate protection if the product is either degraded or damaged during use. A standardized approach for inspection procedures and methods for the identification of CPC items that require maintenance, repair, or disposal should be developed. These methods should also include the development and implementation of self-inspection and auditing to ensure continuing effectiveness of the program.

A routine inspection program should be established and conducted on three levels. The first level should be routine user inspection of the CPC before and after each use. Standardized protocols and training are essential and should be provided to the users to enable them to effectively perform this inspection. The second level should include a scheduled and routine inspection program, conducted by supervisory staff. The third level should include a functional audit by management staff designated as responsible for the CPC program.

A maintenance and repair program for CPC, if applicable, should be established similar to the inspection program. While most CPC is either designed for single use or not repairable, some higher end reusable items may require maintenance and be repairable. Most items of CPC are not user repairable (modifications and unauthorized repair can void any implied warranty and potentially reduce the effectiveness of the CPC). For equipment that may require maintenance or repair, procedures should be developed in conjunction with the inspection program for users to turn in the equipment to their supervisors or other designated persons. The returned equipment should be clearly marked as not operable, or with some other designation to prevent accidental reuse. A fully competent person, qualified to certify that the condition of the repaired equipment meets the original design specifications, should do all major repair work.

Note: Most manufacturers of reusable fully encapsulating suits require that an inflation test (ASTM F 1052-97 or equivalent) be performed at least every six months.

Additional elements and activities may be included in the management system such as purchasing procedures or specifications, medical qualifications and surveillance, and maintenance and storage procedures. Also included are other elements related to the comprehensive approach to management of a CPC, or an overall protective personal equipment program.

Auditing the Program

Management should establish a process to make sure that all elements of the CPC program are effectively performed. This process is best accomplished on at least two levels. First, the manager should conduct internal audits or inspections with designated responsibility for the CPC program on a routine basis (quarterly or semi-annually). Secondly, a team or person that is independent of the daily operations of the CPC program should perform a written audit procedure.

The independent audit team should have at least one individual who is a technically competent professional in the selection and use of CPC. This individual or team should review documentation, standard procedures, and actual use of CPC in the workplace. Following the audit, a written report should be issued with recommendations, as required, for program deficiencies found and improvements required. This peer review is an integral part of maintaining an effective program and of protection of the users of CPC.

Summary

Chemical protective clothing is intended to provide protection for personnel against hazardous chemicals when more effective engineering or administrative controls are not practical or feasible. The actual level of protection provided by CPC will depend on the adequacy and effectiveness of the established management systems to ensure the proper selection, use, and maintenance of CPC. A comprehensive program for CPC should include the following:

- Assessing the need for CPC by reviewing the nature of the job or process hazards,
- Determining the required protection level and performance, based on hazards present,
- Selecting CPC, based on performance and specifications,
- Validating, in the laboratory or field, the effectiveness of the CPC selected,
- Evaluating of potential for contamination and the need for decontamination,
- Providing training in the nature and extent of the hazard, intended use of CPC, limitations of use, inspection, donning and doffing, decontamination, maintenance, and emergency use,
- Developing procedures for inspection, maintenance, and repair of CPC, and
- Establishing a self-inspection and auditing system.

References

1. Baker, R.L., and G.C. Coletta, Eds., *Performance of Protective Clothing*—"Risk Assessment of Chemical Exposure Hazards un the Use of Chemical Protective Clothing-An Overview," ASTM STP 900, American Society for Testing and Materials, West Conshohocken, PA (1986), pp. 207-213.

2. Perkins, J.L., "Chemical Protective Clothing: I, Selection and Use," *Applied Industrial Hygiene Journal*, Vol. 2, 1987, pp. 222-230.

3. Perkins, J.L., "Chemical Protective Clothing: II, Program Considerations," *Applied Industrial Hygiene Journal*, Vol. 3, 1988, pp. 1-8.

4. *PPE Made Easy,* Stull, J., ABS Group, Inc., Rockville, MD, 1998.

5. Alaimo, R.J., Ed., *Handbook of Chemical Health and Safety,* American Chemical Society, Washington, DC (2001).

6. Stern, M. B., and S. Z. Mansdorf, (Eds.) *Applications and Computational Elements of Industrial Hygiene*—"Personal Protective Equipment"(Chapter 11,*)* Lewis Publishers, Inc., Boca Raton, FL., 1998.

7. Gunh, A., P.D. Mellstrom, E. Jan, and M.D. Wahlberg, *Protective Gloves for Occupational Use*—"Industrial Hygiene Assessment for the Use of Protective Gloves," Boca Raton, CRC Press. 1994.

8. *Occupational Safety and Health Guidance Manual for Hazardous Waste Site Activities*, NIOSH Publication 85-115 (1985),

9. Forsberg, K., and S.Z., Mansdorf, 3rd Revised Edition, *Quick Selection Guide to Chemical Protective Clothing*, John Wiley and Sons, New York, NY, 1997.

10. DiNardi, S.R., Ed., *The Occupational Environment-Its Evaluation and Control*—Chapter 35, AIHA, 1997. pp. 956-972.

11. Slater, K., "Comfort or Protection: The Clothing Dilemma", in *Performance of Protective Clothing*, 5th Vol., STP 1237, American Society for Testing and Materials, West Conshohocken, Pa., 1996, pp. 486-497.

12. Occupational Safety and Health Administration, *General Industry Standards*, 29 CFR 1910.132, Occupational Safety and Health Act of 1970 (84 Stat. 1593), Government Printing Office, Washington, DC.

13. ASTM, "Standard Practice for [a] Chemical Protective Clothing Program" F1461-93, Vol. 11.03, 1999 *Annual Book of ASTM Standards*, American Society for Testing and Materials, West Conshohocken, 1999.

Appendix

Certification of the Selection of CPC

(**Note**: For illustrative purposes, editorial liberties were taken on the completed form and the product names are fictitious.) The form is intended as a guide for users and should be altered to suit individual or organizational requirements and situations.

Section 1: **Job or Task Description/Process Description/ Location**

A description of the important health and safety aspects of the job, work task, or process, as well as the location work will be performed.

Section 2: **Potential or Actual Hazards**

An identification of potential or actual hazards observed or anticipated. Include potential hazardous results from process upsets, deviations, and emergencies.

Section 3. **Nature of the potential for contact**

An evaluation of nature of hazard in terms of duration, type, and areas of the body that might potentially be contacted or exposed. Selection of the best CPC ensemble is dependent on the most likely type of contact or exposure. For example, a splash potential might require a "splash suit," whereas a pressurized spray might require a fully-encapsulating suit.

Section 4: Risk level presented by contact or hazards present

An assessment of risk, probably the most important activity. Risk evaluations are typically qualitative rather than quantitative, so the risk usually is expressed as a range from very low to very high. The risk to the user of injury or illness ultimately determines the level or protection required — or if CPC cannot be used.

Section 5: Other control methods considered

A consideration of other control options that might be used in conjunction with or in place of CPC.

Section 6: Operational Considerations and Constraints

A review of operational requirements such as comfort; glove flexibility for manipulating small objects; resistance to abrasion; resistance to puncture or tears by small metal shards from machining operations; soles needed for climbing or walking on wet surfaces; etc.

Section 7: Performance criteria and reference data

A determination and documentation of performance requirements of the CPC drawn from information gathered.

Section 8: CPC Specification/Selection

An evaluation of vendor literature for selection of the best CPC for the intended use, based on specifications. The process may include consideration of cost ,useful life, comfort, and other operational considerations, but in all cases, should be primarily based on risk.

Section 9: Other PPE required/medical surveillance

An identification of CPC worn for other tasks in the same operation. Where practical and feasible, the same CPC should be used for tasks done in a series to avoid any confusion that might result from making frequent changes in PPE.

Also provide an identification of any requirements for other personal protective equipment (PPE) needed in addition to CPC (e.g., respiratory protection, fall protection, head protection, hearing protection, etc.) and any medical surveillance or additional skills or qualifications that are necessary.

Section 10: Decontamination

Documentation of the potential for contamination and proposed methods of decontamination.

Section 11: Training

A description of any special training requirements for the CPC user and others.

Sections 12 and 13: Worksheet Completed By/Checked By

Signature blocks for individual conducting assessment and for individual verifying appropriate assessment performance. Often verification is by the designated individual with overall responsibility and authority for the CPC program.

Example – Certification of the Selection of CPC[1]

Section 1: Job or Task Description/Process Description/Location:
Desmut tank cleaning. Dip tank used for cleaning large aerospace parts prior to other treatment. Tank contains a mix of Nitric (30% v: v), 25% Hydrofluoric+ Water (40% v: v), DM Water (30% v: v) in a 12×20×6 foot tank with ~10,000 gallons of solution, which is drained 1/yr. prior to cleaning of the sludge which forms in the bottom of the tank. In area 42, Job Task # 224. Two maintenance workers assigned with plastic shovels, suction hoses, and a water line. Sludge suctioned out; remainder flushed to drainage.
Section 2: Potential or Actual Hazards (Chemical, Physical, Biological, etc.):
Typical sludge is semi-solid with a high content of acid and metal residues. Potential for contact (direct contact/splash) with the sludge, heat stress from the PPE, confined space hazard, slip/trip hazard in the tank. All CPC must be sealed to prevent contamination during the cleaning.
Section 3: Nature of the Potential for Contact:
Work task takes approximately 4 hours with 2 workers, constant contact of boots with sludge, possible intermittent contact with suit and gloves. Potential full body contact should the workers fall into the sludge.
Section 4: Risk level Presented by Contact or Hazards Present:
Relatively high risk of acid burns and potential severe eye and respiratory hazard. Relatively high risk of skin damage and delayed burns/toxicity from HF. Moderate risk of heat stress from fully enclosing ensembles (will wear cooling vest). Area classified as a confined space.
Section 5: Other Controls Methods
Tank cleaned once per year – mechanical sludge removal system scheduled for installation after tank cleaning.
Section 6: Operational Considerations and Constraints
Abrasion resistant gloves for handling a plastic shovel handle. Some minimal flexibility for manipulating a suction hose and shovel handle. Boots with sure footing needed to climb the ladder into and out of the tank, and to avoid slips and falls on wet surfaces inside the tank. The task will last more than 4 hours and will require decontamination, doffing and donning CPE for breaks and lunch.

Section 7: Performance Criteria and Reference Data:	
Chemical resistance data shows 8-hrs breakthrough resistance for Chem-Tuff suits, Acme Boots, and Special Shield/Neoprene Gloves per the Quick Selection Guide/Manufacturers literature (attached).	
Section 8: CPC Specification/Selection:	
Acme Full Length Neoprene Safety Boots, Model XX-2, Acme Model 24 Fully Encapsulating Chem-Tuff Suit with integrated airline connector, Georgia Cooling Vest, Acme Model YK Neoprene Gloves Over Special Shield Model Gloves (Double Gloving), Safety Guard Model 12 Full Body Harness with lifeline. All suit connections (gloves, boots, hoses) to be sealed with chemically resistant tape. Full ensemble to be inspected prior to donning and immediately after donning and prior to tank entry.	
Section 9: Other PPE Required/Medical Surveillance:	
Full face-piece airline respirators, positive pressure with Neoprene hoses. Full body harness on suits with chemically resistant lines, cooling vests, medically qualified prior to the start of work, attendant topside.	
Section 10: Decontamination:	
All equipment, including hoses, to be decontaminated after use by complexing agent for HF, such as calcium chloride solution, followed by water wash and testing. Decontamination to be performed on personnel prior to doffing of equipment.	
Section 11: Training/Pre-Qualification Required:	
Must have medical approval for use of respirator, full body CPC ensemble and effects of heat stress prior to assignment. Must have had confined spaces training and understand emergency evacuation procedures. Must have demonstrated skills/knowledge of the equipment to be used. Must sign a work permit and be familiar with the hazards presented by this work task as well as emergency procedures and the signs and symptoms of exposure. Must undergo specific work task training before the initiation of work.	
Section 12: Worksheet completed by/date:	**Section 13: Worksheet checked by/date:**
Biff I. Genist, 4/1/00	Johnny Ensemble, 4/1/00

[1] This example is for illustrative purposes only and should not be used to identify appropriate CPC for similar scenarios. In the case of highly hazardous materials, such as hydrofluoric acid, additional precautions or testing (i.e., pinhole testing) should be conducted to ensure that appropriate protection levels are maintained.